T0262223

Forest Ecosystems: Forest Function and Management

Forest Ecosystems: Forest Function and Management

Edited by **Lee Zieger**

New York

Published by Callisto Reference,
106 Park Avenue, Suite 200,
New York, NY 10016, USA
www.callistoreference.com

Forest Ecosystems: Forest Function and Management
Edited by Lee Zieger

© 2015 Callisto Reference

International Standard Book Number: 978-1-63239-342-5 (Hardback)

Printed in the United States of America.

Contents

Preface

This book has been an outcome of determined endeavour from a group of educationists in the field. The primary objective was to involve a broad spectrum of professionals from diverse cultural background involved in the field for developing new researches. The book not only targets students but also scholars pursuing higher research for further enhancement of the theoretical and practical applications of the subject.

Forests are much more than a bunch of trees, contrary to popular belief. They are complicated operating systems of interrelated biological, physical and chemical elements, whose biological features have constantly evolved to preserve itself. These complex combinations of climate, soils, trees and plants result in many varied types of forests throughout the globe. Although trees are the main highlight of forest ecosystems, the vast variety of other creatures and abiotic elements in the forests indicate that there is also a need to study about these elements like soil nutrients and wildlife and consider these compounds when forming management plans. This book does not only provide information about trees, like other books, but also gives an insight into other components related to forests, such as "Forest Function – Energy, Mass and Biological Fluxes", and "Ecosystem-Level Forest Management".

It was an honour to edit such a profound book and also a challenging task to compile and examine all the relevant data for accuracy and originality. I wish to acknowledge the efforts of the contributors for submitting such brilliant and diverse chapters in the field and for endlessly working for the completion of the book. Last, but not the least; I thank my family for being a constant source of support in all my research endeavours.

Editor

Part 1

Forest Function –
Energy, Mass and Biological Fluxes

Fertility, Microbial Biomass and Edaphic Fauna Under Forestry and Agroforestry Systems in the Eastern Amazon

Maria de Lourdes Pinheiro Ruivo[1], Antonio Pereira Junior[2],
Keila Chistina Bernardes[3], Cristine Bastos Amarante[1],
Quezia Leandro Moura[2] and Maria Lucia Jardim Macambira[1]

[1]*Museu Paraense Emilio Goeldi,*
[2]*Program of Post-Graduate in Environmental
Sciences/Federal University of Pará,*
[3]*Post-Graduate Program in Agronomy/
Federal Rural University of Amazonia*
Brazil

1. Introduction

In many countries the rate of deforestation is accelerating. For example, many forest areas of Bangladesh, India, Philippines, Sri Lanka and parts of the rainforest in Brazil could disappear by the end of the century (GLOBAL CHANGE, 2010). The primary forest, especially in the tropics like the Philippines, Malaysia and Thailand such as in Brazil began to be destroyed, because the growth of the agricultural expansion caused a significant decrease in natural resources. Over the past 50 years, the Philippines, there was a loss of 2.4 acres of vegetation every minute, which is attributed to two factors: growth of agriculture and illegal logging

The model of agriculture practiced in Brazil contributes significantly to the expansion of agricultural frontier, increase the production, the productivity and agriculture and the national livestock. However, this performance has led to great reduction of the cover native forest and, consequently, the supply of products of forest origin, besides exposing the lands to loss of fertility, erosion process and water pollution.

In the northeast of Pará state, as in other regions of the Amazon, the intense agricultural activity, with emphasis on removal the primary forest for pasture establishment, agriculture of overthrow and burn, indiscriminate deforestation caused by human activity in terms of economic activities and disorderly logging has been a major factor to accelerate the process of soil alteration.

To mitigate these imbalances, in the Pará state, especially in the northeast region, timber companies, large and small producers located in the cities of Tailândia, Bragança, Igarapé-Açu and Aurora do Pará, began the reforestation on degraded areas in there existent, through use of monocultives and agroforestry systems (ROSA, 2006; RUIVO et al., 2006,

SOUZA et al. 2007; RUIVO et al., 2007, CODEIRO et al. 2009; OLIVEIRA, 2009). It was observed that with the reforestation of these areas the quality and quantity of soil organic matter were slow and continuously recovery .

The fertility of the soil, in edaphoclimatic local conditions, is associated with the content of organic matter in the soil (MOREIRA; COSTA, 2004). However the need to seek a sustainable agriculture, the pressure of national and international society requires techniques that protect the agricultural systems After all, the model of traditional agriculture practiced in the Amazon is unsustainable.

For the reduction of land degradation is necessary the use of conservataive techniques to identify the most profitable activities in the region allowing for a harmonious environment coexistence for agricultural economically viable and environmentally sustainable (SOUSA et al., 2007). The challenge is to identify the correct combinations of species to establish synergistic relationships ideals, so that ensure the key ecological services such as nutrient cycling, biological control of pests and diseases and conservation of soil and water (CARDOSO et al., 2005).

In the state of Para, reforestation with native and exotic species reaches high levels due to the great adaptablity of these species in degraded soils. The answers obtained, either in monoculture agroforestry systems, have been effective in the recovery of deforested areas, providing excellent results both for this action as for commercial use, allowing a decrease in aggression to the primary forest and improving the quality of life of populations where this does occur (CORDEIRO, 1999; MONTEIRO, 2004; RUIVO et al., 2007).

Although there are numerous studies on the growth and development of the species native species (CARVALHO, 2004; CORDEIRO, 2007; JESUS, 2004; LORENZI, 2002), comparative studies with the species subjected to different plantation systems and the nutritional behavior of soil in microbiological and biochemists terms are not commonly found in the literature, like as the influence of coverage with different systems involving the vegetables species and their influence on soil quality are still poorly understood. The addition of organic matter to soil, due to stay vegetables residues leads to creation of an enabling environment for better plant development, enhancing microbial activity and consequently the nutritional conditions of the soil. Based on this assumption this research was conducted in the city of Aurora to identifying the soils modifications under physical, chemical and biological properties in areas under reforestation in forestry cropping systems and agro forestry by antropic actions and their impact on edaphic fauna.

2. Materials and methods

2.1 Localization and characterization of the study area

The study was conducted at the Farm Tramontina Belém S/A, located in the city of Aurora do Pará (Figure 1), which belongs to the Mesoregion of Northeast of Pará state and Microregion Bragantina. This area suffered intense anthropogenic changes in the last 50 years due to high extractivist activity, food production and livestock that decimated almost completely their natural vegetation. Over the years a secondary forest (locally known as capoeira) was developed. Despite being a zone considered in environmental impact, this is an area that food supply the capital, mainly grains, greens and vegetables.

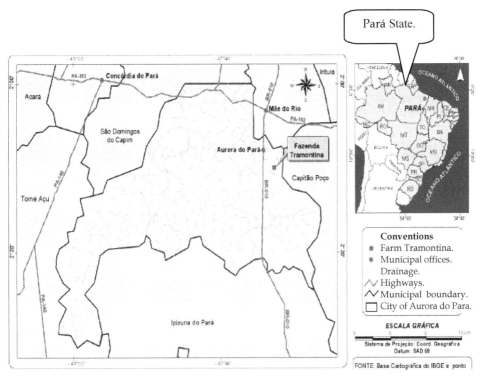

Source: Maps IBGE, 2011.

Fig. 1. Map of geographical location of FarmTramontina Belem S/A, Aurora do Pará.

In the locality where this work was developed, a former cattle ranch acquired by an industry of domestic utensils, that was reforested with purposes sustainable economic exploitation and controlled of forest species. The current vegetation is divided in areas of pasture (livestock) abandoned, predominating such as vegetation   quicuio-da-amazonia (*Brachiaria humidicola*) among other invasive species, beyond agroforest systems consisting of native species, the main Mogno (*Swietenia macrophylla* King), Paricá (*Schizolobium parayba var. amazonicum* Huber ex Ducke), Freijó (*Cordia goeldiana* Huber) and few exotic, such as eucalyptus (*Eucalyptus* sp) and small areas with secondary forest (capoeira) started around 40 years ago, whose seeds have been used for reforestation native species. The selected capoeiras soil was used as a standard for comparison with the reforestations soil.

The in the region from, According to Thorntwaite (1948) the climate classification in the studied area is type Br A'a, ("humid tropical"). The average annual rainfall is 2,200 mm not equally distributed throughout the year. However, the period from January to June is its greatest concentration (OLIVEIRA, 2009). The average temperature and relative atmospheric humidity are 26 ° C and 74% respectively (CORDEIRO 2007, CORDEIRO et al., 2009). Studies conducted in Brazil (CORDEIRO et al., 2009; 2010) allowed to classify the soil in Aurora do Pará area as Yellow Latossol sandy-clay and the occurrence of concretionary laterite levels in some areas, hydromorphic soils along streams and plain relief to gently rolling inserted on the plateau demoted from Amazon. The nutritional characteristics described in these studies show that they have a low supply of available essential nutrients and low tenor in organic matter (CORDEIRO et al., 2010).

2.2 Cropping systems studied

Since the 1990s were planted around 1,043 ha submitted to different types of planting reforestation with the use of species such as Mogno (*Swietenia macrophylla* King) by the great commercial value abroad, Ipê (*Tabebuia heptaphyta* Vellozo), Cedro (*Cedrella fissilis* Vellozo), Jatobá (*Hymenaea intermedia* Ducke var. *adenotricha* (Ducke) Lee & Lang.). Since 1994, the Paricá (*Schizolobium parayba* var. Amazonicum Huber ex Ducke), is the species with high commercial value used in the reforestation because of its applicability in the production of laminates. In 1996 the Freijó (*Cordia goeldiana* Huber) was introduced by the high referential commercial value in Europe. In 2003, was included in the reforestation process in the Tramontina area the Curauá (*Ananas comosus* var. *erectifolius* LBSmith), a bromeliad that in the Amazon, has a higher concentration in the municipality of Santarém, beyond the regions of Xingu River, Tocantins, Maicuru, Trombetas, Paru, Acará and Guamá. In the Pará state, the Curauá stands out in the Bragança and Santarém districts (OLIVEIRA, 2009).

The experimental design was completely randomized, with four systems under reforestation and three repetitions in each system. (S1) Monocultivation system with curauá, (S2) agroforestry system with paricá and curauá, (S3) monocultivation system with paricá, (S4) agroforestry system with paricá, freijó, mogno and (S5) varied capoeira (Table 1). All cropping systems such as the capoeira were subdivided into four parcels 24 x 19 m (456 m²), which totals 1,824 m² of area analyzed.

At the time of planting was performed organic fertilization with manure of corral (500g/pit) and bed chicken (150g/pit) for agronomic and forest species, respectively. In the first year of the forestal planting were performed three fertilizations, at 45, 180 and 300 days, using 150 g/pl of the formula NPK 10-20-20. In the planting Curauá, we used 10 g/pl of the formula NPK 10-10-10 at the beginning and end of the rainy season, in the first two years of planting.

Systems	Forest species	Age (years)
S1 Monocultivation	Curauá (*Ananas comosus var. erectifolius* L.B.Smith)	06
S2 Agroforestry	Paricá/Curauá (*Ananas comosus var. erectifolius* L.B.Smith)	06
S3 Agroforestry	Paricá, Mogno (*Swietenia macrophylla* King), Freijó (*Cordia goeldiana* Huber)	08
S4 Monocultivation	Paricá	06
S5 Capoeira varied		15

Table 1. System of crops, forest species and ages of reforestation, in the Farm Tramontina Belém S/A, Aurora do Pará.

In cropping systems with curauá (S1), agroforestry system with paricá, mahogany, freijó (S3) and in the cropping system with paricá (S4) occured just the cut of the grass, that was left on the soil of these cropping systems, no irrigation in any of them. In agroforestry cropping system with paricá/curauá was made fertilization with manure of corral (500 g/pit) and bed chicken (150 g/pit) (CORDEIRO et al., (2009).

2.3 Collection and preparation of the soil samples for physical, chemical and biological analysis

In all locations were collected soil samples deformed and undeformed in December 2009. Samples were collected by opening mini-trenches where soil samples were extracted from the depths: 0-10, 10-20, 20-40 cm, from transects in areas previously determined.

In each study area were collected 3 composite samples of soil from 5 single samples and were stored in plastic bags, conditioned in cool boxes containing ice for stagnate or decrease the microbial activity. The chemical, physical and biological analysis were made by technicians in the soil laboratory of the Museu Paraense Emílio Goeldi - MPEG.

2.4 Collection, preparation and identification of the soil fauna

The Pedofauna collections was performed using kind traps "pitfall-traps" (Figure 2).

These traps were consisted of plastic containers (08 cm x 12 cm) buried in the soil to a depth of 12 cm, with the leaked extremity leveled with the surface of soil, where they remained for three days (AQUINO et al, 2006).

In each plot of each treatment, the same depth, were placed four (04) traps, and inside each one of them, was added 60 ml of the preservative substance: 70% alcohol, distilled water (ratio 3:1, regarding the use of alcohol); biodegradable detergent (3 drops) and formaldehyde (10 ml). The fall of undesirable objects was prevented with a cover plate of polystyrene, supported by small wooden rods (AQUINO et al, 2006). The edaphic fauna, after collected, was taken to the laboratory where they were sieved (0.2 mm) to remove the fragments of plant and residues of soil. The identification of edaphic fauna was at the level of Order, with the aid of a stereomicroscope and the specific literature (BORROR, DELONG, 1988, BARRETO et al., 2008).

Fig. 2. Trap for capturing the edaphic fauna implanted in the Farms of Aurora do Pará, Bragança and Tailândia.

2.5 Determination of the physical and chemical caracteristics of soil

The granulometric composition was determined by densimeter method (EMBRAPA, 1997) and textural classification of soil in each system was performed using the textural triangle (LEMOS & SANTOS, 2006). The soil density (Ds) was determined by the volumetric ring method type Kopecky.

In the characterization of soil were performed the following measurements: total N, by distillation in semimicro Kjeldahl (BREMNER, MULVANEY, 1982), pH in potentiometer in the relation soil:water 1:2.5, organic C, by volumetric method of oxidation with $K_2Cr_2O_7$ and titration with ammonium ferrous sulphate, Ca, Mg and Al exchangeable in extractor of KCl 1 mol L^{-1} and measured in atomic absorption, exchangeable K and Na in Mehlich-1 extraction solution and determination by flame photometry, P available in Mehlich-1 extraction solution and determination by calorimetry, H + Al were extracted with calcium acetate 0.5 mol L^{-1}, pH 7.0 and determined volumetrically with NaOH solution.

From the values of potential acidity (H + Al), exchangeable bases and exchangeable aluminum, the capacity of total cation exchange (CTC) and cation exchange capacity effective (CTCe) were calculated. Relations were also calculated C/N of soil and the organic carbon stock (EstC), using the formula EstC = Corg x Ds x e/10, according to Freixo et al. (2002).

2.6 Determination of carbon (CBM) and nitrogen (NBM) of the soil microbial biomass

We used the fumigation-extraction method to estimate microbial biomass carbon (CBM) (Vance et al., 1987, Tate et al., 1988). The determination of microbial biomass carbon (C-BMS) of the fumigated and not fumigated extracts was made by titration (dichromatometry) according to De-Polli, Guerra (1999). For CBM calculation , the C content of fumigated samples were subtracted from the values of non-fumigated samples, the difference being divided by the value kc = 0.26 (FEIGL et al., 1995). The estimate of Nmic was made from Kjeldahl digestion. The correction factor (Kn) used for the calculation was 0.54 (BROOKES et al. 1985; Joergensen; Mueller, 1996). From the original values were calculated relations between C_{mic} and C_{org} of soil (C_{mic}/C_{org}), and N_{mic} and N_{total} of soil (N_{mic}/N_{total}), by the following equations: (C_{mic}/C_{org})x100 and (N_{mic}/N_{total})x100, respectively. These indices indicate the fractions of C_{org} and N_{total} that are incorporated in BM, expressing the quality of MOS (GAMA-RODRIGUES, 1999).

2.7 Determination of basal respiration of microbial biomass and the soil metabolic quotient

The basal respiration was estimated by the amount of $C-CO_2$ released within 10 days of incubation (JENKINSON & POWLSON, 1976). This technique allows the determination of the soil microbial activity, being quantified from the evolution of CO2 produced in respiration of the microorganisms in samples free of roots and possible insects. The metabolic quotient (qCO2) is calculated as the ratio between the rate of basal respiration and the microbial biomass carbon (ANDERSON & DOMSCH, 1993).

2.8 Statistical analysis

The two-way ANOVA was used to verify differences between the cropping systems studied. When found a significant (5%), the averages of each variable were tested by the

Tukey test (p <0.05). Additional analysis was also the determination of principal components (PCA) and cluster analysis to determine the degree of correlation between physical, chemical and biological data to be analyzed by soil grouping. Them, according to the variation of its characteristics a multivariate analysis can be use.

3. Results and discussion

3.1 Physical and chemical properties of soil under forestry and agroforestry systems

The production systems studied showed differences in soil physical properties. The type of soil in cropping systems S1 (monocultivation with curauá), S3 (agroforestry system with paricá, mogno, freijó and curauá) and S4 (monocultivation with paricá) is classified as franc sandy loam and only in the S2 (agroforestry system with paricá and curauá), presented soil type franc sandy clay (Table 2). Study conducted in a Latossoil in the Amazon (SILVA JUNIOR et al, 2009) showed that the study of this aspect of the soil is important because it is related to the dynamics of organic matter. This justifies the textural classification made in our study.

NOTATION	DPT (cm)	Ts	Fs	Cl	S	RSC	TC	Ds g/cm³
			g/kg					
S1	0 - 10	729	157	68	46	0,67	S L	1,50
	10 - 20	716	147	83	54	0,65		1,43
	20 - 40	610	160	137	93	0,67		1,54
MEAN	---------	685	155	96	64	0.66		1.47
S2	0 - 10	491	189	184	136	0,73	S C	1,37
	10 - 20	408	169	289	134	0,46		1,62
	20 - 40	362	149	343	146	0,42		1,60
MEAN	---------	420	169	272	139	0.54		1.53
S3	0 - 10	687	148	80	85	1,06	S L	1,42
	10 - 20	672	130	80	118	1,47		1,58
	20 - 40	627	156	122	95	0,77		1,51
MEAN	---------	662	145	94	99	1.10		1.50
S4	0 - 10	651	204	96	49	0,51	S L	1,36
	10 - 20	630	222	80	68	0,85		1,48
	20 - 40	588	170	90	152	1,68		1,57
MEAN	---------	623	199	90	88	1.01		1.47

DPT: Depths; Ts: thick sandy; Fs: fine sand; Cl: clay; S: silt; RSC: relation silt clay; TC: textural classification; SL: Sandy loam; SC:Sandy clay; Ds: density of soil.

Table 2. Granulometry, textural classification and density of soil of cropping systems studied. Farm Tramontina Belém S/A in Aurora do Pará.

With regard to the clay content in the culture system S2 (agroforestry system with paricá and curauá), in the depth from 0 – 10 cm high levels were detected when compared with other systems in this same depth, as shown in Table 2. Study conducted (LAVELLE et al, 1992) in the humid tropics, showed as a result a lateral variation in the soil granulometry and, according to the researchers, this may influence the training capacity of the stocks of exchangeable cations on the surfaces of colloids, this case, clay mineral. Then, the results found in this study for the S2 cultivation system may be indicative of the improved in the capacity of formation of the exchangeable cations.

The lowest content of clay fraction occur in the cropping systems S1 (monocultivation with curauá), S3 (agroforestry system with paricá, mogno, freijó and curauá) and S4 (monocultivation with paricá). The value for the lowest average was found in the cropping system S4 compared with the other cropping systems studied. Freire (1997) reports that the natural fertility of the soil depends on the adsorptive capacity of clay-minerals and organic colloids, with that, it's possible to affirm that in the cropping system S4, despite the low clay content, there are adsorption capacity and organic colloids in balance that allows the maintenance of natural soil fertility of this cropping system.

The relation silt *versus* clay proved to be higher in the cropping system S3 (Table 2), this demonstrates that the degree of weathering occurred in this area decreases with the depth, ie, the degree of weathering of the soil is high, as occur in the Latosoils.

The analysis of soil density showed that, among the systems studied, soil is more dense in the cultivation system S2 (agroforestry system with paricá and curauá). But, as pointed out Santana et al. (2006) density can be an attribute for analysis on the cohesion of the soil horizons, however there is a limitation for such use, ie, the density of the soil suffers from interference of granulometry that can presents high values, this would correspond to cohesive horizons, and this would affect the penetration of root of vegetables.

In this aspect, it was verified that, in the cropping system S2, the density is more pronounced between 10 to 20 cm when compared with other systems. In addition, this cultivation system, based on the exposed by Santana et al. (2006), we can consider this soil as "cohesive" as a results contained in Table 2.

The results indicate very low acidity (pH > 4.5), and as clay-minerals react with water from the soil, absorbing H +, this may explain the variation of acidity occurred in the cropping system S2, although present statistically significant effect. Was verified that in the cropping systems S1 (curauá monocultivation) and S3 (agroforestry system with paricá, mogno, freijó and curauá) the pH value, on average, has no statistically significant effect (Table 3). This can be explained by the content of total clay and sand which is also equivalent between them, as shown in Table 2.

In the cropping system S4 (paricá monocultivation), the pH was found to be constant in the three depths, and this may be due to small variations in the levels of sand and clay that were lower compared to other cropping systems, and this may explain the decrease of pH value in relation to the cropping system S2 and a slight increase in relation to cropping systems S1 and S3 (Table 2).

The tenors of Corg in cultivation system S2 (agroforestry system with paricá and curauá) were superior to other treatments evaluated (Table 3), even with scarce cover vegetation. Study conducted (Silva Junior et al, 2009) in Amazonian Oxisols after transformation to pasture, showed that carbon concentrations are high in clay soils, independent of vegetation cover. But the most plausible explanation is provided by Cordeiro et al. (2009), because the authors report that the area where this system of crops is located was fertilized with cattle manure (500 g/pit) and bed chicken (150 g/pit). This proves that even on degraded land, the use of organic cover helps soil fertility and improving the quality of it (Monteiro, 2004).

In the systems studied, we observed that the Corg content decreases according to depth and clay content (Table 2) in the depth of 0 - 10 cm, the Corg content of the soil surface within the studied systems are high. The research conducted about the relation of Corg contents and soil depth (Dejardins, et al. 1994; Koutika et al., 1997) showed that the trend of content of Corg is in decreasing accordance with increasing depth. This pattern of behavior on the content of Corg was observed in our study. Study carried out in a toposequence in central Amazonia (Marques et al., 200) report that the Carbon content are high in the surface layers to 25 cm (4.48 ± 0.08). Then, the high content of Corg found in cropping systems study corroborates the assertion of those authors.

The highest average of Carbon stock was found in the cultivation system S2 (Table 2), where was found the highest granulometric average, especially the clay and Ds. In descending order, the cultivation system S1 has the second highest average. In the cultivation systems S3 and S4 the average of carbon stock decreases, although the clay content is approximate to that contained in the cropping system S1. However the content of sand has an average ranging from 800 to 840 g/kg, and the variation in average clay content is between 90 and 96 g/kg (Table 1). This may be one explanation for the carbon stocks present a decrease in average of these treatments compared in the cropping system S2 (Table 2).

NOTATION	DPT	pH	Corg g/kg	Cs
	(cm)			Mg C.ha^{-1}
S1	0 - 10	4.5	13,84	20.76
	10 - 20	4.4	11,70	33.46
	20 - 40	4.1	6,90	42.50
MEAN	--------	**4.3**	**10.81**	**32.24**
S2	0 - 10	4.9	24,52	33.61
	10 - 20	4.9	12,96	41.99
	20 - 40	5.0	7,04	45.05
MEAN	--------	**4.9**	**14.84**	**40.22**
S3	0 - 10	4.4	8,59	12.02
	10 - 20	4.3	7,89	24.93
	20 - 40	4.3	7,73	46.68
MEAN	--------	**4.3**	**8.70**	**27.88**
S4	0 - 10	4.4	9,88	13.34
	10 - 20	4.4	7,38	21.84
	20 - 40	4.4	5,83	36.31
MEAN	--------	**4.4**	**7.07**	**23.93**

DPT: Depths; **Corg**: carbon organic; **Cs**: Carbon stock.

Table 3. Values pH (H$_2$O), tenors of carbon organic (Corg in g/kg) and carbon stocks (Mg C.ha^{-1}).

3.2 Relation between physical and chemical attributes

The result of the correlation (r) between the high content of Corg, clay and sand fraction in the cropping system S2, showed an increasing content of Corg in this cropping system (Table 4).

Attributes analyzed	Total clay (g/Kg)	Sand (g/Kg)	Ds (g/cm³)	Relation Silt x clay
pH	0.98 (p < 0.05)	- 0.98 (p < 0.05)	0.77 (ns)	- 0.66 (ns)
Corg (g/kg)	0.81 (ns)	- 0.57 (ns)	0.90 (ns)	- 0.88 (ns)
Carbon stock (Mg C.ha⁻¹)	0.82 (ns)	- 0.75 (ns)	0.86 (ns)	- 0.92 (ns)

Corg: carbon organic; Ds: Density of soil.

Table 4. Analysis of canonical correlation between the physical and chemical attributes of soil in the cropping systems studied. Farm Tramontina Belém S/A in Aurora do Pará.

Study ever done on the relation Corg versus clay content (TELLES, 2002) explained that high clay content allows the formation of macroaggregates and microaggregates that promote physical protection to the Corg, and avoid rapid decomposition of the same. Correlating this result with what was obtained in our study; it is possible to explain the high content of Corg found in S2 culture system, using the same argument. As for the stock of carbon (Cs) results showed that this stock decreases in relation to the increased depth in cropping systems studied.

The determination of the hierarchical clustering (HCA) revealed a approximated relation between physical and chemical attributes of soil subjected to analysis, because, according to Barreiro & Moita Neto (1998) this suggests a correlation between the variables of this data set. The results (Figure 3) show that there is formation of two groups where in the group A there was a split in the A1 and A2. In this group there was a correlation between the pH - Ds (Group A1), Ds - Silt/Clay (group A2) and Silt/Clay - Corg (group A3).

Another group, silt and clay (group B), confirm that these two variables are heterogeneous with respect to those that make up Group A. These results show that there is variability in the functioning of studied soils in the cropping systems SI, S2, S3 and S4. S3 and S4 are more similar across pH and relation attributes of silt/clay. The cultivation system S1 showed lower similarity with cropping systems S3 and S4 for the same attributes. Thus, one can verify which variables that differentiate the systems studied each other and can interfere or not in the edaphic fauna.

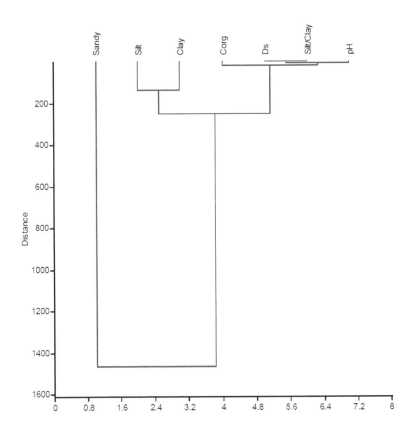

Fig. 3. Grouping of physical and chemical attributes of the cropping systems studied. Farm Tramontina Belém S/A. Aurora do Pará.

3.3 Microbiological attributes

The average content of microbial biomass carbon (CBM) (Figure 4) and the values of the microbial quotient (qMIC) (Figure 5) were higher in the system S4 and S5. In the system S4 the soil was covered with coarse vegetable waste, besides presenting a spontaneous vegetable regeneration between the lines of planting paricá such factors may have favored the maintenance of microorganisms in the soil and therefore increase the microbial carbon content.

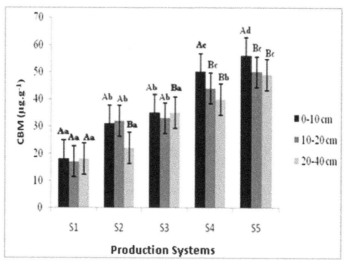

Fig. 4. Values of microbial biomass carbon (CMB) in different production systems.

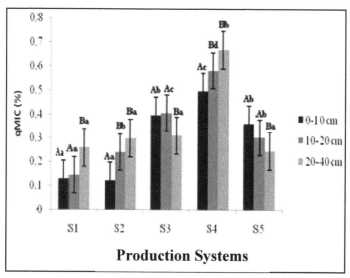

Fig. 5. Values of microbial quotient (qMIC) in different production systems.

For the attribute qMIC very low values were obtained, except in the S4 system, where it was recorded the highest values of qMIC, especially in the layers of the first depth of soil. Jenkinson & Ladd (1981), considered normal that 1-4% of total soil C corresponds to the microbial component, how the collection was done during the dry season, and it is known that water is an important element for microbial activity, it is possible that low values are justified by this fact.

Overall, the results indicate that the incorporation of organic matter is favoring the edaphic conditions and the different systems, especially the S3 and S4 suffering the biggest addition of organic matter, equaling the soils are of capoeira, eventually tending to an equilibrium (OLIVEIRA, 2009; BERNARDES, 2011, PEREIRA Jr, 2011). The Table 5 presents the principal components that shows this relationship. It's possible verify that saturation of bases (V), EC, aluminum saturation (m), Na, and Al were the variables that showed more differences between the systems. In systems S2 and S3, the values of these attributes, in general, are more similar to those found in capoeira.

Attributes	Principal components				
	Comp. 1	Comp. 2	Comp.3	Comp.4	Comp.5
pH	-0,027	0,144	**0,419**	0,195	0,204
Al	**-0,273**	0,003	**-0,314**	0,005	-0,016
H+Al	**-0,303**	0,075	-0,180	-0,018	0,055
Ca	-0,235	**-0,253**	-0,160	0,152	-0,059
Mg	**-0,294**	-0,171	0,119	0,017	-0,008
K	**-0,290**	-0,180	0,095	-0,047	0,032
P	-0,114	**0,371**	-0,206	-0,07	0,071
Na	**-0,254**	0,042	0,091	**0,458**	-0,084
SB	**-0,298**	-0,182	0,074	0,037	-0,024
CTC ef	**-0,312**	-0,095	-0,143	0,021	-0,019
CTC pH7	**-0,324**	-0,03	-0,086	-0,022	0,022
V%	-0,099	**-0,303**	**0,300**	0,243	-0,063
m%	0,111	-0,176	**0,406**	**-0,326**	-0,064
Cot	-0,164	**0,303**	0,231	0,030	-0,199
EC	-0,126	**0,308**	0,048	**-0,421**	0,002
NT	-0,147	0,153	**0,433**	0,008	-0,161
C/N	-0,132	**0,358**	-0,086	0,036	-0,123
CBM	0,158	**0,304**	-0,018	**0,484**	-0,035
NBM	-0,065	-0,011	0,064	0,02	**0,908**
qMIC	**0,279**	-0,078	-0,134	**0,366**	0,031

Table 5. Correlation coefficients between original variables of response and principal components.

The principal component 1 (Comp. 1), that explains 54.00% of the variability of the data, presents the highest correlation for the SB, K, Mg, CEC, Ca and H + Al. The largest negative correlations, which also showed low values occurred with the CBM and qMIC. The principal component 2 (Comp. 2) explains 21.9% of the variability and has the highest correlations for P and COT. The largest negative correlation occurred with the V%.

The principal component 3 (Comp. 3), that explains 12.2% of the variability, presents the higher correlations for P and Ca, and the largest negative correlations were found for pH and COT. In the case of principal component 4 (Comp. 4) explains 5.2% of the variability and was negatively correlated for almost all attributes except for COT. Among the negative correlations stand out CBM qMIC and Ca.

3.4 Edaphic fauna of the forestry system and agroforestry

The faunistic analysis performed on the Farm Tramontina Belém S/A, recorded 2.568 specimens distributed in eighteen (18) taxon of invertebrate and one (01) taxon of vertebrate of the Order Squamata (Gekkonidae) (Table 6) in five culture systems studied, where the prevailing order was Hymenoptera, composed mainly by family Formicidae with 2.151 individuals followed by Coleoptera (78), Collembola (40), Homoptera (67) and Diplopoda (74).

	SYSTEMS						
	S1	S2	S3	S4			
TAXONS	Fi				Total	Fr (%)	(Mean)
INSECTA							
Hymenoptera (Formicidae)	808	667	416	260	2.151	82,82	537,75
Coleoptera	21	15	27	15	78	3.0	19,5
Collembola	6,0	11	21	2,0	40	1.54	9.75
Diptera	13	1,0	11	5,0	30	1.15	7,50
Hemiptera	0,0	1,0	0,0	0,0	01	0.03	0,25
Homoptera	9,0	10	34	14	67	2.57	16,75
Blattariae	0,0	2,0	0,0	2,0	04	0.15	1,00
Odonata	0,0	0,0	1,0	0,0	01	0.03	0,25
Orthoptera	2,0	3,0	5,0	12	22	0.84	5,50
Psocoptera	1,0	0,0	3,0	3,0	07	0.26	1,75
Lepdoptera	0,0	28	0,0	0,0	28	1.07	7,00
Thysanoptera	0,0	0,0	1,0	0,0	01	0.03	0,25
ARACHNIDA							
Acari	8,0	28	14	5,0	55	2.11	13,75
Aranae	3,0	2,0	4,0	6,0	15	0.57	3,75
Opilionida	4,0	3,0	7,0	0,0	14	0.53	3,50
CRUSTACEA							
Isopoda	0,0	1,0	0,0	0,0	01	0.03	0,25
MYRIAPODA							
Chilopoda	1,0	0,0	1,0	0,0	02	0.07	0,50
Diplopoda	17	12	15	30	74	2.84	18,5
SQUAMATA							
Gekkonidae	1,0	3,0	3,0	1,0	08	0.30	1,25
Total	894	785	563	355	2.597	100	

Ordination of taxons second Brusca & Brusca, 2007.

Table 6. Taxons identified, absolute frequency (Fi), relative frequency (Fr) and average of individuals in the cropping systems studied in the Farm Tramontina Belém S/A in Aurora do Pará.

The Diplopoda taxon is present in all treatments, but the highest concentration is in the systems of monocultivation (S1 and S4). In a study of community of invertebrates in litter in agroforestry systems, this order was the second most important (Barros et al., 2006). This importance is due to mobility that they present in the soil, surface and underground, which influences the physical nature of the soil changing porosity, moisture and transport of substances (Correia; Aquino, 2005).

The order Acari is a grouping of vertical habitat in three levels, euedaphics, hemiedaphics and epiedaphics, and the epiedaphics are more tolerant to desiccation (Lavelle & Spain, 2001), although there is low frequency of individuals, it is higher when compared with those obtained in studies in savannas of Pará (Franklin et al., 2007).

Ants have been widely used as biodindicadores in various types of impacts, such as recovery after mining activities, industrial pollution, agricultural practices and other land uses (Smith et al, 2009). In addition, the class Insecta, which belongs to the ant, often grouped according to trophic groups, and the availability of nutrients in the ecosystem (Leivas & Chips, 2008). They are important in below-ground processes, by altering the physical and chemical properties and the environment, its effects on plants, microorganisms and other soil organisms (Folgarait, 1998).

These may be the possible explanations for the faunistic results of this taxons (Hymenopetera) Family Formicidae, with the highest absolute frequency (Fi) in the cropping system S1 (curauá in monocultivation) and lower absolute frequency in the cropping system S4 (paricá in monocultivation) as well as in other culture systems studied (Table 7).

Adult and immature Coleoptera, Collembola, Diplopoda, Diptera and Homoptera adults had higher absolute frequency in the cropping system S3 (paricá + mogno + freijó + curauá) (Table 7). This may show a variety nutritional or of habitat, or still an increase in the prey - predator relation. The same was not found in the monocultivation systems for these taxons.

Macroarthropods of soil has an important role in tropical terrestrial ecosystems, exerting a direct influence on the formation and stability, an indirect influence on the decomposition process through strong participation in the fragmentation of necromass (ARAUJO; BANDEIRA; VANSCONCELOS, 2010). Work already carried out in terra firma forest ecosystems in the state of Pará, also found the presence of Hymenoptera, Coleoptera, Collembola, Homoptera, Acari and Diplopoda, grouped or not (Macambira, 2005; & Garden Macambira, 2007; Ruivo et al. 2007), which corroborates with the edaphic fauna found in Aurora do Pará

The greatest diversity of species occurred at S3 (Table 7), this can be attributed, among other factors, the variety of nutrients, because S3 is an agroforestry system where there is occurrence of paricá, mogno, freijó and curauá. In addition, there may be no natural predators of these species or still the ephemeral life cycle leads to a reproduction in greater numbers, but these factors were not analyzed in this study.

The lowest levels of the Shannon-Wiener diversity, therefore the greater species diversity occurred in the cropping system S1 (curauá in monocultivation) and S2 (agroforestry system paricá + curauá). This may be due to movement of other species of invertebrates through of ecotones that over there exist. It is possible that this displacement has occurred in search of food or place to reproduction, or even to escape from possible predators in areas near to these cultivation systems. It should not be ruled out the hypothesis that the curauá produces some substance that is palatable to invertebrate species diversity and this is a nutritional option for them because this plant is present in both cropping systems that had the lowest diversity indices of Shannon -Wiener.

	Systems			
	S1	S2	S3	S4
Diversity index (H')	0,22	0,26	0,51	0,48
Population density (individuals/m²)	0,69	0,58	0,43	0,26

Table 7. Diversity index Shannon-Wiener (H') and population density found in cropping systems studied at Fazenda Tramontina Belém S/A in Aurora do Pará.

The highest indexes of diversity of Shannon-Wiener occurred in cropping systems S3 (parica + magno + freijó + curauá) and S4 (parica in monocultivation). This shows lower diversity of species, although in S3 there is differents plant species, which would cause nutritionals offerings and different habitats, and this would promote greater diversity of species. Study performed about plant diversity and productivity of plants and the effects on the abundance of arthropods (Perna et al, 2005) showed that productivity in the plant structure, local abiotic conditions, physical disorders of the habitats are factors that interact with the diversity of plants and which also influences the abundance of arthropods. This can be an explanation for the low diversity found in the cropping system S3 in our study.

In S4, occurs monocultivation with paricá. This features a lower nutritional diversity and of habitats, which in turn attracts lower diversity of macrofauna, especially ants. Suggest the hypothesis that this plant produces some substance not palatable or that act on the reproductive cycle in the majority of taxons identified. It is also possible that there are a greater number of predators in relation to other systems cultivation studied, or even occurred edaphic variations that did not allow the diversity of species and greater population density because this cropping system, the highest number of absence (seven taxons) among the eighteen taxons identified (Table 6). When the population density, the highest concentration indivíduos/m², occurs in S1, where there is a monocultivation of curauá. As this culture system does not present sub-forest for vegetal cover and shading that would mitigate the temperature at the soil surface, it is possible that the solar radiation incident directly on the ground promotes an increase in photosynthetic rate, increasing the supply of nutrients, as well as raises the temperature of the same and is one of the factors that contribute to the reproduction of these ants (Harada & Banner, 1994, Oliveira et al., 2009). But do not dismiss as a possible explanation for the results obtained in two cropping systems, the hypotesis of correlated interferences, for example, environmental variables, ephemeral life cycle, presence or absence of predators, temperature, precipitation, brightness, among other variables in this study were not analyzed.

3.5 Edaphic fauna and attributes of soil

The relation between edaphic fauna and soil attributes (pH, Corg, Ds, relation silt/clay) is shown in Figure 6.

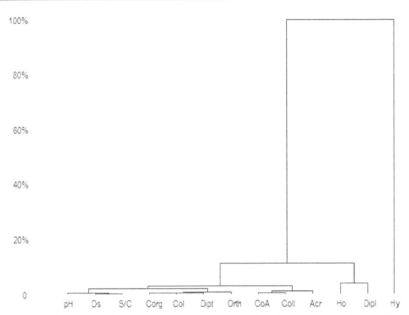

Fig. 6. Grouping between edaphic fauna of greater absolute frequency in the studied treatments, physical and chemical attributes of soil. Campo Experimental of the Tramontina Belém S/A. Corg = carbon organic; S/A = relation silt/clay, Ho = Homoptera; CoA = Coleoptera adult; dipl = Diplopoda; CoI = immature Coleoptera, Coll = Collembola; dipti = Diptera; Orth = Orthoptera; Acr = Acari; Hy = Hymenoptera.

The results showed that (I) the immature Coleoptera taxon has high similarity to the taxon Collembola and constitutes the group 1; (II) the taxon Diptera presents similarity to the taxon Orthoptera and constitute the second group. The group 2 has similarity with group 1, (III) the taxon Acari presents similarities with groups 1 and 2, constituting the third group, (IV) the taxon Coleoptera adult presents similarity to the taxon Diplopoda, constituting the group 4. (V) Groups 1, 2 and 3, have similarity to group 4. (VI) the taxon Homoptera has similarity as group 4, thus constituting the group 5. (VII) Groups 1, 2, 3, 4 and 5 they correlate with the physical attributes (Ds, relation silt/clay), but did not correlate with the chemical attributes (pH, Corg).

The taxon Hymenoptera showed that is not correlated with the physical attributes: Ds and relation silt/clay, and neither with chemical attributes: pH and Corg. Also has no similarity to other taxons analyzed. But it is possible that other edaphic variables correlate with this taxon, but these were not objects of the present study.

Ants have been widely used as biodindicadores in various types of impacts, such as recovery after mining activities, industrial pollution, agricultural practices and other land uses (Smith et al, 2009). In addition, the class Insecta, which belongs to the ant, often grouped according to trophic groups, and the availability of nutrients in the ecosystem (Leivas & Chips, 2008). They are important in below-ground processes, by altering the physical and chemical properties and the environment, its effects on plants, microorganisms and other soil organisms (Folgarait, 1998).

These may be the possible explanations for the faunistic results of this taxons (Hymenopetera) Family Formicidae, with the highest absolute frequency (Fi) in the cropping system S1 (curauá in monocultivation) and lower absolute frequency in the cropping system S4 (paricá in monocultivation).

The order Acari has the major absolute frequency in S2 (Table 3), maybe has better adaptability to the SAF, and the physical attributes shown in Figure 5, which may also is occurring with Homoptera, Collembola and Coleoptera immature. Even in this agroforestry system, the absolute frequency of the taxon Coleoptera, Hymenoptera and the population density decreases, according to data contained in Table 4. This may be related to the increase of soil density and decrease in the silt/clay relation contained in Table 2.

With the modifications imposed by use of soil, particularly by agriculture, the fauna and the microorganisms, in different degrees of intensity, are affected by the impacts caused by agricultural practices, that may alter the composition and diversity of soil organisms.

The addition of new organic matter through the incorporation of waste or the maintenance of forest cover (using the agroforestry system, with or without burning the area) or even a diversified system (as occur in Aurora do Pará), show the importance of maintenance, incorporation and slow decomposition of organic matter on the ground.

So far, the studies show that the types of management adopted did not influence negatively the characteristics of soil and that adding of diversified organic matter in the soil, the retention and incorpration and slow decomposition of these residues led to the creation of an edaph-environment favorable to the maintaining soil quality.

The set of attributes of the soils here studied, especially those related to microbial biomass and chemistry, was adequate to indicate the quality of the substrate. However, the continuation of this kind of work, in the long term, it is necessary, in order to identify differences in biological characteristics of soil between the different management systems, especially taking into account the local climatic variation. Thus, it is necessary to intensify studies of the seasonal variation of soil attributes, the variables listed as indicators of soil quality and intensify the studies in determinated practices of management, such as the tillage and the SAF as potential in the carbon sequestration.

4. Conclusions

The result show that the types of management adopted did not influence negatively the characteristics of soil and adding organic matter to the diverse soil, retention and development and slow decomposition of these residues led to the creation of an edafoambiente of maintaining soil quality. The set of attributes of the soil studied here, especially those related to microbial biomass and chemistry, was adequate to indicate the quality of the substrate. However, the continuation of such work in the long term and different climatic conditions, it is necessary, in order to identify differences in biological characteristics under different soil management systems, especially taking into account the local climatic variation.

The Paricá (*Schizolobium amazonicum* Huber (Ducke)) is a viable native species for recuperation of disturbed areas and with detach in the wood market, nationally and internationally. Its rapid growth and adaptation to areas with low nutrient levels allow it to be optimum in agroforestry, being the second plant species used in reforestation in the Para state , and primarily designed for industry of lams rolled. Thus, it is necessary to intensify studies of the seasonal variation of soil attributes, the variables listed as indicators of soil quality and enhance studies in certain management practices such as tillage and the SAF as a potential in carbon sequestration.

5. Acknowledgements

To INCT/ CNPQ-MPEG and Projetc FAPESPA/MPEG No. 126/2008 by financial support for this project.

6. References

Anderson, J. P. E. & K. H. Domsch. The metabolic quocient (qCO2) as specific activity parameter to assess the effects of environmental conditions, such as pH, on the microbial biomass of forest soils. Soil Biology and Biochemistry, v. 25, n. 3, p. 393-395, 1993.

Aquino, A.M; Aguiar-Menezes, E.L; DE Queiroz, J.M. Recomendações para coleta de artrópodes terrestres por armadilhas de queda (pitffal-traps). Seropédica: Empresa Brasileira de Pesquisa Agropecuária. 18:1-8, 2006.

Barreta, D; Ferreira, C.S; Souza, J.P; Cardoso, E.J.B.N. Colêmbolos (hexapoda: Collembola) como bioindicadores da qualidade do solo em áreas com *Araucaria angustifólia*. R.Bras.Ci. Solo, 32: 2633-2699, 2008. IBGE, Instituto Brasileiro de Geografia e Estatística, 2011 (www.ibge.gov.br)

Barros, E; Mathieu, J; Tapia-Coral, S; Nascimento, A.R. L; Lavelle, P. Soil macrofauna communities in Brazilian Amazonia. In: Moreira, F.M.S; Siqueira, J.O; Bursaard, L. Soil biodiversity in Amazonian and other Brazilian ecosystems. CABI Publishing, 2006. p.43 – 55.

Borror, D.J; Delong, D.M. Introdução ao estudo dos insetos. Rio de Janeiro, Indústrias Gráficas, 1969, 653 p.

Cordeiro, I M C. C. Performance diferencial de crescimento da espécie *Schizolobium amazonicum* Huber (Paricá) em sítios degradados sob diferentes regimes de preparação de área na micro região do Guamá. Aurora do Pará. Apresentado originalmente como monografia de especialização. Universidade Federal do Para. NAEA/UFPA, 1999. 50 p

Cordeiro, I.M.C. C; De Barros, P.L. C; Ferreira, G.C e Filho, A.D.G. Influência da radiação e precipitação no incremento diamétrico de Paricá de diferentes idades e sistemas de cultivo. Disponível em: http://www22.sede.embrapa.br/snt/viicbsaf/cdanais/tema01/01tema39.pdf>. Acesso em 26 out. 2009.

Cordeiro, I.M.C. C; Lameira, O.A; Barros, P.LC; Malheiros, M.A.M. Comportamento do Curauá sob diferentes níveis de radiação fotossinteticamente ativa em condições de cultivo. R. Bras. Ci. Agrárias. 5:49 – 53, 2010

Cordeiro, I.M.C. C; Santana, C; Lameira, O.A; Silva, L.M. Análise econômica dos sistemas de cultivo com *Schizolobium parahyba var. amazonicum* (Huber ex Ducke) Barneby

(Paricá) E *Ananas comosus* var. *erectifolius* (L. B. Smith) Coppus & Leal (Curauá) no município de Aurora do Pará (PA), Brasil. Ver. Fac. Agron. (LUZ), 26: 243-265, 2009.

Cordeiro, I.M.C.C. Comportamento de *Schizolobium var. amazonicum* (Huber ex Ducke) Barneby e *Ananas comosus* var *erectifolus* (L.B.Smith) Copperns & Leal sob diferentes sistemas de cultivo no município de Aurora do Pará (PA). Originalmente apresentada como Tese (Doutorado). Universidade Federal Rural da Amazônia - PA. 2007, 109 p.

Correia, M.R. F; DE Aquino, A.D. Os diplópodes e suas associações com microorganismos na ciclagem de nutrientes. Seropédia – RJ: EMBRAPA, 199:1- 41, 2005.

De-Polli, H.; Guerra, J.G.M. C, N e P na Biomassa microbiana do solo. In: Santos, G. A.; Camargo, F. A. O. Fundamentos da matéria orgânica do solo: ecossistemas tropicais e subtropicais. Porto Alegre – RS. 1999. p. 389-411.

Embrapa (Empresa Brasileira de Pesquisa Agropecuária). Manual de Métodos de Análise de solo. 2. ed. Rio de Janeiro- RJ, Embrapa, 1997. 212 p.

Feigl, B.J.; Sparling, G.P.; Ross, D.J.; Cerri, C.C. Soil microbial biomass in Amazon soils: evaluation of methods and estimates of pool sizes. Soil Biol. Bioch., v. 27, n. 11, p.1467-1472, 1995.

Folgarait, P.J. Ant biodiversity and its relationship to ecosystem functioning: a review. Biodiversity and Conservation, 7: 1221- 1224, 1998.

Franklin, E; Santos, E.M. R; Albuquerque, M.I.C. Edaphic and arboricolous oribatid mites (Acari; Oribatida) in tropical environments: changes in the distribution of higher level taxonomic groups in the communities of species. Braz. J. Biol., 67: 447 – 458, 2007.

Global Change. 2010. Global Deforestation
http:://www.globalchange.umich.edu/globalchange2/current/lectures/deforest/deforest.html

Harada. Y; Bandeira. A.G. Estratificação e diversidade de invertebrados em solo arenoso sob floresta primária e plantios arbóreos na Amazônia Central durante a estação da seca. Acta Amazônica. 24: 103-118, 1994.

Jenkinson, D.S. & Ladd, J.N. Microbial biomass in soil: Measurements and turnover. In: Paul, E.A. & Ladd, J. N., eds. Soil biochem. 5.ed. New York, Marcel Dekker, 1981. p.415-471.

Jenkinson, D.S.; Powlson, D. S. Residual effects of soil fumigation on soil respiration and mineralization. Soil Biology & Biochemistry, v.19, p.703-707, 1976.

Lavelle, P.; Spain, A.V. Soil Ecology. *Kluwer Academic Publishers*, 2001. 654 p.

Lavelle, P; Blanchart, E; Martin, A; Spain, A.V; Martins. Impact of soil fauna on the properties of soil in the Humid Tropics. In LAL, R; Sanches, P.A. Myths and Science of Soils of the Tropics. Madison – Viscont – USA: Soil Science Society of America and American Society of Agronomy, 1992, p157 – 185.

Leivas, F.W. T; Fischer, M.L. Avaliação da composição de invertebrados terrestres em uma área rural localizada no município de Campina Grande do Sul, Paraná, Brasil. Biotemas. 21: 65 – 73, 2008.

Lemos, R.C; Santos, R.D. Manual de descrição e coleta de solo no campo. 4 ed. Viçosa – MG: UFV. 2002.

Marques, J.D.O; Luizão, F.J; Luizão, R.C.C; Souza Neto, A.S. Variação do carbono orgânico em relação aos atributos físicos e químicos do solo ao longo de uma topossequência na Amazônia central. In:Congresso de Ecologia do Brasil. 8. Caxambu. Sociedade de Ecologia do Brasil. 2007.

Moita, G.C; Moita Neto, J.M. Uma Introdução à análise exploratória de dados multivariados. Quim. Nova, 21: 467-469, 1998.

Monteiro, K.F.G. Utilização de resíduos de madeira como cobertura no solo: o estudo de caso de um sistema Agroflorestal no Estado do Pará. Originalmente apresentada como dissertação de mestrado. Universidade Federal Rural da Amazônia. 2004. 102 p: Il.

Moreira A; Costa, D.G. Dinâmica da matéria orgânica na recuperação de clareiras da floresta amazônica. Brasília – DF. Pesq. Agrop. Brasileira, 39: 1013-1019, 2004.

Oliveira, E.A; Calheiros, F.N; Carrasco, D.S; Zardo, C.M.L. Família de Hymenoptera (Insecta) como ferramenta avaliadora da conservação de restingas no extremo sul do Brasil. *EntomoBrasilis*, 2: 64-69, 2009.

Rosa, L. S. Ecologia e silvicultura do Paricá (*Schizolobium amazonicum* Huber ex Ducke), na Amazônia Brasileira. Belém - PA: R. Ci. Agrárias, 45: 135 – 174, 2006

Ruivo, M L P ; Barros, N.F. ; Schaefer, C. E. R. Relações Da Biomassa Microbiana do Solo com Características Químicas de Frações Orgânicas e Minerais do Solo após Exploração Mineral na Amazônia Oriental. Boletim do Museu Paraense Emílio Goeldi. Série Ciências Naturais, v. 2,n2, p. 121-131, 2006.

Ruivo, M L P ; Monteiro, K. F. G. ; Silva, R. M. da; Silveira, I. M.; Quaresma, H. D. A. B. ; SA, L. D. A. ; Prost, Maria Tereza da Costa . Gestão Florestal e Implicações Sócio-Ambientais na Amazônia Oriental (Estado do Pará). Oecologia Brasiliensis, v. 11, p. 481-492, 2007.

Ruivo, M.L. P; Barreiros, J.A. P; Da Silva, R.M; Sá, L.D. A; Lopes. E.L.N. Lba – Esecaflor artificially induced drought in Caxiuanã Reserve, Eastern Amazonia: Soil proprieties and litter spider fauna. Earth Interactions. 11: 1- 13, 2007.

Ruivo, M.L. P; Salazar, E.R. C; Monteiro, K. F.P; Cordeiro, M. I. C; Oliveira, M.L.S. Avaliação do crescimento de Paricá (*Schizolobium parayba var, amazonicum)* em diferentes sistemas agroflorestais no nordeste paraense. 2006.

Santana, M.B; Souza, L.S; Souza, L.D; Fontes, L.E.F. Atributos físicos do solo e distribuição do sistema radicular de citros como indicadores de horizontes coesos em dois solos de tabuleiros costeiros do estado da Bahia. R. Bras. Ci. Solo, 30: 1 – 12, 2006.

Silva Júnior, M.L; Desjardins, T; Sarrazin, M; De Melo, V.S; Da Ilva Martins, F; Santos, E.R; De Carvalho, C.J.R. Carbon content in Amazonian Oxisols after Forest conversion to pasture. R. Bras. Ci. Solo, v. 33: 1602 – 1622, 2009.

Sousa, R. F. de; Barbosa, M. P.; Terceiro Neto, C. P. C. ; Morais Neto, J. M. de; Sousa Junior, S. P. de. Estudo da Degradação das Terras do Município de Boa Vista Paraíba. Engenharia ambiental – Espírito Santo do Pinhal. v.4, n.2, jul/dez 2007. p. 005- 013.

Tate, K.R.; Ross, D.J. & Feltham, C.W. A direct extraction method to estimate soil microbial C: Effects of experimental variables and some different calibration procedures. Soil Biol. Biochem., v. 20, p. 329-335, 1988.

Telles, E.C.C. Dinâmica do carbono no solo influenciado pela textura, drenagem, mineralogia e carvões em florestas primárias na região centro-oriental da Amazônia. Piracicaba. 2002. 92 p. Tese (Doutorado) - Centro de Energia Nuclear na Agricultura, Universidade de São Paulo.

Thornthwaite, C.W. An approach towards a rational classification of climate. Geographical Review, London, v.38,p.55-94, 1948.

Vance, E.D.; Brookes, P.C.; Jenkinson, D.S. An extraction method for measuring soil microbial biomass C. Soil Biol. Biochem., v.19, p.703-707, 1987.

Plant Productivity is Temporarily Enhanced by Soil Fauna Depending on the Life Stage and Abundance of Animals

Ayu Toyota[1,2,*] and Nobuhiro Kaneko[2]

[1]Institute of Soil Biology, Biology Centre,
Academy of Sciences of Czech Republic, České Budějovice
[2]Soil Ecology Research Group,
Yokohama National University, Yokohama
[1]Czech Republic
[2]Japan

1. Introduction

In terrestrial ecosystems, nutrient recycling is driven by the belowground decomposition process because it supplies most of the production to the soil system (Swift *et al.* 1979; Cebrian 1999). Soil links the aboveground plants and the belowground community (Wardle, 2002). Soil nutrient cycling is controlled by interactions among living soil organisms, plant litter quality, soil physical and chemical status, temperature, and water condition. In soil-organism processes, large soil animals rapidly change microbial activity (Hanlon and Anderson 1980) and the litter decomposition rate (Bonkowski *et al.* 1998) and modify soil structure (Barois *et al.* 1993). Consequently, primary decomposition and plant growth are affected not only by microbes but also by soil animals (Wall and Moore 1999; Wardle 2002). The overall faunal contribution to nitrogen (N) mobilization has been estimated as approximately 30% in forest ecosystems (Verhoef and Brussaard 1990). Schröter *et al.* (2003) calculated that the total amount of N mineralized by fauna in European coniferous forests ranged from 11 kg N ha^{-1} a^{-1} in northern Sweden to 73 kg N ha^{-1} a^{-1} in Germany.

The presence of keystone species strongly affects decomposition and nutrient dynamics. In particular, the train millipede *Parafontaria laminata* is widely dominant in soil invertebrate communities of central Japan, with its late-stage (from the 6th to final instar) larvae enhancing N availability by 69% (Toyota *et al.* in press). The enhancement of soil N availability in forest soils by late-stage larvae of this millipede can lead to changes in aboveground plant productivity. Earthworm casts stimulate the growth of most plant species in grassland soil (Zaller and Arnone 1999), but whether soil animals alter plant productivity under natural conditions in forest soils remains poorly understood.

Previous studies on the effects of soil fauna on plant production have mainly been laboratory microcosm experiments, and they have produced mixed conclusions (Brown *et al.* 1999; Scheu 2003). Some studies found that soil animals increase plant production and nutrient utilization (*e.g.*, Setälä and Huhta 1991; Bardgett and Chan 1999), while others reported redundant or negative impacts of soil animals. One possible reason for this discrepancy is that soil animals have different effects on soil nutrient dynamics depend on their feeding type (Lavelle *et al.* 1997). Another possible reason is that differences in soil chemical composition and organic matter input to soil lead to changes in soil animal effects (Tiunov and Scheu 2004).

The rates of nutrient cycling are usually not constant, even in a mature forest ecosystem, due to the effects of processes such as seed masting (Selås *et al.* 2002) and periodic insect outbreaks. The 17-year periodic appearance of cicadas in North America provides a nutrient pulse to the forest soil resulting from the mass of carcasses, in turn increasing foliage N content and seed mass (Yang 2004). Despite these temporal differences in nutrient dynamics, many studies assume links between plant growth and decomposition processes on a forest floor in the short-term.

Populations of *P. laminata* have a synchronized life cycle and undergo simultaneous molts at particular stages according to seasonal temperature rhythms (Fujiyama 1996). The larvae live in the soil for 7 years (Fig. 1a, b) before they reach the adult stage (Fig. 1c). Their population consists of only a single age cohort, so adult swarming is observed at 8-year intervals (Niijima and Shinohara 1988). The huge abundance and synchronized cohort of *P. laminata* provide a good opportunity to investigate the soil-animal contributions to nutrient dynamics at the forest floor (Hashimoto *et al.* 2004; Toyota *et al.* 2006). Due to their heterogeneous abundance within the same region in forest soil, the effects of *P. laminata* on plants can be distinguished, even under field conditions.

The train millipede *P. laminata* can enhance plant production in two major ways. First, changes in soil N availability through millipede activity affect plant production. Late-stage larvae enhance N mineralization in soil when millipede abundance is high (Toyota *et al.* in press). By contrast, adults contribute to transforming plant litter into soil, but do not increase N availability in the adult phase. Due to differences in soil N dynamics with developmental stage, the larvae and adults may affect plant productivity differently. Second, carcasses of adults contain higher concentrations of available phosphorus (P) for plant (Toyota, unpublished data). Since forest production is usually limited by P availability in soil (*e.g.*, Crews *et al.* 1995; Vitousek and Farrington 1997), the P supplied from their carcasses to the forest floor after the death of adults will result in increased plant production and changes in leaf quality.

Here, we studied how aboveground production can be affected by soil animal abundance and developmental stage. Effects of the train millipede were estimated to compare production between natural low-abundance and high-abundance millipede plots in the field. To test whether they temporally have different effects from late stage larvae to after the death of adults, the periods examined for estimating plant productivity were the 2 years before and after the adult active year, over a total of 5 years. Annual herbaceous plants should detect such temporal effects easily. In this paper, we tested the following two predictions: late stage larvae enhance leaf N content and production in the herb layer, and after adult death, the millipede decomposition increases herbaceous annual plant production and leaf P content. Specifically, we examined the effects of the millipedes on the productivity of annual dwarf bamboo, which dominates the undergrowth of temperate forests in Japan.

(a)

(b)

(c)

Fig. 1. The train millipede, *P. laminata*, in the soil as (a) a 6th-instar larva, (b) 7th-instar larvae, and (c) on the forest floor as an adult.

2. Materials and methods

2.1 Study site

We established research plots at four different field sites in larch forests of Mt. Yatsugatake in Yamanashi Prefecture, central Japan (1350–1400 m above sea level, 35°54'12–46"N, 138°20'32"–24'03"E). The plots comparing low- (reference) and high-abundance areas of the train millipede were carefully chosen to avoid differences in soils, vegetation, and topography among plots. This region is characterized by a cool–temperate climate with mean annual precipitation of 1100 mm and mean air temperature of 10.6°C (Japan Weather Association 1998). The vegetation is a plantation forest of Japanese larch [*Larix kaempferi* (Lamb.) Sargent]. The sizes and ages of the larch in each plot are shown in Table 1. The shrub layer was sparse and was dominated by *Quercus crispula* Blume, *Prunus incisa* Thumb., *Ligustrum tschonoskii* Decaisne, *Symplocos coreana* Ohwi, and *Rhododendron obtusum* Planch. Dwarf bamboo (*Sasa nipponica* Makino) dominates the herb layer with *Osmunda japonica* at 50–70 cm height.

| Plots | Latitude and longitude | Altitude (m a.s.l.) | Aspect and slope | Japanese larch | | | |
				Planted in	Canopy density (ind./ha)	Average height (m)	Average DBH (cm)
L1	35° 54' N, 138° 20' E	1350	S16E, 13°	1952	950	17.5	20.4
L2	35° 54' N, 138° 21' E	1360	S42W, 18°	1951	1100	19.4	21.5
H1	35° 54' N, 138° 23' E	1400	S52E, 20°	1961	800	14.6	19.4
H2	35° 54' N, 138° 24' E	1390	N72E, 24°	1961	750	15.6	21.8

Table 1. Location of the study plots and abundance of the larch trees.

The organic layer on the forest floor is mainly composed of larch foliage mixed with larch twigs and bamboo stalks and leaves. The soil type is a well-developed aggregate structure of Andosols (FAO *et al.* 1998). About 50% of the total soil C at 0–100-cm depth was stored in the surface 10 cm, and about 75% was in the upper 0–30 cm (Morisada *et al.* 2002). Volcanic material was deposited after eruptions of Mt. Yatsugatake from 1,300,000 to 10,000 years before present. After that, forests had expanded. In the region around the study plots, natural larch forests remain, and natural larch regeneration at the arly stage follows wildfires. The vegetation was fired artificially in around 1900 (Suka, 2008). In another disturbance, an extraordinarily strong typhoon passed through this region in 1959.

In October 1998, the population of *P. laminata* at this site consisted of only 6th instar larvae (Fig. 1a), which subsequently molted and became 7th instar larvae (Fig. 1b) in August 1999. Adults emerged in late August 2000, and swarming on the forest floor was observed from September (Fig. 1c); the adults died after egg deposition in July 2001 in the foothills of Mt. Yatsugatake (Toyota personal observations). The density of 6th instars ranged from 160 to 1088 m^{-2} (mean ± SD, 485 ± 302 m^{-2}; n = 3) in October 1998 (Toyota personal observation) and from 106 to 2156 m^{-2} (619 ± 208 m^{-2}; n = 10) in June 1999, in the high-abundance area. The density of 7th instar larvae in the high-abundance area ranged from 144 to 720 m^{-2} (469 ± 240 m^{-2}; n = 6) in April 2000 (Toyota *et al.* 2006). The average dry biomass of 7th instar larvae in the high-abundance areas (plots H1 and H2) was 15.7 g m^{-2} and 12.2 g m^{-2}, respectively, in October 1999, and the average dry biomass of adults reached 28.6 g m^{-2} and 15.1 g m^{-2}, respectively, in October 2000 (Hashimoto *et al.* 2004). By contrast, the average dry biomass of

7th instar larvae in the low-abundance areas (plots L1 and L2) was 3.8 g m^{-2} and 1.0 g m^{-2}, respectively, in October 1999, and average dry biomass of adults was 1.0 g m^{-2} and 1.6 g m^{-2}, respectively, in October 2000 (Hashimoto et al., 2004).

2.2 Measurements

Four quadrats (1 × 1 m) were harvested in each plot to estimate the aboveground dwarf bamboo biomass in the period of maximal development of the dwarf bamboo vegetation (from late August to early September 1999, 2000, 2001, 2002, and 2003). The aboveground herb samples were dried at 40°C for 4 days and weighed. After removing the bamboo stalks, samples were homogenized. The 20-g subsamples of dwarf bamboo leaves were ground in a blender and used for N and P analysis. The N content was analyzed using a gas chromatograph (Sumigraph NC-95A; Shimadzu, Kyoto, Japan). The P content was analyzed using a Futura autoanalyzer (Actack, Alliance Instruments, Frépillon, France) for continuous flow analysis.

2.3 Statistical analysis

To test for differences in aboveground biomass, leaf N and P content through the sampling years, multiple comparisons among sampling years within a treatment were performed using the Tukey–Kramer test. Linear regression was used for correlation analysis between millipede biomass and leaf N and P content. All statistical analyses were performed using R 2.8.0 (R Development Core Team 2008).

3. Results

3.1 Temporal changes in aboveground production and leaf quality

The aboveground biomass in high-density train millipede plots varied significantly among years (H1: $P = 0.037$; H2: $P = 0.001$; Fig. 2). No significant differences among years were observed in low-density millipede plots (L1: $P = 0.187$; L2: $P = 0.09$). The observed pattern of the aboveground biomass in high-density plots tended to be high in 1999 (6th instar larvae). In plot H1, aboveground biomass in 1999 (6th instar larvae) was significantly higher than that in 2003 (2 years after adult death). In plot H2, the aboveground biomass in 1999 (6th instar larvae) and 2002 (1 year after adult death) was significantly higher than in 2000 (7th instar larvae) and 2001 (adult) (Tukey–Kramer, $P < 0.05$).

The N content in leaves varied significantly among years in all plots. Temporal dynamics of the N content differed between high- and low-density plots. Similar to the aboveground biomass in high-density plots, the N content tended to be high in 1999 (6th instar larvae). By contrast, in low-density plots, the N content tended to be low in 1999 (6th instar larvae).

The P content in high-density plots varied significantly among years ($P < 0.001$), but not in the low-density plots (L1, $P = 0.45$; L2, $P = 0.08$). The P contents in 2000 (7th instar larvae), 2001 (adult) and 2002 (1 year after adult death) were significantly lower than that in 1999 (6th instar larvae) and 2003 (2 years after adult death) in both plots H1 and H2 (Tukey–Kramer, $P < 0.05$).

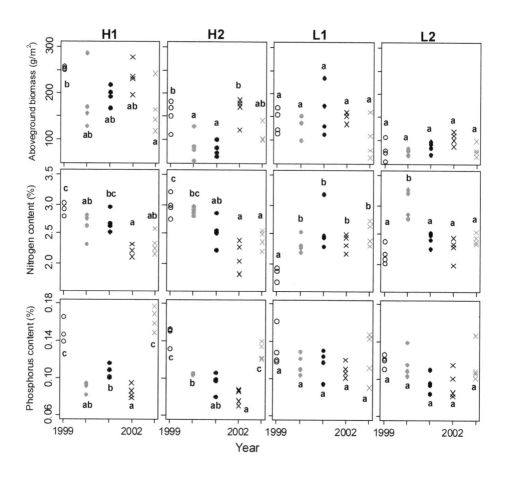

Fig. 2. Temporal variation in the aboveground biomass, nitrogen content, and phosphorus content with developmental stage: open circles, 6th-instar larvae; grey circles, 7th-instar larvae; black circles, adults; and crosses, after the adult period. The same letter indicates no significant difference among years based on the Tukey–Kramer test ($P < 0.05$).

3.2 Effects of the train millipede on leaf quality

The N content of leaves was closely positively correlated with 6th instar larval biomass in 1999 (Fig. 3), but not in other years (Figs. 3, 4). The P content was positively correlated with 6th instar larvae biomass in 1999 (6th instar larvae) and adult biomass in 2003 (2 years after adult death). By contrast, it was negatively correlated with the final instar larvae biomass in 2000 (7th instar larvae) and adult biomass in 2002 (1 year after adult death).

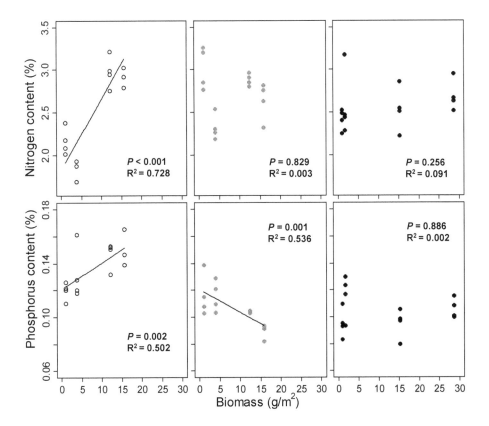

Fig. 3. Developmental stage-dependent effects of the millipede on leaf nitrogen and phosphorus contents: open circles, 6th-instar larvae; grey circles, 7th-instar larvae; and black circles, adults. Solid lines are significant linear regressions.

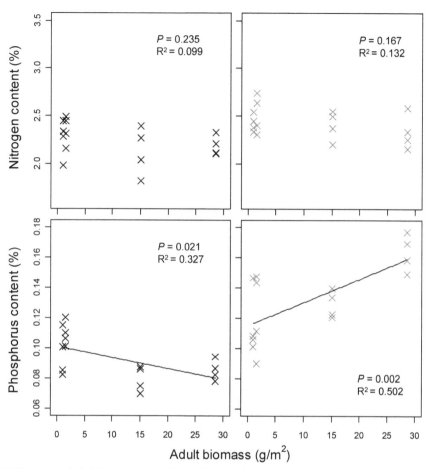

Fig. 4. Effects of adult biomass after adults died on the leaf nitrogen and phosphorus contents. The black crosses indicate 1 year after the adult period. The grey crosses indicate the 2 years after the adult period. Solid lines are significant linear regressions.

4. Discussion

Our results show that temporal changes in dwarf bamboo production and leaf P content occur in plots with high densities of the train millipede, but not in low-density millipede plots. We found that the effects of the millipede on plant leaf N and P content differed with millipede development stage. In agreement with our hypothesis, 6th instar larvae had positive effects on plant production, N and P contents in leaves. In contrast to our hypothesis, the millipede had no effect on dwarf bamboo production and leaf quality during the adult phase, negative effects on the P content for 7th instar larvae and 1 year after the adult phase, and positive effects on the P content in 2 years after the adult phase. The contribution of the millipede to plant growth is therefore twofold. Here, we discuss factors promoting the close relationship between plants and millipedes.

4.1 Temporal differences in production and nitrogen

Observed temporal patterns in production and N content of dwarf bamboo leaves were related to soil N availability in this forest. Increases in soil N availability for plants by 6th instar larval activity (Toyota *et al.* in press) immediately resulted in stimulation of plant biomass and N content in leaves. Under low N availability conditions during the adult phase, plant growth and N content did not increase. These patterns suggest that N limitation may occur in these forests and that millipede larvae can alleviate N limitation in the forest soil system.

4.2 Changes in phosphorus content

Leaf P content increased with the incremental increase in the millipede biomass of 6th instar larvae (Fig. 3). Consequently, the larvae would provide not only available N but also P. However, during the 7th instar, the millipede reduced the leaf P content (Fig. 3). Arbuscular mycorrhizal fungi (AMF), which have a critical role in P uptake by plants, are suppressed by N fertilization (Bradley *et al.* 2006). Similarly, the increase in soil N availability with 6th instar larvae would reduce AMF in the next year, so plants would utilize less P with the incremental increase in millipede biomass. Furthermore, 7th instar larvae may alter soil chemical and physical conditions for the following reasons: (1) excess P consumption by 7th instar larvae might occur to create the adult body, which might result in a reduction of soil P availability, and (2) molting chambers of 7th instar larvae could physically hold P in the soil, lead to a substantial reduction in uptake of P by plants. The train millipedes make molting chambers from their own fecal pellets during summer, for several years the compact structure of the molting chamber of last instar (7th instar) larvae remains in the field at soil depths between 5 and 15 cm (Toyota, unpublished data). Niijima (1984) suggested that the physical structure of soil is greatly altered by 7th instar larvae millipedes due to their large molting chamber (ca. 15 mm in diameter).

After 1 year of the adult phase, the previous adult biomass was negatively correlated with the leaf P content. These decreases in leaf P content can be attributed to the following three reasons. First, P absorption ability in the soil could be high in our forests because of soil engineering by adults (so the P uptake by plants would be low). Second, adult activity might reduce P utilization ability of plants. Adult millipedes were observed in soil to a depth of around 12 cm, near dwarf bamboo roots during the winter (Niijima 1984). This adult activity in soil may damage ectomycorrhizal fungal growth and fine roots of plants due to soil disturbance, which could result in reduced P uptake by plants. Third, there would be fewer adult carcasses on forest floor in high-density millipede plots than adult abundance in the previous year (Hashimoto *et al.* 2004); consequently, phosphate from carcasses would be insufficient to increase the leaf P content.

After 2 years of the adult phase, however, the previous adult biomass was positively correlated with leaf P content. In the previous year, P was probably limited for plants. A lack of P availability leads to the preferential allocation of more resources to root biomass than to aboveground biomass as a plant response (Lynch 1995). Theory predicts a negative correlation between root allocation and aboveground growth rate. Observed low aboveground production in high-density millipede plots after 2 years of the adult phase is consistent with this prediction. Ryan *et al.* (2001) showed that organic acid was supplied from plant roots under low P conditions, and found that roots dissolved iron- or aluminium-bound phosphates by organic acid. Consequently, available P may increase in soil by the activity of plant roots with high-density the millipede. Since plants would allocate carbon to the root system, damaged ectomycorrhizal fungi may recover. Eventually, incremental increases in available P will lead to increases in leaf P content. Further study is required to examine whether P forms and their availability for plants in soil differ according to the developmental stage of train millipedes.

5. Conclusions

By comparing plant production over time between areas with high and low train millipede abundance, we showed that plant growth and leaf quality were differently affected by millipede developmental stage. The plant aboveground biomass and leaf P content varied temporally when millipedes were very abundant but not when their abundance was low. Only 6th instar larvae have a significant role in enhancing both the leaf N and P content with increment in soil mineral N. Last (7th) instar larvae act in the opposite direction; they have a negative effect on the leaf P content. Adults have negative and positive effects after the adult phase probably due to their large impact on rhizosphere and the effects of their carcasses. The roles of the train millipede involve different types of feedback from soil to plant. This different response may lead to changes in competitive ability and in the species composition of the plant community. Although the interactions among plants, microbes, and animals in soil are quite complex, our results indicate that soil animals regulate plant productivity, even in mature forest ecosystems. This insight would be useful for understanding the aboveground–belowground feedback mechanism in the soil community.

6. Acknowledgments

We thank Dr. K. Niijima for valuable advice on the biology of train millipedes and M. Hashimoto and the members of the Soil Ecology Research Group of Yokohama National University for assisting with experimental sampling. We also thank Yamanashi Prefecture for allowing us to use the study site.

7. References

Bardgett, R.D., Chan, K.F., 1999. Experimental evidence that soil fauna enhance nutrient mineralization and plant nutrient uptake in montane grassland ecosystems. Soil Biol. Biochem. 31, 1007–1014.

Barois, I., Villemin, G., Lavelle, P., Toutain, F., 1993. Transformation of the soil structure through *Pontoscolex corethrurus* (Oligochaeta) intestinal tract. Geoderma 56, 57–66.

Bonkowski, M., Scheu, S., Schaefer M., 1998. Interactions of earthworms (*Octolasion lacteum*), millipedes (*Glomeris marginata*) and plants (*Hordelymus europaeus*) in a beechwood on a basalt hill: implications for litter decomposition and soil formation. Applied Soil Ecology 9, 161–166.

Bradley, K., Drijber, R.A., Knops, J., 2006. Increased N availability in grassland soils modifies their microbial communities and decreases the abundance of arbuscular mycorrhizal fungi. Soil Biol. Biochem. 38, 1583–1595.

Brown, G. G., Pashanasi, B., Villenave, C., Patron, J. C., Senapati, B. K., Giri, S., Barois, I., Lavelle, P., Blanchart, E., Blakemore, R. J., Spain, A.V., Boyer, J., 1999. Effects of earthworms on plant production in the tropics. In: Lavelle, P., Brussaard, L., Hendrix, P. (eds) Earthworm management in tropical agroecosystems. CAB International, Wallingford, pp. 87–137.

Cebrian, J., 1999. Patterns in the fate of production in plant communities. American Naturalist 154, 449–468.

Crews, T.E., Kitayama, K., Fownes, J.H., Riley, R.H., Herbert, D.A., Mueller-Dombois, D., Vitousek, P.M., 1995. Changes in soil phosphorous fractions and ecosystem dynamics across a long chronosequences in Hawaii. Ecology 76, 1407–1424.

FAO, ISSS, ISRIC, 1998. World reference bases for soil resources. pp.109. Rome.

Fujiyama, S., 1996. Annual thermoperiod regulating an eight-year life-cycle of a periodical diplopod, *Parafontaria laminata* armigera Verhoeff (Diplopoda). Pedobiologia 40, 541–547.

Hanlon, R.D.G., Anderson, J.M., 1980. Influence of macroarthropod feeding activities on microflora in decomposing oak leaves. Soil Biol. Biochem. 12, 255–261.

Hashimoto, M., Kaneko, N., Ito, M.T., Toyota, A., 2004. Exploitation of litter and soil by the train millipede *Parafontaria laminata* (Diplopoda: Xystodesmidae) in larch plantation forests in Japan. Pedobiologia 48, 71–81.

Japan Weather Association, 1998. Weather yearbook 1998. Ministry of Finance Japan, Tokyo (in Japanese).

Lavelle, P., Bignell, D., Lepage, M., Wolters, V., Roger, P., Ineson, P., Heal, O.W., Dhillion, S., 1997. Soil function in a changing world: the role of invertebrate ecosystem engineers. Eur. J. Soil Biol. 33, 159–193.

Lynch, J., 1995. Root architecture and plant productivity. Plant Physiology 109, 7–13.

Morisada, K., Imaya, A., Ono, K., 2002. Temporal changes in organic carbon of soils developed on volcanic andesitic deposits in Japan. Forest Ecology and Management 171, 113–120.

Niijima, K., 1984. The outbreak of the train millipede. Jpn. J. For. Environ. 26, 25–32 (in Japanese).

Niijima, K., Shinohara, K., 1988. Outbreaks of the *Parafontaria laminata* group (Diplopoda: Xystodesmidae). Jpn. J. Ecol. 38, 257–268 (in Japanese with English summary).

R Development Core Team., 2008. R: A language and environment for statistical computing. R Foundation for Statistical Computing, Vienna, Austria. ISBN 3-900051-07-0, http://www.R-project.org.

Ryan, P.R., Delhaize, E., Jones, D., 2001. Function and mechanism of organic anion exudation from plant roots. Annual Review of Plant Physiology and Plant Molecular Biology 52, 527–560.

Scheu, S., 2003. Effects of earthworms on plant growth: patterns and perspectives. Pedobiologia 47, 846–856.

Schröter, D., Wolters, V., De Ruiter, P.C., 2003. C and N mineralisation in the decomposer food webs of a European forest transect. Oikos 102, 294–308.

Setälä, H., Huhta, V., 1991. Soil fauna increase *Betula Pendula* growth: laboratory experiments with coniferous floor. Ecology 72, 665–671.

Selås, V., Framstad, E., Spidsø, T.K., 2002. Effects of seed masting of bilberry, oak and spruce on sympatric populations of bank vole (*Clethrionomys glareolus*) and wood mouse (*Apodemus sylvaticus*) in southern Norway. J. Zool. Lond. 258, 459–468.

Suka T., 2008. Historical changes of semi-natural grasslands in the central mountainous area of Japan and their implications for conservation of grassland species. Bull. Nagano Environ. Conserv. Res. Inst. 4, 17–31 (in Japanese with English summary).

Swift, M.J., Heal, O.W., Anderson, J.M., 1979. Decomposition in Terrestrial Ecosystems. Blackwell, Oxford.

Tiunov, A.V., Scheu, S., 2004. Carbon availability controls the growth of detritivores (Lumbricidae) and their effect on nitrogen mineralization. Oecologia 138, 83–90.

Toyota, A., Kaneko, N., Ito, M.T., 2006. Soil ecosystem engineering by the train millipede *Parafontaria laminata* in a Japanese larch forest. Soil Biol. Biochem. 38, 1840–1850.

Toyota, A., Kaneko, N., in press. Faunal stage-dependent altering of soil nitrogen availability in a temperate forest. Pedobiologia.

Verhoef, H.A., Brussaard, L., 1990. Decomposition and nitrogen mineralization in natural and agroecosystems: the contribution of soil animals. Biogeochemistry 11, 175–211.

Vitousek, P.M., Farrington, H., 1997. Nutrient limitation and soil development: experimental test of a biogeochemical theory. Biogeochemistry 37, 63–75.

Wall, D.H., Moore, J.C., 1999. Interactions Underground: Soil biodiversity, mutualism, and ecosystem processes. BioScience 49, 109–117.

Wardle, D.A., 2002. Communities and Ecosystems: Linking the Aboveground and Belowground Components. Princeton University Press, Princeton, pp. 7–55.

Yang, L.H., 2004. Periodical Cicadas as Resource Pulses in North American Forests. Science 306, 1565–1567.

Zaller, J.G., Arnone, J.A., 1999. Interactions between plant species and earthworm casts in a calcareous grassland under elevated CO_2. Ecology 80, 873–881.

Carbon Cycling in Teak Plantations in Comparison with Seasonally Dry Tropical Forests in Thailand

Masamichi Takahashi[1], Dokrak Marod[2],
Samreong Panuthai[3] and Keizo Hirai[1]
[1]Forestry and Forest Products Research Institute, Tsukuba
[2]Faculty of Forestry, Kasetsart University, Bangkok
[3]National Parks, Wildlife and Plant Conservation Department, Bangkok
[1]Japan
[2,3]Thailand

1. Introduction

Tropical deforestation has become a significant source of increased atmospheric CO_2 concentration, hence efforts to promote several actions for reducing emissions from deforestation and forest degradation (REDD) in the international society, one important example of which is afforestation in deforested areas (Gibbs et al., 2007). Recently, Pan et al. (2011) estimated that the global average of the gross emission rate of tropical deforestation was 2.9 petagrams of carbon (Pg C y^{-1}) from 1990 to 2007 and that tropical regrowth forests were partially compensated for by a carbon sink of 1.6 Pg C y^{-1} within an area of 557 Mha. In contrast, the carbon sink from intact forests, not substantially affected by direct human activities, was 1.19 Pg C y^{-1} within an area of 1392 Mha, suggesting that tropical plantations acted as strong carbon sinks due to rapid biomass accumulation.

Teak (*Tectona grandis*), as a tall tree species indigenous to India, Myanmar and Thailand and growing in seasonal dry tropical areas in Asia (Bunyavejchewin, 1983), is highly rated among hardwood plantations due to its durability, mellow color, and long straight cylindrical bole. Although natural teak is distributed in relation to productive soils, derived from e.g. limestone (Tanaka et al., 1998), teak is planted over many tropical countries, such as Nigeria, the Côte d'Ivoire, and Sierra Leone in Africa, and Costa Rica, Panama, Colombia, Trinidad and Tobago and Venezuela in central America, as well as Asian countries (Kashio & White, 1998; Pandey & Brown, 2000). By the year 2000, global teak plantation area reached 5.7 million hectares (FAO, 2001) and is still growing for investment by small landholders in agroforestry management as well as industrial wood supply (ITTO, 2004). However, the expansion of teak plantation has been propounding discussion from environmental perspectives, such as reduced biodiversity by mono-cultural plantations involving the clearing of undergrowth vegetation; soil erosion by fire treatment and litter raking; nutrient losses during harvesting; the spread of pests such as defoliators, the bee hole borer, skeletonizer; and the effects of water cycling (Niskanen, 1998; Pandey & Brown, 2000; Hallett et al., 2011).

Nowadays, one of the incentives for planting teak is to meet the demand in terms of carbon sequestration by indigenous tree species, at least in Indochina, with high economical return (Pibumrung et al., 2008; Jayaraman et al., 2010). However, despite several studies on carbon and biomass distribution in teak plantation in many countries, the carbon cycling of teak plantation has rarely been reported (Khanduri et al., 2008; Kraenzel et al., 2003; Viriyabuncha et al., 2002; Pande, 2005). Teak plantation production varies widely among countries and depending on soil conditions (Enters, 2000; Kaosa-ard, 1998). For example, the mean annual increment ranged from 2.0 m^3 ha^{-1} y^{-1} in poor sites in India to 17.6 m^3 ha^{-1} y^{-1} in prime sites in Indonesia with 50 year rotation periods (Pandey & Brown, 2000). Thus, the quantative illustration of carbon cycling in teak plantations is useful for understanding the key carbon sequestration channels, which may serve as the basis for improving forest management.

In this study, we estimated carbon stocks and fluxes of teak plantations in western Thailand, where productive soil was formed by underlying limestone and sandstone geological series (Suksawang et al., 1995). According to carbon allocation and carbon dynamic models (Richter et al., 1999; De Deyn et al., 2008), site productivity usually enhances carbon accumulation in soil by returning litter above and below ground to the soil. To understand the effects of forestry plantation on the carbon sequestration potential, carbon dynamics in adjacent natural forests were compared.

2. Materials and methods

2.1 Study sites, vegetation and soil

The study site was located in the Mae Klong Watershed Research Station (14°35'N, 98°52'E, Fig. 1), Lintin, Thong Pha Phum, Kanchanaburi Province, western Thailand (Takahashi et al., 2009). The annual mean air temperature at the station is about 25°C, ranging from 9.3 to 42.2°C, and the annual mean precipitation is 1,650 mm, most of which falls during the rainy season from April to October (Suksawang et al., 1995). Altitudes at the study sites ranged from 150 to 350 m a.s.l. In the watershed, the CO_2 and water exchange of the forest were monitored by a flux tower (Fisher et al., 2009; Saigusa et al., 2008).

Fig. 1. The location of the Mae Klong Watershed Research Station in Thailand.

Two plots were established in the teak plantation: young aged stand (T1, planted in 1992) and middle aged and gmelina (*Gmelina arborea*) mixed stand (T2, planted in 1977). The T1 plot had 530 trees ha^{-1} and its basal area at a height of 1.3 m was 6.0 m^2 ha^{-1} in 1996 (Photo 1.). In T2, the tree density and the basal area of teak were 384 trees ha^{-1} and 17.4 m^2 ha^{-1}, respectively, while those of gmelina were 181 trees ha^{-1} and 12.3 m^2 ha^{-1}, respectively, in

Photo 1. Photographs of the study sites. Top left: 3-year-old teak (T1), top right: 7-year-old teak (T1), middle left: 16-year-old teak (T2), middle right: natural forest (MDF), bottom left: bamboo undergrowth in the natural forest, and bottom right: aerial photo of the Mae Klong watershed research area (back hills).

1996. The teak plantation was mainly distributed over a lower slope area, which was formerly cultivated land e.g. for upland rice.

The adjacent natural forest was of the mixed deciduous forest (MDF) (Rundel & Boonpragob, 1995). Detailed descriptions of the vegetation were given by Marod et al. (1999). The dominant tree species were *Shorea siamensis, Vitex peduncularis, Dillenia parviflora* var. *Keruii*, and *Xylia xylocarpa* var. *Keruii*, while four bamboo species were mixed as undergrowth vegetation (Takahashi et al., 2007). The forest was spread on a hill behind the teak plantation. Despite being so-called natural forest, it is still thought to have been historically affected by human disturbances such as hunting and collecting forest products, forest fires, and logging, as the other tropical dry forest (Murphy & Lugo, 1986), as well as natural disturbances by winds and storms (Baker et al., 2005).

Soil with a relatively high pH and rich in exchangeable calcium is classified as Alfisols (Soil Survey Staff, 2010) and derived from sedimentary rock, gneiss and limestone. Concisely, the soil pH (H_2O) is 5.7 - 7.1; exchangeable cations (cmol kg^{-1}) are 5.8 - 17.9 for Ca, 1.4 - 3.0 for Mg and 0.5 - 1.3 for K respectively; the cation exchangeable capacity is 10.3 - 16.2 cmol kg^{-1} and the base saturation is 75 – 127%. Soil texture is classified into sandy loam, loam, or clay loam and the soil is well drained. The chemical and physical properties of the soil in the study sites were also reported elsewhere (Takahashi et al., 2009, 2011).

2.2 Tree enumeration, litterfall, and fine root mass

Plots for tree enumeration were 40×60 m for T1 and 30×60 m for T2 in the teak plantation. In the watershed, plantation trees were planted at a spacing of 4×4 m. For the natural forest, enumeration was performed within an area 200×200 m on the slope (Marod et al., 1999). Diameter at the breast height (DBH, 1.3 m above the ground) was measured every year in the plantation and every two years in the natural forest.

The litterfall was measured using 10 litter traps with 1×1 m openings in the teak-gmelina plantation (T2), but not in the young teak plantation (T1). For the natural forest, 100 traps were installed in grids at 20 m intervals. Litter was collected once or twice a month and the oven dry mass was weighed (70 °C). The conversion factor to carbon mass used was 0.47.

Fine and small root biomass (< 2 cm in diameter) was measured by a soil column with an area 15×15 cm and a depth of 15 or 30 cm. The sampling was performed in triplicate in November 1998, the beginning of the dry season. Dead roots were eliminated and bamboo roots were separated by visible inspection. Root diameters (mm) were classified into <1, 1 – 3, 3 – 5, 5 – 10, and 10 – 20 and the oven dry weight was determined. The carbon concentration was assumed to be 0.45 gC g^{-1}.

2.3 Carbon stocks in ecosystem compartments

To calculate the carbon stock in living biomass, biomass conversion equations from basal areas (Kiyono et al., 2010) were used for estimating carbon in leaves, branches, stems, and roots. The wood density of tree species, carbon contents of leaves and woody materials were all collected from the IPCC report (IPCC, 2006).

Soil carbon stocks were determined at the representative soil profile in the plots. For the natural forest, soil pits were set on a slope at different topographical positions: ridge, upper and lower slope positions and a soil sample was taken from each soil horizon described in the soil survey. The soil carbon concentration was analyzed using the dry combustion method (NC analyzer, Shimadzu Co., Kyoto, Japan), while the soil bulk density was measured using a 4×100 cm^2 cylinder core. The soil carbon stock was calculated by multiplying the carbon concentration by the soil bulk density of the soil layer and cumulating to a certain depth. The litter (forest floor) was collected using a 0.5×0.5 m frame with four replications.

2.4 Soil respiration rate

The soil respiration rate was measured using the closed chamber method (Takahashi et al., 2009). The steel chamber used was 30 cm in both diameter and height. About 20 min after the cover had been sealed, the CO_2 concentration in the chamber headspace was determined by an infrared gas analyzer (ZFP5, Fuji Electronics Co., Ltd., Japan), while the soil respiration rate was calculated using a linear model of increasing CO_2 concentration with temperature correction. In this measurement, we manipulated trenching around the chamber and litter removal to separate the respiration sources of the roots, organic layer, and soil. Detailed results were reported elsewhere (Takahashi et al., 2009, 2011).

3. Results and discussion

3.1 Stock and growth of trees

In the teak plantation, plots T1 and T2 showed constant accumulation of carbon in the biomass. The teak biomass was 3.8 MgC ha^{-1} at 3 years, increasing to 28.6 MgC ha^{-1} at 6 years in T1 (Fig. 2). Biomass in T2, where one third of the planted trees were gmelina, was 56 MgC ha^{-1} for teak, 37.5 MgC ha^{-1} for 15-year-old gmelina, increasing to 86.9 MgC ha^{-1} for teak and 59.8 MgC ha^{-1} for gmelina at 20 years. Although plot T2 was a mixed stand and the stand age differed from plot T1, the T1 and T2 growth rates were comparable. Combining these stands, the growth rate of these plantations was estimated as 9.3 MgC ha^{-1} y^{-1} by linear regression ($r = 0.999$, $p < 0.01$). Total biomass consisted of 1.9% for leaves, 14.8% for branches, 69.2% for stems and 14.1% for roots aged 20 years at plot T2.

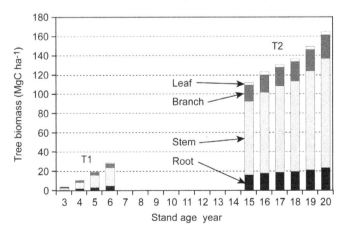

Fig. 2. Leaf, branch, stem and root biomass in teak plantations. Plots T1 and T2 are shown in the same figure according to the stand age.

Previous reports on carbon stocks in teak plantations in several countries were summarized in Table 1. Among the reference data, the site productivity of the teak in this watershed was the highest value. The increase rate of carbon stock of 9.3 MgC ha⁻¹ y⁻¹ was almost equivalent to 24 m³ ha⁻¹ y⁻¹ for the stem volume increment, which would almost represent the upper limit of the mean annual increment of teak plantations under short rotation (< 20 years) system (Enters, 2000; Kiyono et al., 2007). This is probably due to the soil properties: well drained, rich in calcium, and high soil pH (Takahashi et al., 2009), namely ideal soil conditions for teak growth (Kaosa-ard, 1998; Tanaka et al., 1998).

Country, region	AG†	BG‡	Litter	Soil§	Stand age	Tree density	Soil pH	Ref.¶
	(MgC ha⁻¹)				(y)	(ha⁻¹)		
Panama, Boquerón	91.8	13.8	3.6	225 (200cm)	20	586	6.6	1)
Panama, Peñas Blancas	122.2	18.4	3.3		20	566	6.2	
Panama, Tranquilla	117.1	17.6	3.2		20	621	5.9	
Panama, Agua Claras	86.8	13.1	3.5		20	723	6.1	
Nigeria, Oyo	21.2		0.8		5	1184		2)
	57.2		2.2		8	1088		
	57.0		1.4		11	1100		
	67.1		1.5		14	988		
India, Chhindwara	17.8	4.9			16	2500	7.9	3)
India, Kerala	78.7				20	217	5.03	4)
	92.2				20	250	5.36	
	96.0				20	300	5.74	
	70.9				15	233	5.05	
	56.1				15	217	5.36	
	82.5				15	333	5.81	
Thailand, Prachuap Khiri Khan	43.7	13.8		56.77 (50cm)	15			5)
Thailand, Lampang	35.6	9.1		221 (50 cm)	17	844		6)
	41.2	8.2		137 (50 cm)	22	544		
Thailand, Kanchanaburi	24.1	4.4		108 (100cm)	6	530	7.1	7)
	141.0	23.2	2.5	123 (100cm)	20	565	6.2	

†: Aboveground biomass, ‡: Belowground biomass, §: cumulative soil depth in parentheses,
¶:Reference 1) Kraenzel et al., 2003; 2) Mbaekwe & Mackenzie, 2008; 3) Pande, 2005; 4) Chandrashekara, 1996; 5) Meunpong et al., 2010; 6) Hiratsuka et al., 2005; 7) This study.
Note: Dry mass was converted to carbon mass by a factor of 0.5.

Table 1. Comparison of carbon stocks in teak plantations in the seasonally dry tropics.

In the natural forest, the density of trees above 5 cm in DHB was almost stable with slight fluctuation during the 8 years of monitoring, ranging from 174 to 199 trees ha[-1] and 16.9 to 17.7 m[2] ha[-1] for basal areas (Fig. 3). However, the tree density was rather low, compared to the forest in Huai Kha Khaeng, about 120 km north of Mae Klong, where 438 trees (> 10 cm in DBH) ha[-1] (Bunyavejchewin et al., 2001), suggesting a history of high disturbance in this area (Baker et al., 2005). Total tree biomass in 2000 was 139 MgC ha[-1], consisting of 1.5 MgC ha[-1] for leaves, 19.3 MgC ha[-1] for branches, 101 MgC ha[-1] for stems, and 17.8 MgC ha[-1] for roots. Because carbon densities in the Thai forests vary with forest type (Ogawa et al., 1961, 1965), our comparison was limited to the aboveground biomass of trees of the MDF (mixed deciduous forest) type. Similarly, low carbon densities were reported such as 48.14 MgC ha[-1] in Pong Phu Ron station (Terakunpisut et al., 2007) and 24.79 and 50.58 MgC ha[-1] for secondary and primary MDF in Phetchabun province (Kaewkron et al. 2011). As for MDF comparable forests from the early ecological study in Thailand, Ogawa et al. (1965) showed an aboveground dry mass of 157 Mg ha[-1] (77 MgC ha[-1]) for a forest classified as monsoon forest-savanna ecotone in Chiang Mai province and that of 103 Mg ha[-1] (51 MgC ha[-1]) for a mixed savanna forest in Tak province. Although carbon accumulation in MDF varies widely, our result seems typical of this forest type.

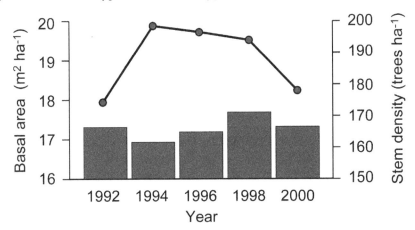

Fig. 3. Changes in stem density (DBH > 5 cm, dots and lines) and basal area (bars) in the natural forest.

Despite the low tree density, no apparent increment of carbon accumulation was detected in the natural forest studied. The natural forest nearby the teak plantation seemed static or in balance with the sequestration of carbon in the forest biomass. Our observation confirmed that most individual trees had increased steadily but some were dead with broken stems and standing dead, probably due to competition during the 8-year monitoring period (Marod et al., 1999 and unpublished data by Marod). In terms of the recruitment of trees, forest fires and drought stress would control the survival rates of tree seedlings and seed germination as well (Marod et al., 2002). In addition, shading by bamboo leaves during the rainy season influences the survival rate of seedlings (Marod et al., 2004). Similar findings showing that abundance in terms of seedlings and species had declined under bamboo in the MDF were reported in the Loei province of Thailand (Larpkern et al., 2011).

3.2 Litterfall

Annual litterfall in the teak-gmelina stand (T2) was 2.22 Mg ha^{-1} y^{-1} on average, consisting of 89.3% for leaf, 2.5% for fruit, and 8.2% for others such as bark. Of the leaf litter, 52.8% was counted for teak, 29.0% for gmelina, and 18.2% for other leaves (Fig. 4). The litterfall in the natural forest was 2.38 Mg ha^{-1} y^{-1}, on average, consisting of 61.5% leaves, 2.3% flowers, 7.7% fruit and 28.5% for others such as branches and bark (Fig. 5). About half the leaf litter (51.7%) was bamboo leaves. The litterfall tended to peak during January to March in both the natural forest and teak plantation: during this quarter, 46% of the litterfall fell in the natural forest and 56% in the teak plantation.

Seasonal patterns of litterfall were reported in several seasonal tropical forests with seasonal drought (e.g. Martínez-Yrízar & Sarukhán, 1990; Bunyavejchewin, 1997), usually as a dry matter basis. For teak plantations, 9.0 Mg ha^{-1} of annual litterfall, 90% of which was leaf litter and 70% or so of which fell during the dry season in Nigeria (Egunjobi, 1974). In India, total litterfall in teak plantations ranged from 3.3 to 4.5 Mg ha^{-1} (Pande et al., 2002). In the natural forests of Thailand, levels of 6.8 Mg ha^{-1} in *Shorea henryana* and 6.4 Mg ha^{-1} in a *Hopea ferrea* were observed (Bunyavejchewin, 1997). Mixed deciduous forest with teak showed total litterfall of 7.98 Mg ha^{-1} (Thaiutsa et al., 1978). The annual litterfall and proportion of the same in this study were comparable to previous studies of seasonal dry tropical forests (Martínez-Yrízar, 1995).

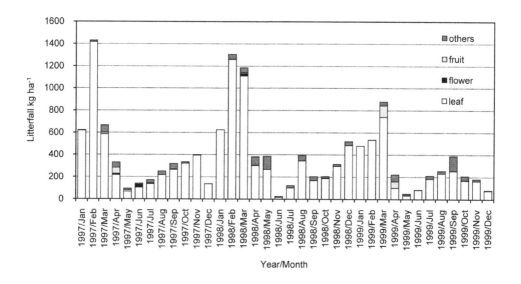

Fig. 4. Dry weight of monthly litterfall in the teak-gmelina plantation (plot T2).

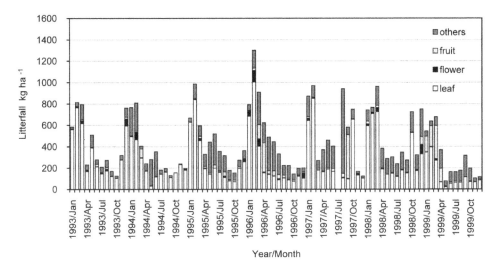

Fig. 5. Dry weight of monthly litterfall in the natural forest.

3.3 Root distribution and biomass

Fine root biomass (< 1 and 1 - 3 mm) was concentrated in the surface soil layer (0 - 15 cm) in all the plots studied (Fig. 6). The teak plantation had a sparse rooting system: fine roots (< 3 mm) with density in the 0 - 15 cm soil layer of 392 and 219 g m^{-3} for T2 and T1 respectively, compared to those in the natural forest (1144 g m^{-3} on average). In the 0 – 30 cm soil layer, fine root biomass (< 3 mm) was 0.47 and 0.32 MgC ha^{-1} for T2 and T1. Total fine and small root biomass (< 2 cm) at depths of 30 cm and 1 m was 2.3 and 2.7 MgC ha^{-1}, respectively, for plot T2 and 0.4 and 1.6 MgC ha^{-1} for plot T1. In the natural forest, the bamboo roots were mostly fine roots (< 3mm) at the surface (725 g m^{-3}), while the fine root biomass (<3 mm) of the 0 – 30 cm layer was 2.8 MgC ha^{-1} on average. Total root biomass (< 2 cm) was 4.0 MgC ha^{-1} at a depth of 30 cm and 8.8 MgC ha^{-1} at a depth of 1 m. These data are included in the root biomass estimated by the allometric equations above.

Although limited data is available for root biomass estimation in seasonal tropical forests, previous research revealed that fine root growth and density in seasonal tropical forests were highly controlled by soil hydrological regime (Cuevas, 1995). Our root biomass measurement, conducted at the beginning of the dry season, was probably lower than that in the rainy season. According to the review data by Cuevas (1995), the root mass in the wet season was about 1.5 times larger than that in the dry season. Singh & Srivastava (1985) found that the dynamics of root tip development, indicating growth and activity of fine roots, occurred mostly in the top 20 cm soil layer in teak plantations and that the highest root tips were observed in the mid rainy season. Similar fluctuations were found in both natural and Mexican dry forests (Kummerow et al., 1990; Kavanagh & Kellman, 1992). We could not clarify the fine root dynamics from the one-time measurement. However, the relative proportion of root biomass between the teak plantation and natural forest would be reasonable because the differences in root respiration as a proportion of total soil respiration between the teak plantation (15%) and the natural forest (19.4 – 33.6%) were comparable to the difference in root biomass of the same (Takahashi et al., 2011). These findings suggest that carbon flux by root necromass and exudates from fine root to mineral soil may be higher in natural forest than teak plantations, although high carbon accumulation by a developing living root system occurred at a rate of 1.3 MgC ha^{-1} y^{-1}; as calculated by total biomass increment and the root biomass proportion of these forests.

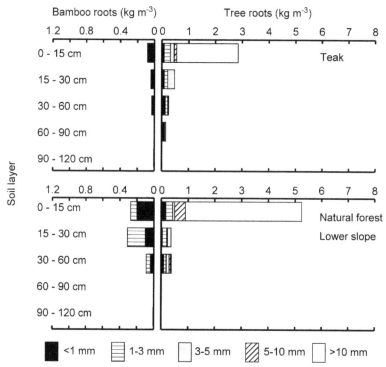

Fig. 6. Vertical distribution of fine and small roots density (< 20 mm) in the teak plantation (T1) and the natural forest (lower slope position). Bamboo roots are shown in the left-hand side and tree and other roots are in the right-hand side.

3.4 Soil carbon stock

In the teak plantation, soil carbon stocks at a depth of 1 m were 108 and 124 MgC ha^{-1} for plots T1 and T2 respectively (Fig. 7). The surface layer 0 – 15 cm had 34 MgC ha^{-1} in both plots T1 and T2. In the natural forest, carbon stocks in soil at a depth of 1 m varied with slope position and tended to have a larger ascribing slope, ranging from 113 to 178 MgC ha^{-1} (Takahashi et al., 2011, and unpublished data by Takahashi). The soil carbon stock in the soil layer of 0 - 15 cm was 32 and 38 MgC ha^{-1} in the lower and middle slope positions respectively.

Tangsinmankong et al. (2007) reported that soil carbon stocks up to 1 m depth under teak plantations of varying age ranged from 78.8 to 157 MgC ha^{-1} and that those in the 0 – 15 cm layer had 26.7 to 37.2 MgC ha^{-1} in the same district of this study. In peninsular Thailand, soil carbon in the 0 – 50 cm layer was 56.8 MgC ha^{-1} in a 15 year-old teak plantation with tree biomass of 57.5 MgC ha⁻¹ (Meunpong et al., 2010). Soil carbon in the 0 – 50 cm soil layer under a teak plantation in Colombia was 54.9 MgC ha⁻¹ (Usuga et al., 2010). These values are comparable to our results. In north Thailand, however, Hiratsuka et al. (2005) reported very high carbon stock in soils in 0 – 50 cm layers, 211 MgC ha⁻¹ in 17 year-old stand, with 2 × 4 m spacing, by accumulating high organic carbon in the top soil layer, and 137 MgC ha⁻¹ in the 22 year-old stand with 4 × 4 m spacing. In Panama, 225 MgC ha⁻¹ was stocked up to the bottom of the soil profile, almost 2 m in depth, under 20 year-old teak plantation (Kraenzel et al., 2003). In MDF of Thailand, the soil carbon stocks at depths of 15 cm and 1 m were 26.7 and 71.0 MgC ha⁻¹ (Tangsinmankong et al., 2007). Soil carbon at a depth of 0 – 30 cm (MgC ha⁻¹) on the sandstone and conglomerate was 27.6 while that on the limestone was 74.9 under MDF in Thailand (summarized by Toriyama et al., 2011).

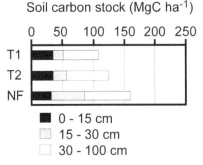

Fig. 7. Vertical distributions of soil carbon stocks in the teak plantations (plots T1 and T2) and natural forest (NF).

It is difficult to compare the results of soil carbon stock between different cumulating depths and different soil groups, which are not identified in some reports. Usuga et al. (2010) pointed out the importance of land-use history for analyzing soil carbon stocks, while Sakai et al. (2010) reported that the accumulation of carbon stock after conversion from arable land to forest plantation was only evident in the organic layer and the surface soil layer with 0 – 5 cm in the Japanese cedar and cypress stands younger than 25 years-old.. With reference to fine root distribution in the top 0 – 15 cm soil layer and litter on the forest floor, comparison of soil carbon stocks should be limited to the upper soil horizons, where the active component of soil organic carbon is concentrated. In addition, Toriyama et al. (2011) showed the importance of parent materials for soil carbon accumulation: basaltic and calcareous parent materials accumulated about twice the carbon stock compared to sedimentary rocks in the forest ecosystems in Cambodia and Thailand. The top soil carbon accumulation, ranging from 32 to 38 MgC ha⁻¹ within the 0 – 15 cm layer, was relatively high, probably due to the influence of the limestone parent material, although no clear differences emerged between the teak plantation and the natural forest.

3.5 Soil respiration

The soil respiration rate (gC m⁻² d⁻¹) showed seasonal fluctuation, peaking in the rainy season and declining in the dry season; ranging from 1.44 to 5.27 for plot T1, 1.71 to 5.46 for plot T2, and 1.85 to 7.21 for plots on the natural forest slope respectively. Applying equations fitting the relationship between soil moisture in the 0 – 30 cm layer monitored by TDR sensors and soil respiration rates, annual carbon efflux from the forest floor was estimated (Takahashi et al., 2009, 2011). Annual carbon efflux in plots T1 and T2 were 11.5 and 10.6 MgC ha^{-1} y^{-1}, respectively. In the natural forest, the soil respiration rate varied with the slope position on the hill (Takahashi et al., 2011) hence the average of the lower and upper slopes was used in this study for annual carbon efflux. In the natural forest, the carbon efflux was 17.7 MgC ha^{-1} y^{-1} on average. To distinguish the CO_2 sources, trenching to separate root respiration and litter removal in the chamber were manipulated (Takahashi et al., 2009, 2011). The proportions of roots, litter, and soil respiration were 15, 17, and 68%, respectively, for plot T2, and 27, 23, and 50 %, respectively, for the natural forest, on average.

Soil respiration is a key channel for returning photosynthesized carbon into the atmosphere, which has been intensively studied in forest ecosystems (e.g. Davidson et al., 2000). In general, annual soil carbon efflux correlates with annual rainfall and the mean annual temperature, as well as with annual litterfall (Raich & Schlesinger, 1992; Raich & Tufekciogul, 2000). For the soil respiration rate, Adachi et al. (2009) reported an average of 6.82 ± 3.55 gC m^{-2} d^{-1} during the rainy season and 2.63 ± 1.35 gC m^{-2} d^{-1} in a forest in Huai Kha Khaeng in northern Thailand. In a dry dipterocarp forest (DDF) in Ratchaburi, western Thailand, the soil respiration rate ranged from 1.26 gC m^{-2} d^{-1} in the dry season to 3.93 gC m^{-2} d^{-1} in the rainy season (Hanpattanakit et al., 2009). Annual carbon efflux from the forest floor was estimated at 13.4 MgC ha^{-1} y^{-1} under a dry evergreen forest (DEF), 12.1 MgC ha^{-1} y^{-1} under an MDF, 10.7 MgC ha^{-1} y^{-1} under a DDF in the Mae Klong watershed basin (Panuthai et al., 2006), and 25.6 MgC ha^{-1} y^{-1} under a hill evergreen forest (HEF) in Chiang Mai, northern Thailand (Hashimoto et al., 2004). As for teak plantations, the soil respiration rate has rarely been measured. In Khon Kaen Province, northeastern Thailand, 15.0 MgC ha^{-1} y^{-1} of CO_2 efflux in a teak plantation on sandy soil was reported (Funakawa et al., 2007). These data, including our results, suggest that soil respiration rates in teak plantations are likely to be similar to those for natural forest vegetation.

3.6 Carbon cycling in teak plantations

The carbon balance in plot T2 of the teak-gmelina plantation was depicted in Fig. 8. To draw this picture, we assumed comparable growth rates for the T1 and T2 plantations, despite the fact T2 were mixed forest with gmelina. We did not measure the litter decomposition of teak and gmelina leaves but, according to the partitioning of soil respiration sources, annual litterfall would be decomposed within a year, which is in accordance with our field observation that leaf litter on the forest floor disappeared during the rainy season. In this watershed, leaves of *Shorea siamensis* and bamboo swiftly decayed; 95 and 85 % of the initial weight having decomposed within a year (Somrithpol, 1997). A rapid decomposition rate was also often reported for teak leaves, usually more than 90 % in a year in several countries (e.g. Sankaran, 1993; Maharudrappa et al., 2000; Pande, 2005). As for forest floor vegetation, this was not measured because in T1, upland rice was initially cultivated between planted teaks for the first two years, whereupon weeding was performed each year. In T2, understory vegetation was sparse, due to bamboo flowering and dying two to three years before our observation commenced. Bamboo recovery was slow compared with the open area outside the forest, hence we ignored forest floor biomass in the teak plantation in this study.

Under these assumptions, the net primary production (NPP), as the gross rate of biomass production (GPP) minus the respiration cost (R): NPP = GPP – R, can be calculated and expressed as an increment of biomass plus necromass, such as litterfall and root litter, while some is consumed by herbivores. In the teak-gmelina plantation, NPP was estimated to be at least 11.5 (=9.3+2.2) MgC ha^{-1} y^{-1}, excluding underground processes. This value would be reasonable and relatively high for plantations in the tropics, referring to the review paper by Pregitzer & Euskirchen (2004). As comparable data, Imvitthaya et al. (2011) estimated the NPP in teak plantations in northern Thailand using BIOME-BGC Model and Spot data and with estimation ranging from 6.06 to 7.76 MgC ha^{-1} y^{-1} with yearly variation.

As for the natural forest, no apparent tree growth was observed from the tree census. If the carbon is balanced, NPP was equivalent to at least over litterfall. Clark et al. (2001) evaluated NPP in tropical primary forests, which ranged from 1.7 to 11.8 MgC ha^{-1} y^{-1} for the lower boundary based on conservative estimates, and featuring close correlation with aboveground biomass increment and fine litterfall. Our data showing the slow carbon sequestration rate by trees seems reasonable. Compared to the carbon cycling in the natural forest, we conclude that teak plantations are certainly valuable for sequestrating carbon, especially the portions above ground, in this area where the soil conditions favor teak.

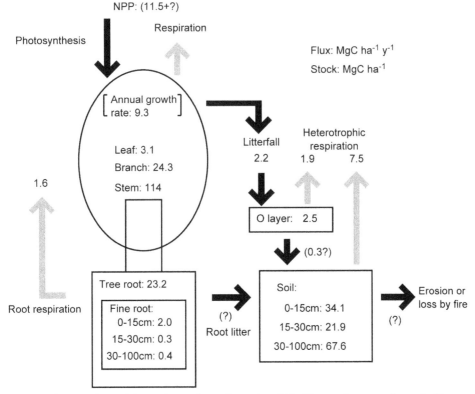

Fig. 8. Carbon cycling of the teak-gmelina plantation (T2). Figures in parentheses indicate uncertain estimates.

3.7 For sustainable teak plantation management

Soil carbon stock usually increased over time after planting trees (Sakai et al., 2010), due to carbon input from litterfall and the turnover of dead roots (Richter et al., 1999), meaning the higher growth of forest plantation would lead to higher soil carbon accumulation. However, despite high production in the plantation studied, there was no apparent difference in the soil carbon between T1 and T2. We speculate that surface soil erosion spoiled the soil carbon sequestrated under the plantation. During our observation, we found that surface soil was eroded due to raindrop splashes in the rainy season, especially in plot T2, which seemed to prevent soil carbon accumulation in the top soil layer. Poor understory vegetation, which induced bamboo flowering and death, dark conditions under the teak canopy, and quick litter decomposition seemed to create soil conditions leading to a bare and exposed surface. A similar observation of erosion under teak plantations was reported by Ogawa et al. (1961), Tangtham (1992), and Boley et al. (2009). This risk of soil erosion in teak plantations, caused by large raindrops falling from broad and large teak leaves, has been pointed by Hall & Calder (1993) and Calder (2002). Another possible risk of preventing carbon accumulation in the soil would be forest fires. Although teak resists fire, litter on the forest floor is lost if fires occur in the dry season, which is considered to be the main cause of low soil carbon accumulation in Myanmar teak plantations (Suzuki et al., 2007).

As well as surface soil, subsurface soil also showed no carbon accumulation in the teak plantation: Soil carbon stocks in the subsurface layer (15 – 30 cm) were smaller than those in the natural forest (Fig. 7). Similarly, in Panama, Kraenzel et al. (2003) also observed that teak plantations on abandoned land promoted no significant increases in soil carbon storage, despite considerable biomass growth. After harvesting, tree stumps remained and decomposed, which may have contributed to the belowground carbon stock to some extent in the short-term. However, for long-term soil carbon storage, undergrowth vegetation with deep rooting systems may help accumulate carbon stock in the subsurface layer.

Apart from soil management, the carbon cycling scheme in Fig. 8 suggests that teak plantations in this watershed are likely to be harvestable with a short rotation cycle, e.g. 20 – 30 years. Such short rotation is also beneficial in terms of carbon sequestration, while the parent material, limestone, of this watershed would promise high future productivity of the soil. However, ideal sites for teak plantation now face competition with agricultural crops and teak is often planted in sites with poor fertility (Enters, 2000), which would thus require longer rotation periods. Appropriate management should be selected in accordance with the site characteristics and management intensity. For sustainable forest management, there is still scope to improve teak plantations from several perspectives, e.g. biodiversity, carbon sequestration, and wood quality (Nair & Souvannavong, 2000).

General criticisms of monoculture plantations in terms of reducing biodiversity were periodically reviewed (e.g. Hartley, 2002; Brocherhoff et al. 2008), with poor undergrowth vegetation in young teak plantations with narrow spacing an example of a serious case. To improve monoculture plantations, mixture with other species, gmelina in our case, would be a live option, although silvicultural prescriptions must be developed. The landscape design of plantations and corridor arrangements may also be helpful (Fischer et al., 2006; Brocherhoff et al., 2008).

Lastly, because teak takes up high levels of nutrients and returns them to the surface soil, the aggrading effect of soil fertility was found on degraded land in Costa Rica (Boley et al., 2009). Similarly, calcium enrichment under teak plantations was observed in Myanmar (Suzuki et al., 2005). If suitable management for top soil conservation is applied, e.g. spacing, weed management, fire control, and mixed planting, teak is likely to represent promising species for land rehabilitation.

4. Conclusion

Teak has been a popular tree species for timber production in commercial and private farmland and remains a promising species for carbon sequestration in the seasonally dry tropics. A carbon cycling scheme obtained in teak (-gmelina) plantations showed a high rate of carbon accumulation in the soil of Alfisols in western Thailand which has high calcium content and high soil pH. In adjacent natural forest, no apparent carbon accumulation was observed, due to slow tree recruitment and disturbance in trees. Based on a comparison of carbon cycling in the natural forest, a teak plantation would represent a reasonable recommendation for tree species when managing plantations with carbon sequestration and high quality timber. However, no soil carbon accumulation is expected, probably due to surface soil erosion caused by raindrop splashes in poor understory vegetation and the ignition of litter by forest fire. Soil erosion control is essential under a teak canopy, which may promote additional carbon sequestration in the soil. Although the results of this study were derived from high productivity sites, teak can be planted as a rehabilitation species on degraded land as well. However, silvicultural prescription must be developed in accordance with economic benefit and ecological services side by side. To achieve this, quantitative measurement of carbon stocks and fluxes are useful for judging forest management appropriately.

5. Acknowledgments

The study was supported by the Science Technology Agency of Japan and the National Research Council of Thailand. The manuscript was summarized for the Global Environment Research Account for National Institutes (Advancement of East Asia forest dynamics plot network), Ministry of the Environment Japan and was partly funded by JSPS KAKENHI 21241010. The authors would like to thank Drs. U. Kutintara, C. Yarwudhi and S. Anusontpornperm at Kasetsart University, P. Limtong, P. Tummakate, C. Leaungvutivirog and V. Sunantapongsuk in the Land Development Department Thailand, S. Suksawang in the National Parks, Wildlife and Plant Conservation Department Thailand, K. Ishizuka, N. Tanaka, T. Kawasaki and H. Tanaka in the Forestry and Forest Products Research Institute, Professors S. Kobayashi at Kyoto University, T. Nakashizuka at Tohoku University, and N. Kaneko at Yokohama National University for their valuable suggestions and supporting research activities.

6. References

Adachi, M., Ishida, A., Bunyavejchewin, S., Okuda, T., & Koizumi, H. (2009). Spatial and temporal variation in soil respiration in a seasonally dry tropical forest, Thailand. *Journal of Tropical Ecology* 25 , 531–539. ISSN 0266-4674

Baker, P. J., Bunyavejchewin, S., Oliver, C. D., & Ashton, P. S. (2005). Disturbance history and historical stand dynamics of a seasonal tropical forest in western Thailand. *Ecological Monographs* 75 (3), 317–343. ISSN 0012-9615

Boley J. D., Drew, A. P. & Andrus, R. E. (2009) Effects of active pasture, teak (*Tectona grandis*) and miced native plantation on soil chemictry in Costa Rica. *Forest Ecology and Management* 257, 2254-2261. ISSN 0378-1127

Brocherhoff, E. G., Jactel, H., Parrotta, J. A., Quine, C. P. & Sayer, J. (2008). Plantation forests and biodiversity: oxymoron or opportunity? *Biodiversity and Conservation* 17, 925-951, ISSN 0960-3115

Bunyavejchewin, S. (1983). Analysis of the tropical dry deciduous forest of Thailand. I. Characteristics of the dominance-types. *Natural History Bulletin of the Siam Society* 31, 109–122. ISSN 0080-9472

Bunyavejchewin, S. (1997). Ecological studies of tropical semi-evergreen rain forest at Sakaerat, Nakhonratchasima, northeast Thailand: II. litterfall. *Natural History Bulletin of the Siam Society* 45, 43–52. ISSN 0080-9472

Bunyavejchewin, S., Baker, P. J., LaFrankie, J. V., & Ashton, P. S. (2001). Stand structure of a seasonal dry evergreen forest at Huai Kha Khaeng wildlife sanctuary, western Thailand. *Natural History Bulletin of the Siam Society* 49, 89–106. ISSN 0080-9472

Calder, I. R. (2001). Canopy processes: implications for transpiration, interception and splash induced erosion, ultimately for forest management and water resources. *Plant Ecology* 153, 203–214. ISSN 1385-0237

Chandrashekara, U. M. (1996) Ecology of *Bambusa arudinacea* (Retz.) Willd. Growing in teak plantations of Kerala, India. *Forest Ecology and Management* 87, 149-162. ISSN 0378-1127

Clark, D. A., Brown, S., Kicklighter, D. W., Chambers, J. Q., Thomlinson, J. R., Ni, J., & Holland, E. A. (2001). Net primary production in tropical forests: An evaluation and synthesis of existing field data. *Ecological Applications* 11, 371–384. ISSN 1051-0761

Cuevas, E. (1995) Biology of the belowground system of tropical dry forests. In: *Seasonally Dry Tropical Forests*, Bullock, S., H., Mooney, H., A., & Medina, E. (Eds), pp. 362– 383, Cambridge University Press, ISBN 0-521-43514-5, New York.

Davidson, E. A., Verchot, L. V., Cattânio, J. H., Ackerman, I. L., & Carvalho, J. E. M. (2000). Effects of soil water content on soil respiration in forests and cattle pastures of eastern Amazonia. *Biogeochemistry* 48, 53–69. ISSN 0168-2563

De Deyn, G. B., Cornelissen, J. H. C. & Bardgett, R. D.(2008). Plant functional traits and soil carbon sequestration in contrasting biomes. *Ecology Letters* 11, 516–531. ISSN 1461-0248

Egunjobi, J. K. (1974). Litter fall and mineralization in a teak *Tectona grandis* stand. *Oikos* 25, 222–226. ISSN 0030-1299

Enters, T. (2000). Site, technology and productivity of teak plantations in southeast Asia. *Unasylva* 201, (51), 55–61. ISSN 0041-6436

FAO (2001). *Global Forest Resources Assessment 2000*, FAO, Forestry Department, Main report 140, Rome. Retrieved from
http://www.fao.org/forestry/fra/2000/report/en/

Fischer, J., Lindenmayer, D. B., & Manning, A. D. (2006). Biodiversity, Ecosystem Function, and Resilience: Ten Guiding Principles for Commodity Production Landscapes. *Frontiers in Ecology and the Environment,* 4, 80-86. ISSN 1540-9295

Fisher, J. B., Malhi, Y., Bonal, D., Da Rocha, H. R., De Araãjo, A. C., Gamo, M., Goulden, M. L., Hirano, T., Huete, A. R., Kondo, H., O'Omi, Loescher, H. W., Miller, S., Nobre, A. D., Nouvellon, Y., Oberbauer, S. F., Panuthai, S., Roupsard, O., Saleska, S., Tanaka, K., Tanaka, N., Tu, K. P., & Von Randow, C. (2009). The land-atmosphere water flux in the tropics. *Global Change Biology* 15, 2694–2714. ISSN 1354-1013

Funakawa, S., Yanai, J., Hayashi, Y., Hayashi, T., Noichana, C., Panitkasate, .T, Katawatin, R., & Nawata, E. (2007.) Analysis of spatial distribution patterns of soil properties and their determining factors on a sloped sandy cropland in Northeast Thailand. P.536 Management of tropical sandy soils for sustainable agriculture. 27th November - 2nd December 2005, Khon Kaen, Thailand. ISBN 978-974-7946-96-3, Retrieved from http://www.fao.org/docrep/010/ag125e/ag125e06.htm

Gibbs, H. K., Brown, S., Niles, J. O., & Foley, J. A. (2007). Monitoring and estimating tropical forest carbon stocks: making REDD a reality. *Environmental Research Letters* 2, 045023. ISSN 1748-9326. Retrieved from http://dx.doi.org/10.1088/1748-9326/2/4/045023

Hall, R. L., & Calder, I. R. (1993). Drop size modification by forest canopies: Measurements using a disdrometer. *Journal of Geophysical Research* 98 (D10), 18 465–18 470. ISSN 0747-7309

Hallett, J. T., Diaz-Calvo, J., Villa-Castillo, J., Wagner, M. R., (2011). Teak plantations: Economic bonanza or environmental disaster? *Journal of Forestry*, 288–292. ISSN 0022-1201

Hanpattanakit, P., Panuthai, S., & Chidthaisong, A. (2009). Temperature and moisture controls of soil respiration in a dry dipterocarp forest, Ratchaburi province. *Kasetsart Journal, Natural Sciences* 43, 650–661. ISSN 0075-5192

Hartley, M. J. (2002). Rationale and methods for conserving biodiversity in plantation forests. *Forest Ecology and Management.* 155, 81-95. ISSN 0378-1127

Hashimoto, S., Tanaka, N., Suzuki, M., Inoue, A., Takizawa, H., Kosaka, I., Tanaka, K., Tantasirin, C., & Tangtham, N. (2004). Soil respiration and soil CO_2 concentration in a tropical forest, Thailand. *Journal of Forest Research* 9, 75–79. ISSN 1341-6979

Hiratsuka, M., Viriyabuncha, C., Peawsa-ad, K., Janmahasatien, S., Sato, A., Nakayama, Y., Matsunami, C., Osumi, Y., Morikawa, Y., (2005). Tree biomass and soil carbon in 17- and 22-year-old stands of teak (*Tectona grandis* l.f.) in northern Thailand. Tropics 14 (4), 377–382. ISSN 0917-415X.

Imvitthaya, C., Honda, K., Lertlum, S., & Tangtham, N. (2011). Calibration of a biome-biogeochemical cycles model for modeling the net primary production of teak forests through inverse modeling of remotely sensed data. *Journal of Applied Remote Sensing* 5, 053516. ISSN 1931-3195

IPCC (2006). *IPCC National Greenhouse Gas Inventories Programme 2006 IPCC guidelines for National Greenhouse Gas Inventories, Volume 4, Agriculture, forestry and other land use.* Technical Support Unit IPCC National Greenhouse Gas Inventories Programme, IGES, Hayama. Japan. Retrieved from http://www.ipcc-nggip.iges.or.jp/public/2006gl/vol4.html

ITTO (2004). *ITTO Tropical Forest Update. Volume 14, Number 1. The prospects for plantation teak.* Alastair Sarre. (Ed), Yokohama, Japan. Retrieved from http://www.itto.int/tfu/id=6660000

Jayaraman, K., Bhat, K.V., Rugmini, P., & Anitha, V. (2010). *Teknet bulletin* Vol. 3 Issue 2, Kerala Forest Research Institute, Peechi-680 653, Thrissur, Kerala, India. Retrieved from http://teaknet.org/files/Teaknet_bulletin%20Vol%203%20Issue%202.pdf

Kaewkron, P., Kaewkla, N., Thummikkapong, S., & Punsang, S. (2011). Evaluation of carbon storage in soil and plant biomass of primary and secondary mixed deciduous forests in the lower northern part of Thailand. *African Journal of Environmental Science and Technology* 5, 8 – 14. ISSN 1996-0786

Kaosa-ard, A. (1998) Overview of problems in teak plantation establishment. In: *Teak for the future, Proceedings of the second regional seminar on teak.* Kashio, M. and White, K. (Eds). RAP Publication. ISBN 974-86342-9-9,

Kashio, M., & White, K. (1998). Teak for the future. *Proceedings of the second regional seminar on teak. TEAKNET Publication: No. 1,* RAP Publication - 1998/05., ISBN 974-86342-9-9, Yangon, Myanmar, 29 May - 3 June 1995.

Kavanagh, T., & Kellman, M. (1992). Seasonal pattern of fine root proliferation in a tropical dry forest. *Biotropica* 24, ISSN 0006-3606.

Khanduri, V.P., Lalnundanga, & Vanlalremkimi J. (2008). Growing stock variation in different teak (*Tectona grandis*) forest stands of Mizoram, *Journal of Forestry Research* 19, 204–208. ISSN 1007-662X

Kiyono, Y., Furuya, N., Sum, T., Umemiya, C., Itoh, E., Araki, M., & Matsumoto, M. (2010). Carbon stock estimation by forest measurement contributing to sustainable forest management in Cambodia. *JARQ. Japan Agricultural Research Quarterly* 44, 81–92. ISSN 0021-3551

Kiyono, Y., Oo, M. Z., Oosumi, Y., & Rachman, I. (2007). Tree biomass of planted forests in the tropical dry climatic zone : Values in the tropical dry climatic zones of the union of Myanmar and the eastern part of Sumba island in the republic of Indonesia. *JARQ. Japan Agricultural Research Quarterly* 41, 315–323. ISSN 0021-3551

Kraenzel, M., Castillo, A., Moore, T., & Potvin, C. (2003). Carbon storage of harvest-age teak (*Tectona grandis*) plantations, Panama. *Forest Ecology and Management* 173, 213–225. ISSN 0378-1127

Kummerow, J., Castillanos, J., Maas, M., & Larigauderie, A. (1990). Production of fine roots and the seasonality of their growth in a Mexican deciduous dry forest. *Plant Ecology* 90, 73–80. ISSN 1385-0237

Larpkern, P., Moe, S.R., & Totland, Ø. (2011). Bamboo dominance reduces tree regeneration in a disturbed tropical forest. *Oecologia* 165, 161–168. ISSN 0029-8549

Maharudrappa, A., Srinivasamurthy, C. A., Nagaraja, M. S., Siddaramappa, R., & Anand, H. S. (2000). Decomposition rates of litter and nutrient release pattern in a tropical soil. *Journal of the Indian Society of Soil Science.* 48,. 92-97. ISSN0019-638X

Marod, D., Kutintara, U., Yarwudhi, C., Tanaka, H., & Nakashisuka, T. (1999). Structural dynamics of a natural mixed deciduous forest in western Thailand. *Journal of Vegetation Science* 10, 777–786. ISSN 1100-9233

Marod, D., Kutintara, U., Tanaka, H., & Nakashizuka, T. (2002). The effects of drought and fire on seed and seedling dynamics in a tropical seasonal forest in Thailand. *Plant Ecology* 161, 41–57. ISSN 1385-0237

Marod, D., Kutintara, U., Tanaka, H., & Nakashizuka, T. (2004). Effects of drought and fire on seedling survival and growth under contrasting light conditions in a seasonal tropical forest. *Journal of Vegetation Science* 15, 691–700. ISSN 1100-9233

Martínez-Yrízar, A. (1995). Biomass distribution and primary production of tropical dry forests. In: *Seasonally Dry Tropical Forests*, SH Bullock, HA Mooney, & E Medina (Eds), pp. 326—345, ISBN-0-521-43514-5, Cambridge University Press, ISBN 0-521-43514-5, New York.

Martínez-Yrízar, A., & Sarukhán, J. (1990). Litterfall patterns in a tropical deciduous forest in mexico over a five-year period. *Journal of Tropical Ecology* 6, 433–444. ISSN 0266-4674

Mbaekwe, E. I., & Mackenzie, J.A. (2008). The use of a best-fit allometric model to estimate aboveground biomass accumulation and distribution in an age series of teak (*Tectona grandis* L.f.) plantations at Gambari Forest Reserve, Oyo State, Nigeria. *Tropical Ecology* 49, 259-270, ISSN 0564-3295

Meunpong, P., Wachrinrat, C., Thaiutsa, B., Kanzaki, M., & Meekaew, K. (2010). Carbon pools of indigenous and exotic trees species in a forest plantation, Prachuap Khiri Khan, Thailand. *Kasetsart Journal. Natural Sciences* 44, 1044–1057

Murphy, P. G., & Lugo, A. E. (1986). Ecology of tropical dry forest. *Annual Review of Ecology and Systematics* 17 , 67–88. ISSN 0066-4162

Nair, C. T. S. & Souvannavong, O. (2000). Emerging research issues in the management of tesk. *Unasylva* 201, (51) 45–54. ISSN 0041-6436

Niskanen, A. (1998). Value of external environmental impacts of reforestation in Thailand. *Ecological Economics* 26, 287–297. ISSN 0921-8009

Ogawa, H., Yoda, K., & Kira, T. (1961). A preliminary survey of the vegetation in Thailand. *Nature and Life in Southeast Asia* 1. 22 – 157. Fauna and flora research society, Kyoto, Japan.

Ogawa, H., Yoda, K., Ogino, K., & Kira, T. (1965). Comparative ecological studies on three main types of forest vegetation in Thailand. II. Plant biomass. *Nature and Life in Southeast Asia* 4, 49 – 80. Fauna and flora research society, Kyoto, Japan.

Pan, Y., Birdsey, R. A., Fang, J., Houghton, R., Kauppi, P. E., Kurz, W. A., Phillips, O. L., Shvidenko, A., Lewis, S. L., Canadell, J. G., Ciais, P., Jackson, R. B., Pacala, S. W., McGuire, A. D., Piao, S., Rautiainen, A., Sitch, S. & Hayes, D. (2011) A large and persistent carbon sink in the world's forests. *Science*, 333, 988–993, ISSN 0036-8075

Pande P. K., Meshram P. B., & Banerjee S. K. (2002). Litter production and nutrient return in tropical dry deciduous teak forests of Satpura plateau in central India. *Tropical Ecology* 43, 337-344. ISSN 0564-3295

Pande, P. K., (2005). Biomass and productivity in some disturbed tropical dry deciduous teak forests of Satpura plateau, Madhya Pradesh. *Tropical Ecology* 46, 229–239. ISSN 0564-3295

Pandey, D., & Brown, C. (2000). Teak: a global overview. an overview of global teak resources and issues affecting their future outlook. *Unasylva* 201, (51) 3–13. ISSN 0041-6436

Panuthai, S. Junmahasatein, S., & Diloksumpun, S. (2006). Soil CO_2 emission in the Sakaerat dry evergreen forest and Maeklong mixed deciduous forest. In: *Research paper collections on study of carbon cycle in Sakaerat dry evergreen forest and mixed deciduous forest paper in Maeklong watershed basin.* pp.237 – 242. Forest and Plant Conservation Research Office, National Park, Wildlife and Plant Conservation Department, Ministry of Natural Resources and Environment, Bangkok, Thailand. (in Thai).

Pibumrung, P., Gajaseni, N., & Popan, A., (2008). Profiles of carbon stocks in forest, reforestation and agricultural land, northern Thailand. *Journal of Forestry Research* 19, 11–18. ISSN 1007-662X

Pregitzer, K.S., & Euskirchen, E.S. (2004) Carbon cycling and storage in world forests: biome patterns related to forest age. *Global Change Biology* 10, 2052–2077. ISSN 1354-1013

Raich, J. W., & Schlesinger, W. H. (1992). The global carbon dioxide flux in soil respiration and its relationship to vegetation and climate. *Tellus B* 44, 81–99. ISSN 0280-6509

Raich, J. W., & Tufekciogul, A. (2000). Vegetation and soil respiration: Correlations and controls. *Biogeochemistry* 48, 71–90. ISSN 0168-2563

Richter, D. D., Markewitz, D., Trumbore, S. E., & Wells, C. G. (1999). Rapid accumulation and turnover of soil carbon in a Re-Establishing forest. *Nature.* 400, 56-58. ISSN 0028-0836

Rundel, P.W., & Boonpragob, K. (1995). Dry forest ecosystems of Thailand. In: *Seasonally Dry Tropical Forests*, Bullock, S., H., Mooney, H., A., & Medina, E. (Eds), pp. 93 – 123, Cambridge University Press, ISBN 0-521-43514-5, New York.

Saigusa, N., Yamamoto, S., Hirata, R., Ohtani, Y., Ide, R., Asanuma, J., Gamo, M., Hirano, T., Kondo, H., Kosugi, Y., Li, S.-G., Nakai, Y., Takagi, K., Tani, M., & Wang, H. (2008). Temporal and spatial variations in the seasonal patterns of CO_2 flux in boreal, temperate, and tropical forests in east Asia. *Agricultural and Forest Meteorology* 148, 700–713. ISSN 0168-1923

Sakai, H., Inagaki, M., Noguchi, K., Sakata, T., Yatskov, M. A., Tanouchi, H., & Takahashi, M. (2010). Changes in soil organic carbon and nitrogen in an area of Andisol following afforestation with Japanese cedar and Hinoki cypress. *Soil Science and Plant Nutrition* 56, 332–343. ISSN 0038-0768

Sankaran, K. V. (1993). Decomposition of leaf litter of albizia (*Paraserianthes falcataria*), eucalypt (*Eucalyptus tereticornis*) and teak (*Tectona grandis*) in Kerala, India. *Forest Ecology and Management* 56, 225-242. ISSN 0378-1127

Singh, K. P., & Srivastava, S. K. (1985) Seasonal variations in the spatial distribution of root tips in teak (*Tectonia grandis* Linn. f.) plantations in the Varanasi Forest Division, India. *Plant and Soil* 84, 93-104. ISSN 0032-079X

Soil Survey Staff. (2010). *Keys to Soil Taxonomy, 11th ed.* USDA-Natural Resources Conservation Service, Washington, DC. Retrieve from http://soils.usda.gov/technical/classification/tax_keys/

Somrithipol, S. (1997) Decomposition of bamboo and Rang (*Shorea siamensis* Miq.) leaf litters in mixed deciduous forest and their decomposition fungi. Retrieve from http://agris.fao.org/agris-search/search/display.do?f=2000/TH/TH00010.xml;TH2000001591

Suksawang, S. (1995). Site overview: Thong Pha Phum study site. *Proceedings of the international workshop on the changes of tropical forest ecosystems by El Nino and others*, JSTA, NRCT, and JISTEC, pp. 33−37, 7-10 February 2005, Kanchanaburi, Thailand.

Suzuki, R., Takeda, S., & Thein, H. M. (2007). Chronosequence changes in soil properties of teak (*tectona grandis*) plantations in the Bago mountains, Myanmar. *Journal of Tropical Forest Science* 19, 207–217. ISSN 0128-1283

Takahashi, M., Furusawa, H., Limtong, P., Sunanthapongsuk, V., Marod, D., & Panuthai, S. (2007). Soil nutrient status after bamboo flowering and death in a seasonal tropical forest in western Thailand. *Ecological Research* 22, 160–164. ISSN 0912-3814

Takahashi, M., Hirai, K., Limtong, P., Leaungvutivirog, C., Suksawang, S., Panuthai, S., Anusontpornperm, S., & Marod, D. (2009). Soil respiration in different ages of teak plantations in Thailand. *JARQ, Japan Agricultural Research Quarterly* 43, 337–343. ISSN 0021-3551

Takahashi, M., Hirai, K., Limtong, P., Leaungvutivirog, C., Panuthai, S., Suksawang, S., Anusontpornperm, S., & Marod, D. (2011). Topographic variation in heterotrophic and autotrophic soil respiration in a tropical seasonal forest in Thailand. *Soil Science and Plant Nutrition* 57, 452–465. ISSN 0038-0768

Tanaka, N., Hamazaki, T., & Vacharangkura, T. (1998). Distribution, growth and site requirements of teak. *JARQ, Japan Agricultural Research Quarterly* 32, 65–77. ISSN 0021-3551

Tangsinmankong, W., Pumijumnong, N., & Moncharoen, L. (2007). Carbon stocks in soil of mixed deciduous forest and teak plantation. *Environment and Natural Resources Journal* 5, 80–86. ISSN 1686-5456

Tangtham, N. (1992). Soil erosion problem in teak plantation. In: *Proceeding of the Seminar on 50 Anniversary of Huay-Tak Teak Plantation: 60th Birthday Celebration of Her Majesty the Queen of Thailand*, Lampang, Thailand, 5-8 Aug 1992. pp.247–259. (In Thai with English summary). Retrieve from http://agris.fao.org/agris-search/search/display.do?f=1998%2FTH%2FTH98008.xml%3BTH1997020402

Terakunpisut, J., Gajaseni, N., & Ruankawe, N. (2007) Carbon sequestration potential in aboveground biomass of Thong Pha Phum National Forest, Thailand. *Applied Ecology and Environmental Research*. 5. 93–102. ISSN 1589-1623

Thaiutsa, B., Suwannapinunt, W., & Kaitpraneet, W. (1978). Production and chemical composition of forest litter in Thailand. *Forest Research Bulletin*. 52. ISSN 0216-4620, Kasetsart University, Bangkok, Thailand.

Toriyama, J., Ohta, S., Ohnuki, Y., Ito, E., Kanzaki, M., Araki, M., Chann, S., Tith, B., Keth, S., Hirai, K., & Kiyono, Y. (2011). Soil carbon stock in Cambodian monsoon forests. *JARQ, Japan Agricultural Research Quarterly* 45, 309–316. ISSN 0021-3551

Usuga, J. C., Toro, J. A., Alzate, M. V., & de Jesús Lema Tapias, A. (2010). Estimation of biomass and carbon stocks in plants, soil and forest floor in different tropical forests. *Forest Ecology and Management* 260, 1906–1913. ISSN 0378-1127

Viriyabuncha, C. Chittachumnonk, P., Sutthisrisinn, C., Samran, S. & Peawsa-ad, K. (2002). Adjusting equation to estimate the above-ground biomass of teak plantation in

Thailand, In. *Proceedings of the 7th of silvicultural seminar: Silviculture for commercial plantations*, Kasetsart University. Bangkhen Campus, Bangkok (Thailand). ISBN 974-537-141-6. pp. 239-260 (In Thai with English summary)

4

Ecohydrology and Biogeochemistry in a Temperate Forest Catchment

Su-Jin Kim[1], Hyung Tae Choi[1],
Kyongha Kim[2] and Chunghwa Lee[1]
[1]Division of Forest Water & Soil Conservation
[2]Division of Forest Disaster Management
Department of Forest Conservation
Korea Forest Research Institute
South Korea

1. Introduction

The assimilated carbon stored in terrestrial ecosystems is exported with water movement in both organic and inorganic forms, which are defined as particulate organic carbon (POC), dissolved organic carbon (DOC), and dissolved inorganic carbon (DIC). The transport of terrestrial carbon into streams, rivers and eventually the oceans is an important link in the global carbon cycle (Ludwig et al., 1996; Warnken and Santschi, 2004). The Committee on Flux of Carbon to the Ocean estimated that of the organic carbon entering rivers globally, around 50% is transported to the ocean, 25% is oxidized within the system and 25% stored as POC in the system as sediment (Hope et al. 1994). As compared to the terrestrial carbon sinks (1.9 Gt-C/yr; Prentice et al., 2001), the organic carbon transport from terrestrial ecosystems to oceans is 0.4–0.9 Gt-C/yr (Meybeck, 1982; Hope et al., 1994; Prentice et al., 2001), representing a substantial component of the ecosystem carbon balance.

The water and carbon cycles in forest catchments are important elements for understanding the impact of global environmental changes on terrestrial ecosystems. Various theories have been suggested to better understand water discharge (Horton, 1933; Betson, 1964; Kirkby, 1978; Anderson and Burt, 1991; Kim et al., 2003) and its effect on carbon efflux processes from forest catchments (McGlynn and McDonnell, 2003; Kawasaki et al., 2005; Schulze, 2006; Kim et al., 2007b; Kim et al., 2010). Most of the results indicated that the hydrological flowpaths are important in carbon dynamics within the forest catchments.

Data from major results show export of organic carbon to be highly correlated with annual river discharge and watershed size (Table 1; Fig. 1). Hydrological processes strongly affect organic carbon discharge from terrestrial ecosystems, especially in monsoon climate zone of East Asia, and 60-80% of annual organic carbon export to the ocean during summer rainy season (Tao, 1998; Liu et al., 2003; Kawasaki et al., 2005; Zhang et al., 2009; Kim et al., 2010).

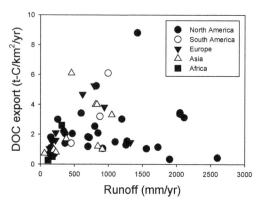

Fig. 1. Annual DOC export by rivers from watersheds (modified from Ittekkot and Laane (1991), Hope et al. (1994), and Table 1)

Ecosystem type, location	Annual precipitation (mm)	Annual runoff (mm)	Watershed size (km²)	Loss of organic carbon		Reference
				DOC (t-C/ha/yr)	POC (t-C/ha/yr)	
Huanghe, Semi-arid area, Central China	-	59	745,000	0.007	-	Gan et al. (1983)
Yichun, Humid temperate area, North-eastern China	500-650	-	2500	0.03	-	Tao (1998)
Luodingjiang River, Subtropical mountainous, Southern China	1534	844	3164	0.012	0.011	Zhang et al. (2009)
Guandaushi, Subtropical forest, Taiwan	2300-2700	-	0.47	0.025	-	Liu et al. (2003)
Tomakomai, Cool temperate mixed forest, Northern Japan	1200	-	9.4	0.0052	0.0076	Shibata et al. (2005)
Kiryu, Temperate conifer, Central Japan	1645	911	0.006	0.01	-	Kawasaki et al. (2005)
Gwangneung, Temperate deciduous, Central Korea	1332	809	0.22	0.04	0.05	Kim et al. (2010)
Han River, Temperate area, Central Korea	1244	-	26,018	0.04	0.02	Kim et al. (2007a)

Table 1. Export of organic carbon East Asian watersheds

Forests are the major terrestrial biome, in which soils and vegetation are the primary sources of DOC and POC in streamwater. Within the forest soil profile, concentrations of DOC typically are highest in the interstitial waters of the organic-rich upper soil horizons (McDowell and Likens, 1988; Richter et al., 1994; Dosskey and Bertsch, 1997). Both column experiments and field observations have indicated that significant transport of DOC occurs by preferential flow, given that the state of adsorption equilibrium cannot be reached, owing to the reduction of the contact time between DOC and the soil surface (Jardine et al., 1989; Hagedorn et al., 1999). Understanding the flow paths of DOC discharge from forested catchments to streams is important because DOC provides a source of energy to microorganisms in water systems (Stewart and Wetzel, 1982) and carbon fixation in the soil (Neff and Asner 2001; Kawasaki et al., 2005).

Since the 1970s 2.3 million hectares of land have been planted with coniferous species such as *Pinus Koraiensis, Abies holophylla* and *Larix leptolepis* in South Korea. Because coniferous forests lose and consume water resources much more than deciduous forests due to higher leaf area index (LAI) and year-round transpiration, theses planted coniferous forests may deteriorate the physical properties of the topsoil owing to the water repellence of the soil surface and decrease the availability of water resources by high evapotranspiration rates. To conserve soil and water resources, densely planted coniferous forests must be managed using silvicultural techniques (e.g. pruning, thinning) that could influence water quantity and quality. In South Korea, various studies have shown that forest management practices in coniferous forests decreased the amount of interception loss and increased discharges during the dry season. Clear cutting resulted in catastrophic augmentation of runoff and soil loss. The water quality of stream headwaters improved after thinning and pruning because these techniques tended to ameliorate soil physical properties and increase soil ion exchange capacity.

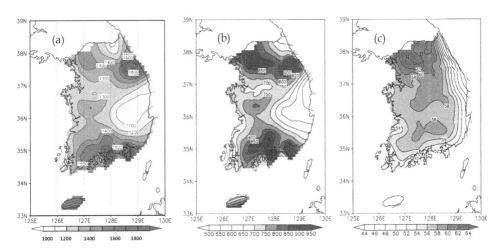

Fig. 2. Accumulated climatological precipitation (mm) at 60 stations for (a) annual total precipitation, (b) summer (from June to August) in South Korea. (c) Percentage (%) of accumulated climatological precipitation with respect to annual total precipitation summer. (Seo et al., 2011)

Hydrological circumstances in South Korea are unfavourable to manage water resources. Temporal and spatial variations of rainfall are very large (Fig. 2). Annual rainfall ranges from 754 to 1,683 mm. In South Korea, more than 50% of the annual precipitation falls in the summer monsoon season (Fig. 2(b, c)), which quickly discharges to the ocean due to the steep slopes and short river lengths (<500 km). Therefore, the water regime in the catchment undergoes drastic changes with recurring wet and dry seasons, which makes it difficult to interpret and predict hydrological processes and subsequently their effect on nutrients cycling (Kim et al., 2009).

The amount of water storage capacity in a forest stand increases with the forest aging, when a forest stand grows, the amount of litter falls and roots also increases. The mineral soils and humus materials tend to aggregate into their structure. The aggregation may change the distribution of pore sizes and often increase the total porosity of soil. Forests have been called 'Green Dam' or 'Reservoir' because of its function of controlling the flood and drought through a litter layer and topsoil like a sponge filter. Net infiltration rate of a well-developed forest soil has been estimated 76 mm/hr in comparison with 8mm/hr of a bare land (Brooks et. al. 1991). Generally, infiltration capacity of soil in a deciduous forest is higher than that in a coniferous forest because the litter fall of the former is easily decomposed and incorporated with mineral particles compared to that of the latter. Most of stream water in South Korea comes from mountain headwaters as mountains occupy 65% of the total area. Stream in forested headwaters yields clean water. Forest soils hold the water like sponge, which is 3.3 times more than the soil in a bare land. Water holding capacity of forest soil in Korea is estimated about 18 billion tons, as shown in Table 2 (Ministry of Science and Technology, 1992).

Bed rock	Igneous	Metanorphic	Basalt	Sedimentary I	Sedimentary II	Lime	Total
Maximum storage	A 34.3	40.1	32.4	36.1	33.9	39.8	
capacity(%)	B 39.5	44.4	35.1	39.0	35.5	41.1	
Total storage	A 15.2	20.0	5.0	4.0	1.5	2.1	
(0.1 billion tons)	B 36.5	60.3	8.5	14.2	4.6	7.9	
Sub-total	51.7	80.3	13.5	18.2	6.1	10.0	179.8

Table 2. Water holding capacity of forest soils depending on bed rock in South Korea (Ministry of Science and Technology, 1992)

Despite decades of dedicated scientific efforts on these fundamental questions, it is still difficult to find a robust interpretation even for some basic hydrological processes such as discharge and runoff. The up to date results showed that the geophysical and meteorological conditions greatly affect the hydrological processes (Hooper et al., 1990; Elsenber et al., 1995; Katsuyama et al., 2001; McGlynn and McDonnell, 2003; Kim et al., 2010).

In this chapter, we have implemented a comprehensive ecohydrological measurement system at the temperate forest catchment in South Korea. Most importantly, high quality long-term data of hydrological and meteorological conditions have been collected, which may be also important in monitoring global environmental changes and their effects. The study was also designed based on a nested watershed concept (smaller catchments are nested in successively larger catchments) to investigate how catchment processes change as scale varies. In this chapter, we introduce the concepts and techniques that were implemented to investigate the movement of water and carbon in a forest catchment. We also briefly discuss preliminary results and their implications for the interactions between hydrological and biogeochemical processes in a temperate forest catchment.

2. Hydrological cycle of forested catchments in South Korea

2.1 Interception loss and evapotranspiration

The differences in the amount and process of interception loss and evapotranspiration depend on the factors of the forest structure and local climate. The forest structure includes the forest type, age and density. Generally, coniferous forests intercept rain and snowfall more than deciduous because the former has higher LAI and longer leaf-period than the latter.

The first research on interception loss by tree canopy and stem in Korea had conducted in 1935 for determining the total and net precipitation at the forest stand in Korea Forest Research Institute (Kim and Jo, 1937). The experiment was conducted during 23 months on the natural 50-year-old red pine (*Pinus densiflora*) with the tree height of 12 m and DBH of 18 cm. It showed that the annual total and net rainfall were 1,194.2 mm and 1,066.3 mm, respectively. The percentage of interception loss from the total rainfall varied from less than 10% in the season to more than 26% in the growing dormant season.

To clarify the effects of forest types on interception, three types of forest were chosen in Gwangnung experiment station during the period of 1982 to 1988, namely natural matured-deciduous, planted young-coniferous and rehabilitated mixed forest. The results of the research are shown in Table 3 (Lee et al. 1989). Among the three forest types, the planted young-coniferous forest showed the most interception loss of 32.6%, compared to 29.1% in natural matured-deciduous and 18.5% in rehabilitated mixed forest. Even though the naturally matured deciduous has the largest forest structure of 80 years old, its amount of interception loss resulted in less percentage compared with the planted young-coniferous for the cause mentioned above.

Forest type	Precipitation (mm)	Intensity (mm/hr)	Throughfall (mm)	Steamflow (mm)	Interception (mm)	Interception (%)
Mixed	1733.2	6.7	1312.6 (75.8)	99.2 (5.7)	321.4	18.5
Coniferous	1477.3	6.0	945.2 (64.0)	50.4 (3.4)	481.5	32.6
Deciduous	1172.4	6.2	758.7 (64.7)	72.2 (6.2)	341.5	29.1

( ) means % for precipitation

Table 3. The amount of interception loss by three forest types in South Korea

In other results, the young coniferous forest of 26-year-old *Pinus rigitaeda* and deciduous forest of 16-year-old *Quercus mongolica* intercepted 17.4 and 13.9% of the total rainfall, respectively, during the period of July 1986 to September 1987 in Seoul National University's Gwanak arboretum (Kim and Woo 1988a, b).

It is difficult to measure the exact amount of interception loss due to the large variations of forests and climate factors. Several forest hydrologists have tried to predict the amount by using an interception model. There are three kinds of interception model for the estimation of the processes and amount of interception; the dynamic, analytical and regression-methods. The dynamic-interception model was developed using the forest stand structure and Penman-Monteith model to predict the amount of evaporation under saturation condition (Kim and Woo 1997).

Another loss component of hydrological cycle in forested catchment is evapotanspiration from tree canopy during a period of no rainfall. The amount of evapotranspiration can be estimated by using the water budget method in a short term or a calculating method like penman or Thornthwaite method. The amount of evapotanspiration estimated by Thornthwaite method in three forest types is shown in Table 4 (Kim 1987).

(unit: ton/day/ha)

Type \ Month	Jan	Feb	Mar	Apr	May	Jun	Jul	Aug	Sep	Oct	Nov	Dec
Yangju[a]	0.0	0.0	2.7	15.7	31.0	38.4	48.5	47.7	31.2	15.2	3.6	0.0
Mixed[b]	3.5	4.1	8.7	11.5	16.3	18.4	25.7	26.1	20.6	10.5	10.7	7.3
Gwangnung[a]	0.0	0.0	3.2	17.8	31.4	40.3	50.6	49.9	32.4	17.0	4.0	0.2
Coniferous[b]	6.3	6.6	15.7	14.0	33.3	34.0	44.4	56.4	40.3	17.8	12.0	8.2
Deciduous[b]	6.0	4.2	11.5	10.8	20.9	35.7	17.9	29.3	24.6	17.8	10.7	8.2

[a] means the amount of evaportranspiration by Thornthwaite method.
[b] means the amount of evaportranspiration by short-term water budget method.

Table 4. Monthly evaportranspiration determined by means of Thornthwaite and short-term water budget methods.

2.2 Discharge and soil loss variations depending on land cover

In the 1960s and 70s, the main theme on forest hydrology was to evaluate soil and water conservation at the different land cover types. Because in those periods, most of the lands in South Korea were completely devastated all over the country, development of techniques on the erosion control was urgent, especially in the fields (Fig. 3).

Lee et al. (1967) clarified that land cover type influence discharge in the small plot. They concluded that the coniferous plot produces the least discharge (26%) while bare land produces most (76%). They also found that the discharge in the bare land plot started at the rainfall of 10 mm and increased radically at the rainfall more than 80 mm, whereas in the coniferous plot started at the rainfall of 30 mm.

Kim (1987) estimated effects of floods, direct flow reduction and long-term yields on forest by utilizing the measured rainfall-runoff data from the three above-mentioned experimental catchments. He found that the flood peak discharge of the young coniferous and mature deciduous stands were 49% and 36% of the devastated-mixed stand respectively. Also, the direct flow dropped to 53% in the young coniferous stand and to 55% in the mature deciduous stand, compared to that of the devastated-mixed catchment.

Lee et al. (1989) analyzed the runoff rate and soil at the the natural deciduous, the planted coniferous and the rehabilitated mixed forests, using the data from 1980 to 1988. The runoff rates of the three forest types in an order above were 61.9%, 48.5% and 71.3%, respectively. They concluded that the natural deciduous forest mitigated the peak of flow during the rainy season while it discharged more of low-flow during the dry season, in comparison with the rehabilitated mixed forest. The amount of soil loss during the rainy season was the highest in the rehabilitated mixed forest (2.2 ton/ha/yr) and the least in the deciduous one (0.7 ton/ha/yr).

Fig. 3. Before and after the construction of a hillside planting work in South Korea

Several techniques for analyzing the discharge components include surface runoff, interflow and groundwater, for different forest types in a long-term. The recession coefficient represents the rate of runoff that is released from a soil and streamside. If the recession coefficient of an independent event in the hydrograph changes statistically with the lapse of time, the hydrological characteristics of the forested catchment would be changed. Korea Forest Research Institute (1998) studied the hydrological variation of discharge, soil loss and recession coefficient in three small, forested catchments, using a long-term hydrological data from 1983 to 1992. This study included the naturally matured deciduous, planted coniferous and erosion-controlled mixed forest. The amount of discharge and soil loss varied with the rainfall and forest type. Fig. 4 (up left) shows the variation of the recession coefficient of surface runoff ($\alpha 1$) for 10 years. $\alpha 1$ gradually decreases in the coniferous forest while it does not show the tendency in the others. This may be caused by the change of the forest structure in the coniferous forest after the planting. The amount of the initial loss by the interception and transpiration has been greatly increased since 1976 as the coniferous trees grow. However, the forest structures in others have not much changed since 1983.

The recession coefficient of interflow ($\alpha 2$) decreased in the coniferous and mixed forests with time Fig. 4 (up right). This can be interpreted by an increase in the soil storage capacity after the planting and erosion control work. As the amount of evapotranspiration increases, the storage opportunity of rainfall in the soil improves. Increment of the storage capacity may result in delaying the releasing time of interflow from the soil.

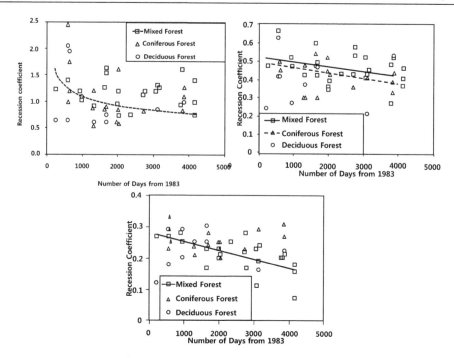

Fig. 4. Variation of the recession coefficient of surface runoff (up left), interflow (up right), and groundwater (down) for 10 years

In the case of the recession coefficient of groundwater ($\alpha 3$), only mixed forest showed a gradual reduction for 10 years (Fig. 4. down). The mixed forest has been a devastated land until erosion control work had finished in 1974. After the work, the soil layer rapidly formed and the soil's physical properties improved.

2.3 Residence time of water in a forest catchment

Various radioactive tracers have provided valuable information regarding hydrological processes, such as mean residence time of water, flowpaths during storm events, groundwater movement, and biogeochemical reactions occurring along the flowpaths (Michel and Naftz, 1995; Shanley et al., 1998; Sueker et al., 1999). For example, ^{3}H and ^{14}C have been widely used for determination of time scale of hydrological processes (Matsutani et al., 1993). However, these tracers are inadequate for studying hydrological processes in small and headwater catchments with expected time scales of a year or less because of their long half lives (decadesto thousands of years). In this study, we will introduce a short-lived cosmogenic radioactive isotope of [35]S (half life = 87 days) for measuring the mean residence time of water in the Gwangneung catchment.

The measured activity of ^{35}S in water can be expressed as an equation:

$$C = C_o e^{-\lambda t}$$ (1)

where C_o is the initial ^{35}S activity, λ is the decay constant (0.0079655), t is the number of days from the start of decay, and C is the measured ^{35}S activity. The ^{35}S activity in water provided information of the residence time of atmospherically deposited sulfate. Biogeochemical reactions such as adsorption/desorption in soil and groundwater are also important in affecting the calculated residence time of water in a forested catchment. Assuming a conservative response of sulfate in streamwater, the mean residence time of water was < 40 days during the summer monsoon period in the natural deciduous forest catchment. However, the mean residence time of water increased to around 100 days in the dry season with increasing contribution of the base flow to the stream water (Fig. 5). These results demonstrate that ^{35}S is useful in estimating the age of water exiting a small catchment where the time scales of hydrologic processes are on the order of 1 year or less.

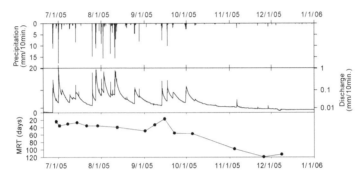

Fig. 5. Temporal variation in mean residence time (MRT) calculated from ^{35}S based method along with the precipitation and stream discharge (Kim et al., 2009)

From this MRT estimate, the existence of substantial, long-term subsurface water storage is not supported in the studied catchment. The assumed rapid turnover of water in the catchment indicates that the hydrological conditions will respond to the change in precipitation directly and immediately. Therefore, surplus (flooding) and shortage (drought) of water supply may alternate at a relatively short time scale (even within a year) depending on the seasonal distribution of precipitation. A secure water resource planning in catchments of this type will require a reliable prediction and efficient management of precipitation and surface water bodies (Kim et al., 2009).

2.4 Flow paths of water during storm events

The identification of flow paths in forested catchments has been elusive because of difficulties in measuring subsurface flow. Forested catchments are spatially complex and subsurface flow is invisible. Hence, one can only infer the movement and mixing of water from the natural tracer elements that the water carries (Pinder and Jones, 1969). Using various tracers, the end-member mixing analysis (EMMA) has been used to elucidate flow paths and hydrological processes in several catchments (e.g. Hooper et al., 1990; Christophersen et al., 1990; Elsenbeer et al., 1995; Katsuyama et al., 2001). Numerous conceptual models have adopted the flow path dynamics proposed by Anderson et al. (1997), i.e., both pre-event soil water and bedrock groundwater contribute to the formation of a saturated zone in the area adjacent to the stream (e.g., McGlynn et al., 1999; Bowden et al., 2001; Uchida et al., 2002).

The EMMA can be applied for individual storm events to quantitatively evaluate the contribution of each solutions component. The source waters are called 'end members'. The tracer concentrations of end members are more extreme than stream water since streamwater is a mixture of these sources (Fig. 6). In order to apply EMMA, (1) tracers should be conservative, (2) sources should be significantly different in tracer concentrations, (3) unmeasured sources must have same concentration with known sources or don't contribute significantly, and (4) the sources should maintain a constant concentration. Typical source waters are those from organic rich soil horizon, hillslope groundwater, valley bottom groundwater, throughfall, and precipitation.

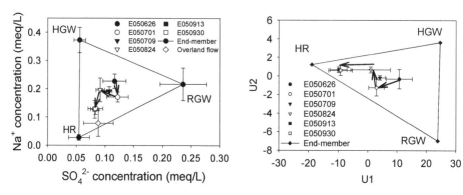

Fig. 6. Three-component mixing diagram for each storm event (left) and mixing diagram showing stream water evolution and end-member composition in U space during six storm events (right)

The hydrological characteristics of the six storm events observed during the summer of 2005 are summarized in Table 5. The maxima of precipitation intensity and discharge intensity were observed on 1 July 2005, which were 17.7 mm/10min. and 1.0 mm/10min., respectively. Stream discharge as a proportion of total precipitation ranged from 15 to 60% with an average of 30%. The maximum discharge rate also was observed in E050701, but associated with 5 days' antecedent precipitation.

The end-member mixing analysis (EMMA) with principal components analysis (PCA) was applied to each storm event to evaluate quantitatively the contribution of each water component (Christophersen and Hooper, 1992; Burns et al., 2001). Three-component mixing diagrams are shown in Fig. 6. Stormflow in E050626 lay near the groundwater end-member, and moved to soil water in E050701. Stormflow were also closer to that of groundwater through E050709 and E050824. After moving near groundwater, stormflow were closer to that of throughfall in E050913 and E050930. Stormflow solutes in E050913 and E050930 were not significantly different from overland flow.

In E050913 and E050930, the values of water-filled porosity in the surface layer (0–0.1 m) was about 5% higher than the maximum observed during the previous storm events. This higher water-filled porosity (as compared to prior storm events) led to a low water infiltration rate and an increase in the contribution of surface discharge. Previous studies suggested that a maintained precipitation expands the saturation zone and increases macropore flows in the forested catchment (e.g., McDonnell, 1990). Such macropore flows deliver new water in which dissolved ion concentrations are low because of the short contact time with soil and bedrock (Burns et al., 1998). The calculated mean residence time of water based on the ^{35}S analysis varied with changing water regime in the study area, ranging from 20 to 40 days during the summer monsoon period (Kim et al., 2009). Especially, for the stream water sample taken on 15 September when the surface runoff increased due to the storm event, the mean residence time of water also decreased abruptly (Kim et al., 2009; Fig. 5).

	E050626	E050701	E050709	E050824	E050913	E050930
Observed period	26–28, Jun.	1–3, Jul.	9–10, Jul.	24–26, Aug.	13–15, Sep.	30, Sep.–2, Oct.
Total precipitation (mm)	160.5	104.0	40.5	83.5	85.5	87.0
Max. precipitation intensity (mm/10min)	11.1	17.7	2.5	4.5	7.5	2.5
Total discharge (mm)	23.6	61.5	11.5	22.8	18.1	29.1
Max. discharge intensity (mm/10min)	0.32	1.05	0.06	0.14	0.28	0.16
Total discharge / Total precipitation (%)	15	60	28	27	21	33
Antecedent precipitation (5 days)	0.0	161.9	1.3	1.5	7.0	1.0
Antecedent precipitation (10 days)	1.3	161.9	154.3	19.5	7.0	43.5

Table 5. Hydrological characteristic of storm events in 2005

3. Dynamics of water and dissolved materials in forest soils

The dynamics of water in the soil layer are important for the understanding of water storage and dissolved material fluxes in a forest catchment. In the field measurement, an intensive monitoring is useful using a precise multiplex Time Domain Reflectometry system to capture and characterize variation patterns of soil moisture on a steep hillslope. Here, we introduce the methods for estimating the water and dissolved material flux in soils with tensiometer and water table fluctuations.

3.1 Estimation of soil water and dissolved material flux using a tensiometer

Tensiometer consists of a pressure transducer which measures the pressure (when saturated) or tension (when unsaturated) that the soil moisture exerts on a column of water, a porous cup which is in contact with the soil water at the measurement level, and a water body with a PVC pipe. According to Kim (2003), the one-dimensional, vertical water flow equation for unsaturated soil in a compartment can be written as:

$$Q_{in} = Q_{out} - E + \Delta W \tag{2}$$

where Q_{in} and Q_{out} are input and output of water to and from the compartment, respectively, E is the evapotranspiration, and ΔW is the change of water content in the compartment during the period. For example, Q_{in} in the 0-10 m soil compartment can be obtained from the throughfall measurement, and ΔW, E by direct observations. The calculated Q_{out}, in turn, becomes Q_{in} for the 0.1-0.2 m soil compartment. Therefore, the equation can be used to calculate the water flux through a series of compartments up to 1.0 m soil depth.

E can be calculated from temporal variations of evapotranspiration (Suzuki, 1980).

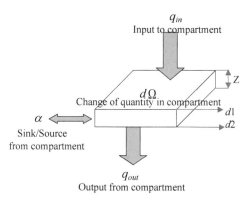

Fig. 7. Calculation of dissolved material flux in soil compartment (Kim, 2003)

$$E_{d1-d2} = cE \tag{3}$$

where E_{d1-d2} is evapotranspiration in soil depth from $d1$ to $d2$, E is the total evapotranspiration from the entire soil column, and c is the proportion of E_{d1-d2} to E. For example, c in the 0-0.1 m soil compartment (if the total soil depth is 1.0 m) during time t is calculated from the change of water content by using equation (4).

$$c = \frac{\left(\theta_{0-10}^{t+\Delta t} - \theta_{0-10}^{t}\right)}{\left(\theta_{0-10}^{t+\Delta t} - \theta_{0-10}^{t}\right) + \left(\theta_{10-20}^{t+\Delta t} - \theta_{10-20}^{t}\right) + \left(\theta_{20-30}^{t+\Delta t} - \theta_{20-30}^{t}\right) + \left(\theta_{30-50}^{t+\Delta t} - \theta_{30-50}^{t}\right) + \left(\theta_{50-100}^{t+\Delta t} - \theta_{50-100}^{t}\right)} \tag{4}$$

ΔW can be calculated from the change of water content, which is derived from the relationship between θ and ψ (Kosugi, 1994; Kosugi, 1996).

$$\Delta W = \left(\theta_{(d1+d2)/2}^{t+\Delta t} - \theta_{(d1+d2)/2}^{t}\right) \cdot Z \tag{5}$$

where θ_d^t is water content during time t at soil depth $(d1+d2)/2$, and Z is soil thickness.

Dissolved ions and compounds in soils move with water infiltration processes. Therefore, dissolved material flux is calculated by multiplying dissolved material concentration with the water flux. The calculation method of dissolved material flux is described in Fig. 7. The dissolved material flux is calculated from the change of quantity in a compartment. The sink/source (a) property of the compartment can be estimated from q_{in}, q_{out} and the change of quantity in the compartment ($d\Omega$), such as:

$$a = d\Omega - (q_{in} - q_{out}) \tag{6}$$

where $d\Omega$ is calculated from the concentration of dissolved materials and water content.

$$d\Omega = \left(\theta_{(d1+d2)/2}^{t+\Delta t} \cdot S_{(d1+d2)/2}^{t+\Delta t} - \theta_{(d1+d2)/2}^{t} \cdot S_{(d1+d2)/2}^{t} \right) / Z \tag{7}$$

where $S_{(d1+d2)/2}^{t}$ is the dissolved material concentration during time t at soil depth $(d1+d2)/2$. The equation (7) indicates the change of dissolved material budget in the soil compartment during time t. Moreover, q_{in} and q_{out} at depth d can be described as:

$$q_{in} = \left(f_{d1}^{t} + f_{d1}^{t+\Delta t} \right) / 2 \cdot \Delta t \tag{8}$$

$$q_{out} = \left(f_{d2}^{t} + f_{d2}^{t+\Delta t} \right) / 2 \cdot \Delta t \tag{9}$$

where f_{d1}^{t} is dissolved material flux at soil depth $d1$ during time t.

3.2 Estimation of water infiltration rate using a water table fluctuation

The water infiltration rate can be calculated indirectly from the groundwater recharge rate. To estimate the water infiltration rate, the groundwater recharge rate from the water table fluctuation can be calculated as follows (Moon et al., 2004):

$$\alpha = \frac{\Delta h}{\sum P} \times S_y \tag{10}$$

where α is the recharge rate, h is the change of groundwater level, P is precipitation, and S_y is the specific yield. On specific conditions, groundwater recharge rate may practically represent the infiltration rate. We can also estimate the dissolved material flux, such as dissolved organic carbon (DOC) by multiplying groundwater recharge rate with the measured concentration. This technique has been applied to the headwater region in the Gwangneung catchment, and its reliability has been critically evaluated by comparing with other methodologies. The uncertainty of this technique is largely due to the measurement error of specific yield (S_y) caused by the heterogeneity of geologic materials, and other factors influencing the water table fluctuation such as changes in atmospheric pressures, air entrapment during the infiltration of water, irrigation, and pumping (Choi et al., 2007).

According to the results from the water infiltration rates, 0.44 t-C ha[-1] DOC was infiltrated into the soil from late June to early October in 2005, which represented approximately 8% of the stored carbon in the forest floor (5.6 t-C ha[-1]; Lim et al., 2003) and 30 to 50% of NEE (-0.84 to 1.56 t-C ha[-1] yr[-1]; Kwon et al., 2010) (Fig. 8). These results indicate that a considerable amount of decomposed organic matter is stored in the soil through water movement processes. If most of the infiltrated DOC were to accumulate as soil organic carbon in the shallow soil and to be decomposed in the deep soil, then 0.5% of the soil carbon (92.0 t-C ha[-1]; Lim et al., 2003) would be retained from DOC during the summer monsoon (Fig. 8).

While these values seem to be relatively small, soil organic carbon can be accumulated in the mineral soil for an extended period (e.g., Michalzik et al., 2003); potentially making the 0.5% of soil carbon retained from DOC during the summer monsoon an important component of the forest carbon budget to consider (e.g., Battin et al., 2009).

Based on these estimates of NPP ranging from 4.3 to 5.8 t C ha⁻¹ yr⁻¹, the observed amount of total DOC and POC effluxes is roughly 2% of the annual NPP – a small but non-negligible amount in terms of net ecosystem carbon exchange (NEE). Considering the averaged NEE of -0.84 t C ha⁻¹ yr⁻¹ (negative sign indicates net uptake of carbon by the forest; Kwon et al., 2010), approximately 10% of NEE would escape from this forest catchment as DOC and POC (Fig. 8). Our results further indicate that 50 and 80% of the respective annual DOC and POC effluxes were transported out of this forest catchment during the summer monsoon period.

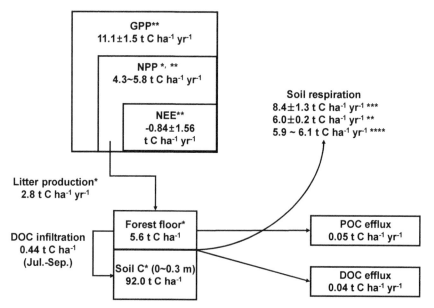

Fig. 8. The contribution of DOC and POC to the carbon budget in the Gwangneung deciduous forest catchment. * Lim et al. (2003; observation periods 1998 to 1999), ** Kwon et al. (2010; observation periods 2006 to 2008), *** Chae (2008; observation periods 2001 to 2004). The difference of soil respiration is due to difference of observation periods and methods. Modified from Kim et al., 2010.

3.3 Adsorption of DOC in forest soil

Many field studies have shown that the concentration of DOC in soil water significantly decreases with increasing soil depth (Fig. 9). It is generally assumed that adsorption of DOC to the surface of mineral soil is important than decomposition in reducing DOC concentrations. Various sorption mechanisms have been reported, including anion exchange, cation bridging, physical adsorption, etc. (Jardine et al., 1989; Gu et al., 1994; Edwards et al., 1996; Kaiser and Zech, 1998a; Kaiser and Zech, 1998b). These DOC sorptions are irreversible under natural soil conditions (Gu et al., 1994). Because Fe and Al oxides are the most important sources of variable charge in soils (Jardine et al., 1989; Moore et al., 1992; Kaiser and Zech, 1998a), DOC adsorption can be related quantitatively to the Fe and Al oxide contents of soils (Moore et al., 1992). The proportion of clay in mineral soil is also an important factor for DOC adsorption. DOC concentrations in catchment runoff are negatively correlated with the clay contents of soils in the catchment. The adsorption process is relatively rapid, which completed within 2 to 12 hours (Kaiser and Zech, 1998b). The effect of pH on the adsorption of DOC in forest soil is also important. Tipping and Woof (1990) calculated that an increase in soil pH by 0.5 units would lead to an increase by about 50% in the amount of mobilized organic matter. Nodvin et al. (1986) also calculated the reactive soil pool of DOC under various pH conditions.

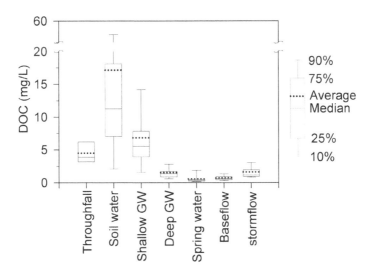

Fig. 9. Spatial variations in the concentrations of DOC of throughfall, soil water, shallow groundwater (0.5 m), deep groundwater (0.8-1.0 m), spring water, and baseflow, with respect to and stormflow (Kim et al., 2007b)

3.4 Temporal and seasonal change of DOC export from temperate forest catchment

Typical temporal variations in DOC concentrations during storm events are shown in Fig. 10. With the onset of heavy precipitation, DOC concentration in streamwater increases significantly, and after the precipitation ceased, DOC concentrations returned to pre-storm levels. The results from the hydrograph separation during storm events indicated that a large amount of water discharged through surface and subsurface soil layers (Fig. 6). DOC concentration in the surface soil is higher than the deep soil and the groundwater (Fig. 9). The Storm event leads to the increase in the surface runoff with a high DOC concentration. During the baseflow period, most stream waters flow out from the groundwater with a low DOC concentration (Fig. 11). These results indicate that hydrological processes strongly affect the DOC export and thereby the carbon budget in the catchment.

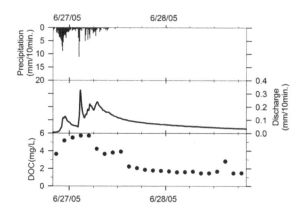

Fig. 10. Precipitation, stream discharge and temporal variations of DOC concentration in streamwater during storm event (Kim et al., 2007b)

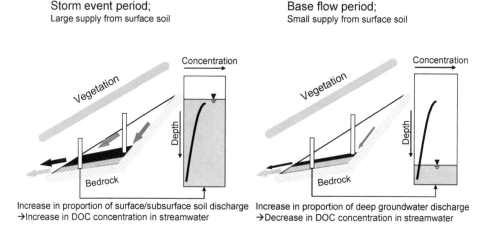

Fig. 11. Precipitation, stream discharge and temporal variations of DOC concentration in streamwater during storm event (Kim et al., 2007b)

Forest treatments		Throughfall (mm)	Stemflow (mm)	Interception loss (mm)	Interception loss (%)	Rainfall (mm)
Pinus	Not practice	85.8	6.5	74.0	44.6	166.0
Koraiensis	Practiced	100.2	7.5	58.3	35.1	
Abie	Not practice	90.1	7.8	68.1	41.0	
holophylla	Practiced	104.8	8.2	53.0	31.9	
	Deciduous forest	118.6	33.9	12.0	7.3	
	Mixed forest	115.3	17.6	33.1	19.9	

Table 6. Percentage of interception loss by forest practices and vegetation types

4. Effects of forest managements on water cycle and quality

Change of the forest stand structure causes the modification of the hydrological characteristics because the components of water loss by interception and evapotranspiration change immediately, owing to the reduction of LAI. Moreover, physical and chemical properties of forest soil change in a long term.

Kim et al. (1993) conducted a research on the effects of site conditions in headwater stream on water storage of reservoirs on small-forested watersheds. The result shows that the water storage of the reservoirs during the dry season is positively correlated with the tree height, DBH, stand ages and crown closure, but negatively with understory coverage and drainage density.

The first change in hydrological components after forest practices is the interception loss from the tree canopy surface during rainfall. The amount of interception loss decreases after thinning and cutting. The rate of reduction for interception loss is correlated positively with the percentage of thinning and cutting of forest types. Table 6 represents the effects of forest practices and types on the percentage of interception loss. In the coniferous stand, the percentage of interception loss decreased to about 10% after forest practices. The mixed forest intercepted the rainfall in about half the amount of the coniferous, whereas the deciduous stand did in about one-sixths of that. Forest treatments increase not only interception loss but also discharge due to the reduction of loss components such as evapotranspiration. The amount of discharge after the forest practices during the dry season was increased by two and three-tenths of that, respectively, before the treatments.

Generally, forest soil has a filtering property like sponge and conserves soil and water resources. If forests, regardless of the types, are cut clearly, catastrophic amounts of soil and water are produced. Fig. 12 explains the effects of clear cutting on the peak flow in a small-forested catchment. The amount of peak flow in the clear-cut site increased to 78.3 mm compared to the controlled site during the rainfall of 400 mm.

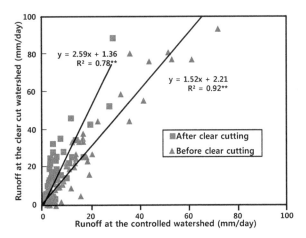

Fig. 12. Change of peak flow after clear cutting

Jeong et al. (1997) analyzed the influential factors on the electrical conductivity of stream and soil water in a small-forested watershed. They concluded that the electrical conductivity was correlated with the total amount of a cation and an anion in stream and soil water. Their results proposed that the amounts of NO_3- and Na^+ in the stream have a statistical significance for the electrical conductivity in streams and the amounts of K^+ and Ca^{2+} and pH in soil water for the electrical conductivity in soil water.

Jeong et al.(1999a, b) further clarified the effect of forest management practices (thinning and pruning) on soil physical properties and water quality to obtain fundamental information on the facility of purifying water quality after forestry practices. They investigated the water quality of rainfall, throughfall, stemflow, and soil and stream water at the coniferous stands that consisted of *Abies holophylla* and *Pinus koraiensis*. The seasonal variation of water qualities of throughfall, stemflow and soil water were decreased after practices. Some researches supported the mesopore ratio on pore geometry of surface soil to be used as an index of the water retention capacity of forestlands.

Jeong et al. (2001a) investigated 23 parameters, including site conditions and soil properties to analyze the influencing factors of mesopore ratio on pore geometry of surface soil in coniferous stands. They found that the factors influencing the mesopore ratio (pF2.7) on the surface soil were macropore ratio (pF 1.6), slope, crown-cover rates, and thickness of F-layer, organic matter contents, and the growing stock. They concluded that crown-cover rates of stands should be controlled to be less than 80% for enhancing the water resource retention capacity in coniferous stands.

Jeong et al. (2001b) investigated fifteen factors, including site conditions and soil properties to analyze the influencing factors of mesopore ratio on a pore geometry of surface soil in deciduous stands. The factors influencing the mesopore ratio (pF2.7) on the surface soil were found to be the tree height, under vegetation coverage and organic matter contents of soil in deciduous stands. Hence, they concluded that the water resource retention capacity would be improved when under vegetation coverage was increased from 30 to 80%.

5. Upscaling of observation data through hydrological modeling

The rainfall-runoff process, which is an important component of the hydrological system, is very complex considering the large number of factors involved and their temporal and spatial distribution. Hydrological modelling is a suitable technique to represent the rainfall-runoff process in various symbolic or mathematical forms using known or assumed functions expressing the various components of a rainfall-runoff response (Ndiritu and Daniell, 1999). In the last half-century there have been hundreds of hydrological response models, each with their own attributes and shortcomings, developed by many different researchers. Furthermore, with the current rapid developments within computer technology and hydrology, the application of computer based hydrologic models is only likely to increase in the near future (Loague and Van der Kwaak, 2004).

The distributed hydrological models aim to better represent the spatio-temporal variability of hydrological characteristics governing the rainfall-runoff response at the catchment scale. One of the distributed hydrological models used commonly is TOPMODEL, which is a quasi-physically based semi-distributed hydrological model (Beven and Kirby, 1979; Beven et al., 1995; Beven, 1997; Beven, 2001; Beven and Freer, 2001a, b).

Most physically based distributed models have parameters which are effective at the scale of the computational elements. In order for a rainfall-runoff model to have practical utility or be useful for hypothesis testing, it is necessary to select appropriate values for the model parameters. Unfortunately, it is not normally possible to estimate the effective values of parameters by either prior estimation or measurement, even given intensive series of measurements of parameter values. Therefore, parameter values must be calibrated for individual applications (Refsgaard and Knudsen, 1996; Refsgaard, 1997; Freer, 1998; Beven, 2001).

In general, the process of parameter calibration has involved some form of determination of a parameter set that gives a simulation that adequately matches the observation. However, many calibration studies in the past have revealed that while one optimum parameter set could often be found, there would usually be a multitude of quite different parameter sets that can produce almost equally good simulation results. Recognition of multiple acceptance parameter sets results in the concept of equifinality of parameter sets (Beven and Freer, 2001b; Beven, 2002; Freer et al., 2003). In addition, in the general case for rainfall-runoff modelling with multiple storm sequences, it might be difficult to assess model performance using a single likelihood measure, because the form of the distribution of uncertain predictions varies markedly over the range of streamflow and the appropriate error structure might vary with both of type of data and the model parameter set (Freer et al., 2003). It may often be the case that the available data are not adequate to allow identification of complex models and/or that a single performance measure (objective function) is not adequate to properly take into account the simulation of all the characteristics of a system used. Thus, the multi-criteria or multi-objective methods using multiple objective functions or other data in addition to rainfall-runoff data may allow more robust analyses of models, and aid hypothesis testing of competing model structures (Gupta et al., 1999; Beven, 2001; Madsen et al., 2002; Freer et al., 2003).

The multi-criteria performance measures based on the concept of equifinality of behavioral model simulations were used for calibration of the rainfall-runoff model, TOPMODEL at natural deciduous forest in South Korea. Totally 100,000 parameter sets uniformly sampled by Monte Carlo Simulations from the ranges for each TOPMODEL parameters, and hourly stream flow and rainfall data observed from April to October, 2005 i n the deciduous forest catchment located in the Gwangnung experimental forests were used for model calibration.

The performance of each parameter set was evaluated and identified with 6 different performance measures against behavioral acceptance thresholds defined for each performance measure, and the results were analyzed focused on the variability and relationship between the behavioral parameter distributions according to the definitions of performance measures.

The results demonstrate that there are many acceptable parameter sets scattered throughout the parameter space, all of which are consistent in some sense with the calibration data, and the range of model behavior for each parameter varied considerably between the different performance measures. Sensitivity was very high in some parameters, and varied depending on the kind of performance measure (Fig. 13). Compatibilities of behavioral parameter sets between different performance measures also varied, and a very small minority of parameter sets could produce reliable predictions regardless of the kind of performance measures (at least, for the performance measures used in this paper).

Especially, the results indicate that using a single performance measure for the calibration of a hydrological model may lead to an increase in model uncertainty. Therefore, careful consideration should be given to the choice of performance measure appropriate to the characteristics of used model and data and the purpose of study.

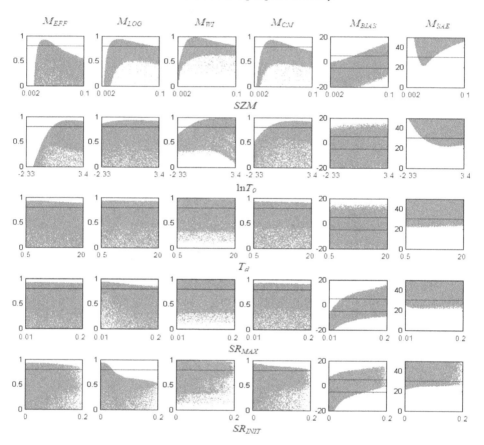

Fig. 13. Scatter plots of likelihood values for TOPMODEL parameters from Monte Carlo simulations of the deciduous forest catchment conditioned on the 2005 discharge period using six performance measures. Each dot represents one simulation with a likelihood weight calculated by a given performance measure, and horizontal lines mean thresholds identifying behavioural parameter sets for each performance measure; dots over the line (in cases of M_{EFF}, M_{LOG}, M_{WI} and M_{CM}), dots between both lines(in case of M_{BIAS}) and dots below the line (in case of M_{SAE}) are classified as behavioural simulations.

Differences in the behavioral parameter distributions according to the performance measures may be directly caused by the definitions of performance measures. However, it also should be considered that the effects of model nonlinearity, covariation of parameter values and errors in model structure, input data or observed variables may be taken into account in the nonlinearity of the response of acceptable model.

The performance of the parameter set can be used to produce the likelihood-weighted marginal parameter distributions for individual parameters, and the likelihood weighted model simulations can be used to estimate prediction quantiles in a way that allows that different models may contribute to the ensemble prediction interval at different time steps and that the distributional form of the predictions may change from time to time step (Fig. 14).

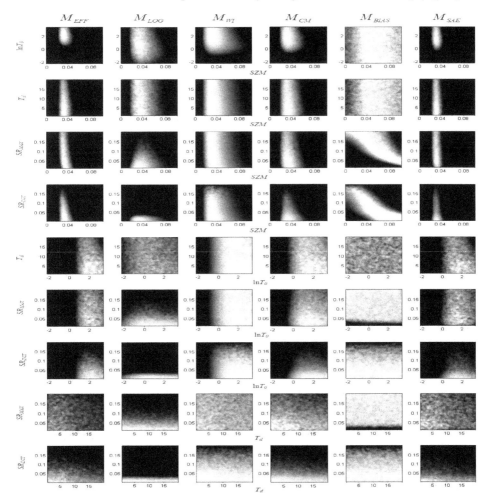

Fig. 14. Likelihood response surfaces between the major parameters of TOPMODEL, conditioned on the 2005 discharge period of the deciduous forest catchment (Behavioural parameter sets with higher model performance are in the white zone.)

6. Conclusions

The ecohydrological and biogeochemical studies have proposed a major scientific question: What is the role of hydrology in the carbon budget of complex forest catchment and how will it change in the hydrologic cycle in monsoon Asia and influence the forest carbon budget? (Kim et al., 2006) To properly answer this question, some of the most fundamental aspects in catchment hydrology need to be clarified i.e., (1) How much water is stored in the catchments? (2) What flowpaths does water take to the stream? (3) How long does water reside in catchments? (4) How can we scale or transfer our observations to other catchments? Despite decades of dedicated scientific efforts on these fundamental questions, it is still difficult to find a robust interpretation even for some basic hydrological processes such as discharge and runoff. The up to date results showed that the geophysical and meteorological conditions greatly affect the hydrological processes (Hooper et al., 1990; Elsenber et al., 1995; Katsuyama et al., 2001; McGlynn and McDonnell, 2003; Kim et al., 2010).

To understand carbon cycling in this catchment better, it is necessary to estimate the annual accumulation and movement of water and DOC in the soil. The organic carbon has been continuously discharged from terrestrial ecosystems of river basin. This organic carbon will contribute for an important sink for carbon through burial in coastal sea sediments or floor. These missing values have to consider for estimation of carbon budget in terrestrial ecosystems. Our results suggest that storm events during summer monsoon (including the typhoon season) are important to estimate flow paths of water and carbon budget in a Korean forested catchment and East Asia. The seasonally concentrated precipitation increases the surface runoff, when the infiltration capacity of the soil decreases during summer monsoon. The outbreak of surface runoff reduced the mean residence time of water in the catchment, and increased DOC export from the surface soil layer. The precipitation also plays an important role in infiltration processes of dissolved material. The precipitation patterns and hydrological processes strongly affect the carbon cycling in the Korean temperate forest during summer monsoon. The increasing occasions of heavy precipitation may not lead to the simultaneous increase of available water resources in the catchment due to the shortening of the water residence time. However, the heavy precipitation will clearly increase material discharge such as DOC. Therefore, the effect of monsoon climate on water and carbon cycling in forest catchment should be critically evaluated on the basis of improved understanding of catchment hydrological and biogeochemical processes.

Our understandings in water and carbon cycling obtained from the hydro-biogeochemical approaches are limited due to the prescribed spatial scale of the measurements. The scaling issues are implicitly built into our field measurements and model representations (Kim et al., 2006). The information provided in this chapter should be carefully considered in modelling formulations at the hydrologic catchment and grid scales of ecohydrological/biogeoche-mical models and satellite image analyses. Such efforts should provide insights as to how various information is transferred across scales, and hence on how to simplify and aggregate measurements, models and satellite products. Future research must be focused on how to make measurements at scales that are appropriate for parameterization and model validation, and how to make the scales of modeling and satellite algorithm converge with those of field measurements (Kim et al., 2006).

7. Acknowledgment

This research was supported by the Development of the Integrated Forest Water Resources Management Technology for Climate Change Mitigation and Adaptation Project of Korea Forest Research Institute. Our thanks go out to the editor who provided valuable comments.

8. References

Anderson, M. G. & Burt, T. P. (1991). *Process Studies in Hillslope Hydrology.* 539, John Wiley and Sons

Anderson, S. P.; Dietrich, W. E.; Montgomery, D.R.; Torres, R.; Conrad, M.E. & Loague, K. (1997) Subsurface flow paths in a steep unchanneled catchment. *Water Resources Research*, Vol.33, pp. 2637–2653

Battin, Y. J.; Luyssaert, S.; Kaplan, L. A.; Aufdenkampe, A. K.; Richter, A. & Tranvik, L. J. (2009). The boundless carbon cycle. *Nature Geoscience*, Vol.2, pp. 598-600

Betson, R. P. (1964). What is watershed runoff? *Journal of Geophysical Research*, Vol.69, pp. 1541-1551

Beven, K. J. (1997). TOPMODEL: a critique. *Hydrological Processes*, Vol.11, No.3, pp. 1069-1085

Beven, K. J. (2001). *Rainfall-Runoff Modelling.* pp. 360, The Primer, Wiley, Chichester

Beven, K. J. (2002). Towards a coherent philosophy for modelling the environment. *Proceedings of the Royal Society of London Series A – Mathematical Physical and Engineering Sciences*, Vol. 458, No.2026, 2465–2484.

Beven, K. J. & Freer, J. (2001a). A dynamic TOPMODEL. *Hydrological Processes*, Vol.15, pp. 1993-2011

Beven, K. J. & Freer, J. (2001b). Equifinality, data assimilation, and uncertainty estimation in mechanistic modeling of complex environmental systems using the GLUE methodology. *Journal of Hydrology*, Vol.249, pp. 11–29

Beven, K. J. & Kirkby, M. J. (1979). A physically-based variable contributing area model of basin hydrology. *Hydrological Science Bulletin*, Vol.24, No.1, pp.43–69

Beven, K. J.; Lamb, R.; Quinn, P.; Romanowicz, R. & Freer, J. (1995). TOPMODEL, In: *Computer Models of Watershed Hydrology*, V. P. Singh, (Ed.), 627–668, Water Resource Publications, Colorado

Bowden, W. B.; Fahey, B. D.; Ekanayakem, J. & Murray, D. L. (2001). Hillslope and wetland hydrodynamics in a Tussock grassland, south island, New Zealand. *Hydrological Processes*, Vol.15, pp. 1707–1730

Brooks, K.N.; Ffollitt, P. F; Gregersen, H. M. & Thames, J. L. (1991). *Hydrology and the management of watersheds.* Iowa State University Press.

Burns, D. A.; Hooper, R. P.; McDonnell, J. J.; Freer, J. E.; Kendall, C. & Beven, K. (1998). Base cation concentrations in subsurface flow from a forested hillslope: The role of flushing frequency. *Water Resources Research*, Vol.33, pp. 3535–3544

Burns, D. A.; McDonnell, J. J.; Hooper, R. P.; Peters, N. E.; Freer, J. E.; Kendall, C. & Beven, K. (2001). Quantifying contributions to storm runoff through end-member mixing analysis and hydrologic measurements at the Panola Mountain Research Watershed (Georgia, USA). *Hydrological Processes*, Vol.15, pp. 1903–1924

Chae, N. (2008). *Soil CO_2 in a temperate forest ecosystem under monsoon climate in northeast Asia.* pp. 26-80, PhD thesis, Yonsei University, Seoul

Choi, I.-H; Woo, N. C.; Kim, S. J.; Moon, S. K. & Kim, J. (2007). Estimation of the groundwater recharge rate during a rainy season at a headwater catchment in Gwangneung. *Korean Journal of Agricultural and Forest Meteorology*, Vol. 9, pp. 75-87, ISSN 1229-5671

Christophersen, N.; Neal, C.; Hooper, R. P.; Vogt, R. D. & Andersen, S. (1990). Modelling streamwater chemistry as a mixture of soilwater end-members – a step towards second-generation acidification models. *Journal of Hydrology*, Vol.116, pp. 307–320

Christophersen, N. & Hopper, R. P. (1992). Multivariate analysis of stream water chemical data: The use of principal components analysis for the end-member mixing problem. *Water Resources Research*, Vol.28, pp. 99–107

Dosskey, M. G. & Bertsch, P. M. (1997) Transport of dissolved organic matter through a sandy forest soil. *Soil Science Society of American Journal*, Vol.61, pp.920–927

Edwards, M.; Bejamin, M. M. & Ryan, J. N. (1996). Role of organic acidity in sorption of natural organic matter (NOM) to oxide surfaces. *Colloid Surfaces A*, Vol. 107, pp. 297-307

Elsenbeer, H.; Lorieri, D. & Bonell, M. (1995). Mixing model approaches to estimate storm flow sources in an overland flow-dominated tropical rain forest catchment. *Water Resources Research*, Vol.31, pp. 2267-2278

Freer, J. (1998). *Uncertainty and calibration of conceptual rainfall runoff models.* Ph.D. Thesis, Lancaster University

Freer, J. E., Beven, K. J. & Peters, N.E. (2003). Multivariate seasonal period model rejection within the generalised likelihood uncertainty estimation procedure, In: *Calibration of watershed models*, Q., Duan; H., Gupta; S., Sorooshian; A. N., Rousseau & R., Turcotte, (Ed.), 346, AGU, Water Science and Application Series, Washington

Gan W.; Chen H.-M. & Han Y.-F. (1983). Carbon transport by the Yangtze (at Nanjing) and Huanghe (at Jinan) Rivers, People's Republic of China. In: *Transport of Carbon and Minerals in Major World Rivers*, E. T. Degens; S. Kempe & H. Soliman, (Ed.), 459-470. SCOPE/UNEP Sonderbd. 55

Gu, B.; Schimitt, J.; Chen, Z.; Liang, L. & McCarthy, J. F. (1994). Adsorption and desorption of natural organic matter on iron oxides: Mechanisms and models. *Environmental Science & Technology*, Vol.28, pp. 38-46

Gupta, H. V.; Bastidas, L. A.; Sorooshian, S.; Shuttleworth, W. J. & Yang, Z. L. (1999). Parameter estimation of a land surface scheme using multicriteria methods. *Journal of Geophysical Research*, Vol.104, D16, doi:10.1029/1999JD900154

Hagedorn, F.; Mohn, J.; Schleppi, P. & Fluhler, H. (1999). The role of rapid flow paths for nitrogen transformation in a forest soil: A field study with micro suction cups. *Soil Science Society of American Journal*, Vol.63, pp. 1915–1923

Horton, R. E. (1933). The role of infiltration in the hydrologic cycle. *American Geophysical Union. Transaction*, Vol,14, pp. 446-460

Hooper, R. P.; Christophersen, N. & Peters, N. E. (1990) Modeling streamwater chemistry as a mixture of soilwater end-members–an application to the Panola Mountain Catchment, Georgia, U.S.A. *Journal of Hydrology*, Vol.116, pp. 321-343

Hope, D., Billett, M. F. & Cresser, M. S. (1994) A review of the export of carbon in river water: Fluxes and processes. *Environmental Pollution*, Vol.84, pp. 301–324

Ittekkot, V. & Laane R. W. P. M. (1991). Fate of riverine particulate organic matter. In: *Biogeochemistry of major world rivers*, E. T. Degens; S. Kempe & J. E. Richey, (Ed.), pp. 233–243, Wiley, Chichester

Jardine, P. M.; Wilson, G. V.; Luxmoore, R. J. & McCarthy, J. F. (1989). Transport of inorganic and natural organic tracers through an isolated pedon in a forest watershed. *Soil Science Society of American Journal*, Vol.53, pp.317–323

Jeong, Y. H.; Won, H. K.; Kim, K. H.; Park, J. H. & Ryu, J. H. (1997). Influences of electrical conductivity on stream and soil water quality in small forested watershed. *Korea FRI Journal of Forest Science*, Vol.55, pp. 125-137

Jeong, Y. H.; Park, J. H.; Kim, K. H. & Lee, B. S. (1999a). Influence of forest management on he facility of purifying water quality in *Abies holophylla* and *Pinus koraiensis* watershed (I). *Journal of Korean Forestry Society*, Vol.88, No.3, pp. 364-373

Jeong, Y. H.; Park, J. H.; Kim, K. H., Youn, H. J. & Won, H. K. (1999b). Influence of forest management on he facility of purifying water quality in *Abies holophylla* and *Pinus koraiensis* watershed (II). *Journal of Korean Forestry Society*, Vol.88, No.4, pp. 498-509

Jeong, Y. H.; Park, J. H.; Kim, K. H. & Youn, H. J. (2001a). Analysis of the factors influencing the mesopore ratio on the soil surface to investigate the site factors in a forest stand(I) -With a special reference to coniferous stands-. *Journal of Korean Forestry Society*, Vol.90, No.3, pp. 314-323

Jeong, Y. H.; Park, J. H.; Kim, K. H. &Youn, H. J. (2001b). Analysis of the factors influencing the mesopore ratio on the soil surface to investigate the site factors in a forest stand(II) -With a special reference to deciduous stands-. *Journal of Korean Forestry Society*, Vol.90, No.4, pp. 450-457

Kaiser, K. & Zech, W. (1998a). Soil dissolved organic matter sorption as influenced by organic and sesquioxide coatings and sorbed sulfate. *Soil Science Society of America Journal*, Vol.62, pp.129-136

Kaiser, K. & Zech, W. (1998b). Rates of dissolved organicmatter release and sorption in forest soils. *Soil Science*, Vol.62, pp.129-136

Katsuyama, M.; Ohte, N. & Kobashi, S. (2001). A three-component end-member analysis of streamwater hydrochemistry in a small Japanese forested headwater catchment. *Hydrological Processes*, Vol.15, pp. 249-260

Kawasaki, M.; Ohte, N. & Katsuyama, M. (2005). Biogeochemical and hydrological controls on carbon export from a forested catchment in central Japan. *Ecological Research*, Vol.20, pp. 347-358.

Kim D. S. & Jo, C. H. (1939). The measurement of total and net rainfall. *Report of Forestry Research institute*, Vol.20, pp. 19-37

Kim, J. -K.; Shin, M.; Jang, C.; Jung, S. & Kim, B. (2007a). Comparison of TOC and DOC distribution and the oxidation efficiency of BOD and COD in several reservoirs and rivers in the Han River systems. *Journal of Korean Society on Water Quality*, Vol.23, pp. 72–80 (in Korean with English abstract)

Kim J. S. (1987). Forest effects on the flood discharges and the estimation of the evapotranspiration in the small watershed. *Korea Forest Research Institute Reports*, Vol.35, pp.69-78

Kim, J.; Lee, D.; Hong, J.; Kang, S.; Kim, S. J.; Moon, S. K.; Lim, J. H.; Son, Y.; Lee, J.; Kim, S.; Woo, N.; Kim, K.; Lee, B.; Lee, B. L. & Kim, S. (2006). HydroKorea and CarboKorea: cross-scale studies of ecohydrology and biogeochemistry in a heterogeneous and complex forest catchment of Korea. *Ecological Research*, Vol. 21, pp.881-889

Kim, K. H. & Woo B. M. (1988(a)). Study on rainfall interception loss from canopy in forest(I). *Journal of Korean Forestry Society*, Vol.77, No.3, pp. 331-337

Kim, K. H. & Woo B. M. (1988(b)). Study on rainfall interception loss from canopy in forest(II). *Research Bulletin of the Seoul National University Forests*, Vol.24, pp. 29-37

Kim, K. H. & Woo B. M. (1997). Measurement and modeling of rainfall interception in Quercus mongolica and Pinus rigitaeda during the rainy season. *Journal of Korea Water Resources Association*, Vol.30, No.5, pp. 503-513

Kim, S. J. (2003). *Hydro-Biogeochemical Study on the Sulfur Dynamics in a Temperate Forest Catchment*. Ph. D. Dissertation, Kyoto University, Kyoto

Kim, S. J.; Ohte, N.; Kawasaki, M.; Katsuyama, M.; Tokuchi, N. & Hobara, S. (2003). Interactive responses of dissolved sulfate and nitrate to disturbance associated with pine wilt disease in a temperate forest. *Soil Science and Plant Nutrition* Vol.49, pp. 539-550

Kim, S. J.; Lee, D.; Kim, J & Kim, S. (2007b). Hydro-Biogeochemical Approaches to Understanding of Water and Carbon Cycling in the Gwangneung Forest Catchment. *Korean Journal of Agricultural and Forest Meteorology*, Vol. 9, No. 2, pp. 109-120, ISSN 1229-5671

Kim, S. J.; Lee, D. & Kim, S. (2009). Use of isotope data to determine mean residence time (MRT) of water in a forest catchment: ^{35}S and ^{3}H-based estimates. *Asia-Pacific Journal of Atmospheric Sciences*, Vol.45, No.2, pp. 165-173, ISSN 1976-7633

Kim, S. J.; Kim, J & Kim, K. (2010). Organic carbon efflux from a deciduous forest catchment in Korea. *Biogeosciences*, Vol. 7, 1323–1334, doi:10.5194/bg-7-1323-2010

Kim, T. S.; Lee, C. Y.; Kim, K. H. & Jeong, Y. H. (1993). *Forest practices for increasing water retention capacity*, 493-501, Annual Report of Korean Forest Research Institute (4-1) (In Korean)

Kirkby, M. J. (1978). *Hillslope Hydrology*, 389, John Wiley and Sons

Korea Forest Research Institute (1998). *Annual Research Report for Forest Resources*. Seoul, Korea

Kosugi, K. (1994). Three-parameter lognormal distribution model for soil water retain. *Water Resources Research*, Vol.30, pp. 891-901

Kosugi, K. (1996). Lognormal distribution model of unsaturated soil hydraulic properties. *Water Resources Research*, Vol.32, pp.2697-2703

Kwon, H.; Kim, J.; Hong, J. & Lim, J.-H. (2010). Influence of the Asian monsoon on net ecosystem carbon exchange in two major ecosystems in Korea. *Biogeosciences*, Vol. 7, pp. 1493-1504

Lee J. H.; Kim, T. H.; Lee, W. K.; Choi, K.; Lee, C. Y. & Joo, J. S. (1989). A study on regulating discharges by forest. *Korea Forest Research Institute Reports*, Vol.38, pp. 98-111

Lee I. H.; Kim, K. D. & Kwon, H. M. (1967). The effect of vegetation cover on head water control. *Forest Genetics Reports*, Vol.4, pp. 139-149

Lim, J. H.; Shin, J. H.; Jin, G. Z.; Chun, J. H. & Oh, J. S. (2003). Forest stand structure, site characteristics and carbon budget of the Gwangneung natural forest in Korea, Kor. *Korean Journal of Agricultural and Forest Meteorology*, Vol. 5, pp. 101-109, ISSN 1229-5671

Liu, C. P. & Sheu, B. H. (2003). Dissolved organic carbon in precipitation, throughfall, stemflow, soil solution, and stream water at the Guandaushi subtropical forest in Taiwan. *Forest Ecology and Management*, Vol.172, pp. 315-325

Loague, K. & Van der Kwaak, J. E. (2004). Physics-based hydrologic response simulation: platinum bridge, 1958 Edsel, or useful tool. *Hydrological Processes*, Vol.18, pp.2949-2956

Ludwig, W.; Probst, J. L. & Kempe, S. (1996). Predicting the oceanic input of organic carbon by continental erosion. *Global Biogeochemical Cycles*, Vol.10, pp. 23-41

Madsen, H.; Wilson, G. & Ammentorp, H. C. (2002). Comparison of different automatic strategies for calibration of rainfall-runoff models. *Journal of Hydrology*, Vol.261, pp. 48-59

Matsutani, J.; Tanaka, T. & Tsujimura, M. (1993). Residence times of soil, ground, and discharge waters in a mountainous headwater basin, central Japan, traced by tritium, In: *Tracers in Hydrology*, N. E. Peters, E. Hoehn, Ch. Leibundgut, N. Tase., & D. E. Walling (Ed.), No. 215, 57-64, IAHS Publication

McDonnell, J. J. (1990). A rationale for old water discharge through macropores in a steep, humid catchment. *Water Resources Research*, Vol.26, pp. 1821-2832

McDowell, W. H. & Likens, G. E. (1988). Origin, composition, and flux of dissolved organic carbon in the Hubbard Brook valley. *Ecological Monographs*, Vol.58, pp. 177-195

McGlynn, B. L.; McDonnell, J. J.; Shanley, J. B. & Kendall, C. (1999). Riparian zone flow path dynamics during snowmelt in a small headwater catchment. *Journal of Hydrology*, Vol.222, pp. 75-92

McGlynn, B. L. & McDonnell, J. J. (2003). Role of discrete landscape units in controlling catchment dissolved organic carbon dynamics. *Water Resources Research*, Vol.39, No.4, 1090, doi:10.1029/2002WR001525

Meybeck, M. (1982). Carbon, nitrogen, and phosphorus transport by the world rivers. *American Journal of Science*, Vol.282, pp. 401-450

Michalzik, B.; Tipping, E.; Mulder, J.; Gallardo Lancho, J. F.; Matzner, E.; Bryant, C. L.; Clarke, N.; Lofts, S. & Vicente Esteban, M. A. (2003). Modelling the production and transport of dissolved organic carbon in forest soils. *Biogeochemistry*, Vol.66, pp. 241-264

Michel, R. L., & Naftz, D. L. (1995). Use of sulfur-35 and tritium to study runoff from an alpine glacier, Wind River Range, Wyoming, In: *Biogeochemistry of Seasonally Snow-Covered Catchments*, K. A. Tonnessen; M. W. Williams & M., Tranter (Ed.), Publication No. 228, 441-443, International Association of Hydrological Sciences, Boulder, CO

Ministry of Science and Technology (1992). *Studies on to quantification of welfare functions of forests(II)* (in Korean)

Moon, S. -K.; Woo, N. C. & Lee, K. S. (2004). Statistical analysis of hydrographs and water-table fluctuation to estimate groundwater recharge. *Journal of Hydrology*, Vol.292, 198-209

Moore, T. R.; Desouza, W. & Koprivnijak, J. F. (1992). Controls on the sorption of dissolved organic carbon in soils. *Soil Science*, Vol.154, pp. 120-129

Neff, J. C. & Asner, G. P. (2001) Dissolved organic carbon in terrestrial ecosystems: synthesis and a model. *Ecosystems*, Vol.4, pp. 29-48

Ndiritu J.G. & Daniell, T.M. (1999). Assessing model calibration adequacy via global optimization. *Water SA*, Vol.25, No.3, pp. 317-326

Nodvin, S. C., Driscoll, C. T. & Likens, G. E. (1986). Simple partitioning of anions and dissolved organic carbon in a forest soil. *Soil Science*, Vol.142, pp. 27-35

Pinder, G. F., & Jones, J. F. (1969). Determination of the groundwater component of peak discharge from the chemistry of total runoff water. *Water Resources Research*, Vol.5, No.2, pp. 438-445

Prentice, I. C.; Farquhar, G. D.; Fasham, M. J. R.; Goulden, M. L.; Heimann, M.; Jaramillo, V. J.; Kheshgi, H. S.; Le Quéré, C.; Scholes, R. J. & Wallace, D. W. R. (2001). The carbon cycle and atmospheric carbon dioxide, In: *Climate Change 2001: The Scientific Basis. Contribution of Working Group I to the Third Assessment Report of the Intergovernmental Panel on Climate Change*, J. T. Houghton; Y. Ding; D. J. Griggs; M. Noguer; P. J. van der Linden; X. Dai; K. Maskell & C.A. Johnson (Ed.), 190, Cambridge University Press

Refsgaard, J. C. (1997). Parameterisation, calibration and validation of distributed hydrological models. *Journal of Hydrology*, Vol.198, pp. 69-97.

Refsgaard, J. C. & Knudsen, J. (1996). Operational validation and intercomparison of different types of hydrological models. *Water Resources Research*, Vol.32, No.7, pp. 2189-2202

Richter, D. D.; Markewitz, D.; Wells, C. E.; Allen, H. L.; April, R.; Heine, P. R. & Urrego, B. (1994). Soil chemical change during three decades in an old-field loblolly pine (*Pinus taeda* L.) ecosystem. *Ecology*, Vol.75, pp. 1463-1473

Seo, K. H.; Son, J. H. & Lee, J. Y. (2011). A New Look at Changma. *Atmosphere*, Vol.21, No.1, pp. 109-121, ISSN 1598-3560 (in Korean with English abstract)

Schulze, E.-D. (2006). Biological control of the terrestrial carbon sink. *Biogeosciences*, Vol.3, pp. 147-166

Shanley, J. B.; Pendall, E.; Kendall, C.; Stevens, L. R.; Nichel, R. L.; Phillips, P. J.; Forester, R. M.; Naftz, D. L.; Liu, B.; Stern, L.; Wolfe, B. B.; Chamerlain, C. P.; Leavitt, S. W.; Heaton, T. H. E.; Mayer, B.; Cecil, L. D.; Lyons, W. B.; Katz, B. G.; Betancourt, J. L.; McKnight, D. M.; Blum, J. D.; Edwards, T. W. D.; House, H. R.; Ito, E.; Aravena, R. O. & Whelan, J. F. (1998). Isotope as indicators of environmental change, In: *Isotope tracers in catchment hydrology*, C. Kendall & J. J. McDonnell (Ed.), 761-816, Elsevier

Shibata, H.; Hiura, T.; Tanaka, Y.; Takagi, K. & Koike, T. (2005). Carbon cycling and budget in a forested basin of southwestern Hokkaido, northern Japan. *Ecological Research*, Vol.20, pp. 325-331

Stewart, A. J. & Wetzel, R. G. (1982). Influence of dissolved humic materials on carbon assimilation and alkaline phosphatase activity in natural algal-bacterial assemblages. *Freshwater Biology*, Vol.12, pp. 369-380

Sueker, J. K.; Turk, J. T. & Michel, R. L. (1999). Use of cosmogenic ^{35}S for comparing ages of water from three alpine–subalpine basins in the Colorado Front Range. *Geomorphology*, Vol.27, pp. 61-74

Suzuki, M. (1980). Evapotranspiration from a small catchment in hilly mountain (I) Seasonal variations in evapotranspiration, rainfall interception and transpiration. *Journal of Japanese Forest Society*, Vol.62, pp. 46-53

Tao, S. (1998). Spatial and temporal variation in DOC in the Yichun River, China. *Water Resource Research*, Vol.32, pp. 2205-2210

Tipping, E. & Woof, C. (1990). Humic substances in acid organic soils: Modeling their release to the soil solution in terms of humic charge. *Journal of Soil Science*, Vol.41, pp. 573-585

Uchida, T.; Kosugi, K. & Mizuyama, T. (2002). Effect of pipeflow and bedrock groundwater on runoff generation at a steep headwater catchment, Ashiu, central Japan. *Water Resources Research*, Vol.38, pp. 1119, doi:10.1029/2001WR000261

Warnken, K. W. & Santschi, P. H. (2004). Giogeochemical behavior of oganic carbon in the Trinity River downstream of a large reservoir lake in Texas, USA. *Science of the Total Environment*, Vol.329, pp. 131-144.

Zhang, S.; Lu, X. X.; Sun, H.; Han, J. & Higgitt. D. L. (2009). Geochemical characteristics and fluxes of organic carbon in a human-disturbed mountainous river (the Luodingjiang River) of the Zhujiang (Pearl River), China. *Science of The Total Environment*, Vol.407, pp. 815-825

5

Entomopathogenic Fungi as an Important Natural Regulator of Insect Outbreaks in Forests (Review)

Anna Augustyniuk-Kram[1,2] and Karol J. Kram[2]
[1]Institute of Ecology and Bioethics,
Cardinal Stefan Wyszyński University, Warszawa,
[2]Polish Academy of Sciences Centre for Ecological
Research in Dziekanów Leśny, Łomianki
Poland

1. Introduction

With over 1 million species insects are not only the largest group of animals, but also a group that causes the most damage in forest management. Hence it is important to understanding the biology of their natural enemies. Among them are entomopathogenic fungi. Entomopathogenic fungi are a very heterogeneous group. Belong to different systematic groups and even their biology is often very different. However, all of them are pathogenic in relation to insects, and actually all arthropods, and their effectiveness in infecting their hosts is so large that it can become a factor regulating the abundance of insects. Importantly, the harmful insects (from the human point of view) include of course the forest pests.

In this paper we would like to introduce biology, systematics, geographical distribution, and give examples of natural and man-stimulated biocontrol of forest pests by entomopathogenic fungi.

2. Geographical and ecological distribution of entomopathogenic fungi

Entomopathogenic fungi are an important and widespread component of most terrestrial ecosystems. It seems they are not only in places where there are no victims – insects nor other arthropods. Of course spread of individual species of entomopathogenic fungi are different. However some of them can be found practically throughout the world. An example of such species may be *Beauveria bassiana* which is reported from tropical rainforest (Aung et al., 2008), and has been found in Canada as far north as latitude 75° (Widden & Parkinson, 1979). Entomopathogenic fungi have been also recorded north of the Arctic Circle. They have been *Tolypocladium cylindrosporum, B. bassiana* and *Metarhizium anisopliae* in Norway (Klingen et al., 2002), and *B. bassiana, M. anisopliae* and *Isaria farinosa* (=*Paecilomyces farinosus)* in Finland (Vänninen, 1995). What more, entomopathogenic fungi have been reported also from Arctic Greenland (Eilenberg et al., 2007) and Antarctica. In the latter location including endemic Antarctic species *Paecilomyces antarctica* isolated from the Antarctic springtail *Cryptopygus antarcticus* in the peninsular Antarctic (Bridge et al., 2005).

Also cosmopolitan fungi belonging to the genus *Beauveria, Lecanicillium, Conidiobolus* and *Neozygites* have been found on Antarctic sites, but without their arthropod hosts (Bridge et al., 2005).

Studies of Quesada-Moraga showed that altitude has no influence on presence of entomopathogenic fungi in range up to 1608 m, what more altitude was found to be predictive for the occurrence of *B. bassiana* (Quesada-Moraga et al., 2007). However, studies made on wider range of altitudes (up to > 5200 m) made by Sun & Liu showed great importance of this factor on the species diversity of insect-associated fungi (Sun & Liu, 2008).

There are different groups of entomopathogenic fungi in different habitats. Different insect pathogenic mycofloras could be found in the soil and different in the overground environment. Sosnowska found in Poland that in the Białowieża Forest soil litter and soil surface layer dominated *Hypocreales*, but in understory trees and in the canopy – *Entomophthorales*, and in the meadow and rush communities species of spider pathogenic fungi of the genus *Gibellula* (Sosnowska et al., 2004). The *Entomophthorales* are commonly reported as pathogens of forest pests in temperate forest habitats (Burges, 1981), but are rare in tropical forests (Evans, 1982). Humid tropical forests had a rich and varied insect pathogenic fungal species and the great majority of species belong in the genus *Cordyceps* (*Ascomycota: Hypocreales*) (Evans, 1982; Aung et al., 2008). While other species of *Hypocreales* such as *Beauveria, Metarhizium* and *Isaria* were the dominant fungi found on soil insects (Samson et al., 1988; Keller & Zimmerman, 1989).

Despite the fact that both *B. bassiana* and *M. anisopliae* are common everywhere there is known that *B. bassiana* seems to be very sensitive to the disturbance effects of cultivation and thus restricted to natural habitats. The ability of *M. anisopliae* to persist in cultivated soils is well established. Therefore the first is more frequent in forest, and second in arable soils (Rath et al., 1992; Vänninen, 1995; Quesada-Moraga et al., 2007; Sánchez-Peña et al., 2011).

Most reports show that frequency of entomopathogenic fungi in intensively cultivated soils is lower than in forest soils (Vänninen et al., 1989; Miętkiewski et al., 1991; Vänninen, 1995; Chandler et al., 1997; Bałazy, 2004). However, there were some exceptions from this rule, e.g. higher frequency of bait insect infections in pasture soils than in soils from either forest or cropland (Baker & Baker, 1998).

Entomopathogenic fungi are commonly found in soil and leaf litter of worldwide forests, however in temperate forests the diversity of entomopathogenic fungi is relatively low in comparison with tropical habitats (Evans, 1982; Grunde-Cimerman et al., 1998; Aung et al., 2008). However, compared to agricultural areas the diversity of entomopathogenic fungi in the temperate forests is quite high (Sosnowska et al., 2004). The differences in their prevalence and diversity of species were also found between the different types of forests (Miętkiewski et al., 1991; Chandler et al., 1997).

3. Systematics and biology of entomopathogenic fungi

3.1 Systematics of entomopathogenic fungi

Entomopathogenic fungi are a very heterogeneous group of insect pathogens. It is known nearly 700 species belonging to approximately 100 orders. Although only a few of them have been studied well. Most of them belong to the order *Entomophthorales* of the phylum

Glomeromycota and to *Hypocreales* of the phylum *Ascomycota* (Hibbett et al., 2007; Sung et al., 2007). Recent phylogenetic studies within entomopathogenic fungi resulted in significant revision of many species of entomopathogenic fungi. For example, such species as *Paecilomyces farinosus* and *P. fumosoroseus* currently belong to the genus *Isaria* (Luangsa-ard et al., 2004), and species *Verticillium lecanii* to the genus *Lecanicillium* (Zare & Gams, 2001). There are many species in the *Ascomycota* in which the sexual phase (teleomorph) is not known and which reproduce entirely asexually (anamorphic fungi). Sexually reproducing hypocrealean fungi occur in the genera *Cordyceps* and *Torrubiella*. These fungi are important natural control agents of many insects in tropical forests. Genus *Cordyceps* has many anamorphs, of which *Beauveria*, *Lecanicillium* and *Isaria* are the best known and described (Blackwell, 2010). Recent phylogenetic studies have demonstrated that the genus *Beauveria* (so far known to be only anamorphic fungus) is monophyletic within the *Cordycipitaceae* (*Hypocreales*), and has been linked developmentally and phylogenetically to *Cordyceps* species (Rehner et al., 2011). Despite recent interest in the genetic diversity of many groups of entomopathogens, the genus *Beauveria,* in contrast to other species, has not received critical taxonomic review (Rehner et al., 2011).

3.2 The life cycles of *Hypocreales* and *Entomophthorales*

The life cycles of *Hypocreales* and *Entomophthorales* are slightly different. Nevertheless, the survival and spread in the environment of both groups is dependent on the infection of the host that invariably leads to its death. The life cycle of entomopathogenic fungus consists of a parasitic phase (from host infection to its death) and a saprophytic phase (after host death) (Fig. 1 and Fig. 2).

In contrast to other entomopathogens (bacteria and viruses), which enters the insects with food, entomopathogenic fungi infect their host through the external cuticle. The process of infection involves: adhesion of the spore on the insect cuticle, penetration of the cuticle by the germ tube, development of the fungus inside the insect body and colonization of the hemocoel by fungal hyphae. The spores of the entomopathogenic fungi are usually covered with a layer of mucus composed of proteins and glucans, which facilitates their attachment to the insect cuticle. Germinating spores of several entomopathogenic fungi produce specialized structures called appressoria. The appressorium is responsible for attachment of germinating spore to the epicuticular surface. The process of penetration of the insect cuticle is a result of mechanical pressure and enzymatic activity of the germ tube. The major role in the penetration plays the secretion of sequential lipases, proteases and chitinases. Inside the insect body most entomopathogenic fungi grow as yeast-like propagules (blastospores), hyphal bodies or protoplasts lacking a cell wall. These structures are spread through the hemocoel. Death of an insect is usually a result of mechanical damage caused by growing mycelia inside the insect (mummification), or toxins produced and released by the pathogen. *Beauveria*, *Metarhizium, and Tolypocladium* are known that secrete a whole range of toxins. Some of them like destruxin, bavericin, and efrapeptins are fully described chemically, and is known their action and contribution in the process of pathogenesis (Roberts, 1981; Hajek & St. Leger, 1994). For *Entomophthorales* there are limited data about the release of toxins (Boguś & Scheller, 2002). In this case, death is the result of the total colonization of host tissues by the fungus.

After host death, the fungus colonizes the cadaver and during 2-3 days forms aerial hyphae and then sporulates (Fig. 1 and Fig. 2). Whereas *Hypocreales* produce only asexual spores, species of *Entomophthorales* produce two types of spores: asexual (primary conidia) and sexual (zygo- or azygospores) called resting spores (Fig. 1 and Fig. 2). Conidia of *Hypocreales* and primary conidia of *Entomophthorales* are produced externally on the surface of an insect after its colonization and death. *Entomophthorales* and *Hypocreales* differ in the way dispersal of spores. The first of these are actively discharged from cadavers by hydrostatic pressure, while the latter are spread by wind. If primary conidium from cadavers does not land on a new host, it germinates and forms secondary conidia (some species can also produce tertiary and quaternary conidia). The majority of *Entomophthorales* produce resting spores (internally within cadavers). Cadavers containing resting spores (azygospores) initially attach to the branches of trees, and then fall to the ground and then azygospores are leached into the soil. Under favourable conditions, azygospores begin to germinate to form germ conidia and infect new hosts. Resting spores allow entomophthoralen species to survive unfavourable periods or the temporary lack of hosts. In this way many species of *Entomophthorales* synchronize their development with the development of insects. Hypocrealen fungi can also survive in the environment (if do not land on a new host), as mummified cadavers or as conidia in soil (Hajek & St. Leger, 1994; Hajek & Shimazu, 1996).

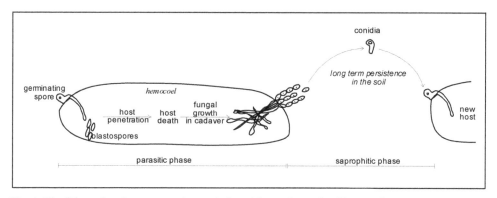

Fig. 1. The life cycle of entomopathogenic fungi from the order *Hypocreales*

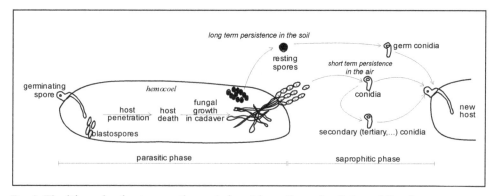

Fig. 2. The life cycle of entomopathogenic fungi from the order *Entomophthorales*

3.3 Intraspecific variation of entomopathogenic fungi

Entomopathogenic fungi can attack insects from different orders: Lepidoptera, Coleoptera, Hemiptera, Diptera, Orthoptera, Hymenoptera, as well as non-insect arthropods. But while some species of fungi (who belongs mainly to the *Hypocreales*) have a very wide spectrum of potential victims, others (mainly *Entomophthorales*) are pathogens only one particular species of insect.

The question arises whether the specialization of the pathogen, and thus its virulence towards particular hosts, is associated with genetic diversity of the pathogen. In the case of entomopathogenic fungi with a wide range of hosts such as *Beauveria bassiana* numerous studies have confirmed that strains of diverse geographical origin but isolated from the same host species showed a greater similarity in the genetic structure than strains from the same area but isolated from different host species (Poprawski et al., 1988; Neuvéglise et al., 1994; Cravanzola et al., 1997; Couteaudier & Viaud, 1997; Maurer et al., 1997; Castrillo & Brooks, 1998). For example, Maurer et al. (1997) by analysing the RFLPs (restriction fragment length polymorphism) and RAPDs (random amplification of polymorphic DNA) of 38 strains of *Beauveria bassiana* isolated from various geographical sites and from diverse species of Lepidoptera (*Ostrinia nubilalis, Diatraea saccharalis, Maliarpha separatella*) and Coleoptera (*Sitona humeralis, S. discoideus*), found a few homogenous groups of strains with similar genetic structure. The first group consisted of all strains isolated from *O. nubilalis*, and the second one included strains from *D. saccharalis*. Strains isolated from various species of *Sitona* formed the third group of strains with similar genetic structure. Furthermore, in laboratory tests, strains from *O. nubilalis* were highly virulent towards this host and less (or not) virulent against the other hosts. Similarly, Neuvéglise et al. (1994) have shown the relationship between the genetic structure of *Beauveria brongniartii* strains and their biological origins (host species). The results of PCR-RFLP revealed a perfectly homogenous group of strains isolated from *Hoplochelus marginalis*. All the strains isolated from *H. marginalis* were more virulent against this host (30-100% mortality) than the strains isolated from different insects (10% mortality). Castrillo & Brooks (1998) also found a high similarity between *B. bassiana* isolates obtained from *Alphitobius diaperinus*.

Literature data on *Metarhizium anisopliae* do not confirm such a clear relationship between genotype and specialization in relation to a particular host (Tigano-Milani et al., 1995; Fungaro et al., 1996; Bridge et al., 1997). A lot of studies indicate the crucial importance of the geographical origin of isolates (St. Leger et al., 1992; Cobb & Clarkson, 1993; Leal et al., 1994; Leal et al., 1997) and habitat type (Bidochka et al., 2001). Studies of Leal et al. (1994; 1997) showed that among 40 isolates of *M. anisopliae*, strains from the same country were more similar in the genetic structure than those from different countries despite the same host. Interesting observations to support a relationship between the genetic structure of *M. anisopliae* isolates and habitat type (agricultural and forested areas) along with abiotic factors (temperature and exposure to UV radiation) provided Bidochka et al. (2001). On the basis of various genetic markers (allozymes, RAPD, RFLP), Bidochka et al. (2001) divided 83 strains of *M. anisopliae* into two distinct groups, each associated with different habitat type. The group from forested areas showed ability for growth at low temperatures (at 8°C), while the group from the agricultural areas showed ability for growth at high temperatures (37°C) and resilience to UV exposure. The association of habitat and thermal preferences was also found for *B. bassiana* (Bidochka et al., 2002). Recently, more sensitive and reliable molecular methods also indicate a certain association between *B. bassiana* isolates and their geographical origins and not between the genetic structure of the fungus and host systematic position (Wang et al., 2005; Fernandes et al., 2009). There is a hypothesis that the saprophytic phase has an evolutionary impact on genetic structure of many species of entomopathogenic fungi including *B. bassiana* (Bidochka et al., 2002; Ghikas et al., 2010; Ormond et al., 2010; Garrido-Jurado et al., 2011). In studies conducted by Ormond et al. (2010) in a conifer forest, molecular analyses (ISSR-PCR) indicate that below-ground and above-ground isolates of *B. bassiana* are genetically diverse.

Relatively little is known about genetic diversity of *Entomophthorales*. The host range within *Entomophthorales* is generally narrow. Therefore, it would seem that the genetic diversity of isolates from the same host may be small. One of the better-studied species in this respect is *Entomophaga maimaiga*. In North America where it produces numerous epizootics in a population of *Lymantria dispar*, *E. maimaiga* characterized by relatively low genetic diversity. *E. maimaiga* was introduced to the USA from Japan, and such low genetic diversity proves that it spread from a small number of individuals (Hajek et al., 1995). Nielsen et al. (2005a) comparing by AFLPs 30 *E. maimaiga* isolates originating from the USA, Japan, China and Russia found that native populations from Asia were more diverse than the USA populations. The authors hypothesize that the population now present in the USA came from Japan of a result of accidental introduction rather than the deliberate release. In contrast to the Asian isolates, no correlation between geographical location and clade was found among the US isolates. The authors explain this by the fact that *E. maimaiga* was introduced into the USA relatively recently; therefore genetically distinct subpopulations may not have evolved yet.

Another specialized entomophthoralen species *Erynia neoaphidis* – major aphid pathogen - shows greater genetic diversity. Rohel et al. (1997) using PCR-ITS and RAPD methods, identified four separate groups with high genetic variability among 30 isolates originating from diverse countries and hosts. Only in some cases RAPD groupings could be related with geographical origin and there was no apparent relationship between host and ITS or RAPD pattern. Similar results have received Tymon & Pell (2005) using ISSR, ERIC and RAPD techniques.

Very little intraspecific variation has been demonstrated for *Entomophthora muscae*, a common pathogen of flies (Jensen et al., 2001; Jensen & Eilenberg, 2001). These studies show low genetic variation within isolates from the same host taxon. In a study conducted by Jensen et al. (2001) several different genotypes within *E. muscae s. str.* have been documented, and each of the genotypes was restricted to a single host taxon, suggesting high host specificity.

Molecular studies of entomopathogenic fungi are very important in the context of choosing the appropriate species or strains for biological control, as well as distinguish wild strains from those introduced artificially in order to monitor and to track isolates after field application.

4. Epizootiology of insect disease

4.1 Fungal epizootic as a limiting factor of insect outbreaks

Insects are essential part of forest ecosystems and at low density have negligible impact on tree growth. Although occasionally, some insect species quickly increase their numbers giving catastrophic impacts on trees and, in some cases this can lead to the complete destruction of large areas of natural or planted forests. Insect outbreaks are often the result of disturbance in biocenotic balance caused for example by sudden events such as fire or hurricane, or as a result of human activities, such as changes in the planting structure (monocultures), but also by more global processes like climate change (Hunter, 2002). In many specific cases, the initiation of insect outbreaks is the result of many factors, and mechanisms of their occurrence are not fully understood and explained. Similarly, there are many factors causing the collapse of outbreaks: depletion of food resources, natural enemies, and unfavourable weather.

Among the natural enemies of insect, infections (epizootics) caused by entomopathogenic fungi are one of the frequently observed causes of collapse of outbreaks. In insect pathology epizootic is defined as an unusually large number of cases of disease in a host population. Epizootic diseases are sporadic, limited in time, and in a given area, and characterized by a sudden change in prevalence (Fuxa & Tanada, 1987). Entomopathogenic fungi are constantly present in populations of insect hosts but when density of the host population is normal infections occur sporadically (enzootic phase of insect diseases). However, during insects outbreaks fungi that infect insects can increase their numbers enough to spread in the environment and contribute to the reduction of insect's population (epizootic phase) (Fuxa & Tanada, 1987). It is very difficult to predict the occurrence of epizootic and not always in different pathogen-host systems the same factors initiate its development in pest population. Furthermore, epizootics caused by entomopathogenic fungi in forested habitats are less numerous than those in other habitats, particularly in agricultural areas because the forest ecosystem is more complex and more stable compared with "agroecosystem". Forest ecosystems also have many different mechanisms for regulating the number of pests.

Spatial and temporal spread of the epizootic depends on the effective transmission of the pathogen in the population of the target insect and insect susceptibility to infection. These two main factors are closely related to climatic factors (mainly temperature and humidity) and biotic environments (other pathogens, parasitoids and predators). All these factors act simultaneously on both sides of pathogen-host system by modifying the growth and development of both pathogen and insect population. Transmission is the transfer of infective propagules between individuals through direct contact. Entomopathogenic fungi in the insect population can be transmitted in three ways: horizontally (from infected insects on healthy individuals within a single generation), vertically (between generations), and be moved by vectors. This third method plays an important role in the transmission of fungi to new habitats (Fuxa & Tanada, 1987). It is supposed that natural epizootic in a population of insects do not have to be initiated by a highly virulent strain of the fungus. In the laboratory, the virulence of the strain can be increased with passaging (repeated in-vitro transmission of the tested strains of fungi from infected to healthy individuals of target insect) (Hyden et al., 1992; Hughes & Boomsma, 2006). Most likely, under natural conditions there is a natural passage during which the pathogen increases its virulence.

Fungal epizootics occur naturally in many insect populations during outbreaks and are frequently the primary cause of the collapse of many insect populations (Table 1). Hicks & Watt (2000) reported that in 1998 outbreak of the pine beauty moth (*Panolis flammea*) in lodgepole plantations in Scotland collapsed by a fungal epizootic. *Entomophaga aulicae*, *Nomuraea rileyi* and *Beauveria bassiana* were recorded from infected larvae. Together they caused 88% mortality in the population of *P. flammea*.

Epizootics *Lymantria dispar* are found in areas of natural occurrence of the pest (from Europe to Asia), but also in new areas where *L. dispar* is an invasive species (North America). Shimazu & Takatsuka (2010) in Japan, during outbreak of *L. dispar* found many dead and living larvae on the surface of boles of the host trees. Laboratory investigation of the gypsy moth cadavers revealed that most of them were infected with the nuclear polyhedrosis virus (NPV), the fungal pathogen *Entomophaga maimaiga*, or a mixed infection of those two pathogens. They also found larvae infected by *Isaria javanica*. Earlier Aoki (1974) reported *Paecilomyces canadensis* and *Entomophthora aulicae* (=*Entomophaga maimaiga*) from *L. dispar* larvae in the same Prefecture as the study of Shimazu & Takatsuka (2010). Aoki (1974) found that 80% of collected larvae were killed by *Entomophthora aulicae* and 19% by a mixed infection with *P. canadensis*.

In North America *Lymantria dispar* is an invasive species. It was imported into the USA from France that was accidentally introduced in Massachusetts in 1869, while *E. maimaiga* – natural enemy causing widespread epizootics in the areas of natural occurrence – was first recorded in North America in 1989. *E. maimaiga* probably comes from the introductions that took place in 1910 and 1911 when diseased gypsy moth cadavers collected in the Tokyo area were released near Boston, Massachusetts. However, no fungal infections were recovered and establishment was presumed to have failed. And until 1989, despite several attempts of introduction, neither *E. maimaiga* nor any other entomophthoralean fungus has ever been observed in the North American gypsy moth populations (Andreadis & Weseloh, 1990; Hajek et al., 1995). Since 1989 almost every year epizootics caused by *E. maimaiga* in the USA gypsy moth populations have been observed and *E. maimaiga* is considered the main factor suppressed outbreaks of *L. dispar* (Hajek et al., 1995; Hajek, 1997). During outbreaks of *L. dispar*, epizootics of nuclear polyhedrosis virus (NPV) are as common as fungal infections caused by *E. maimaiga* (Table 1). However, NPV requires high-density populations for development of epizootics, whereas *E. maimaiga* can cause high level of infection in low-density as well as high-density populations (Hajek et al., 1995). Furthermore, it was found that climatic conditions especially moisture is critical in the development of epizootics of *E. maimaiga*. Hajek (1997) found that in the year of low rainfall prevailed viral infections, whereas in year with more rainfall dominated fungal infections. In other surveys it was found that the greatest production of conidia of *E. maimaiga* by cadavers took place on the day of rainfall, whereas maximum germination of resting spores occurred 1-2 days after significant precipitation (Weseloh & Andreadis, 1992).

Mixed infections of entomopathogenic fungi and viruses are also common in other pathogen-host systems (Table 1). Ziemnicka (2008a) has confirmed that in the population of *Leucoma salicis* nuclear polyhedrosis virus acts as a density-dependent mortality factor and most epizootics were recorded at temperature above 15°C and low humidity. Some laboratory studies and field observations also indicated that fungi and viruses that occur within one insect's population might act in a synergistic manner (Malakar et al., 1999; van Frankenhuyzen et al., 2002; Ziemnicka, 2008b). For example, artificially induced viral epizootic in the population of *L. salicis* not interfere with the occurrence of the epizootic caused by *Beauveria bassiana*, which extended the population decline phase of 6 to 8 years (Ziemnicka, 2008b).

In Japan, outbreaks of *Syntypistis punctatela* in beech stands are known to occur synchronously among different areas at intervals of 8–11 years. Kamata (2000) has found that this periodicity is the result of several factors (predators, parasitoids, pathogens, delayed induced defensive response), which act as time-delayed and density-dependent factors. Among them fungal disease caused by predominantly *Cordyceps militaris*, was considered to be the most plausible factor for generating cycles of the beech caterpillar population. Kamata (1998, 2000) observed the most severe infections caused by *C. militaris* at the population peak and in the period of its decreasing phase. His field observations also showed that infections caused by *C. militaris* affected the beech caterpillar population both in outbreak and non-outbreak areas. During outbreaks mortality of *S. punctatela* caused by *C. militaris*, *B. bassiana* and *P. farinosus* (= *I. farinosa*) ranged from 96 to 100% suggesting important roles of these parasites in natural control of this pest (Kamata, 1998).

Quite unique epizootic loci in the summer-autumn complex of the multispecies communities of forest lepidopterans and sawflies in Siberia had been discovered by Kryukov et al. (Kryukov et al., 2011). In this survey at least 30 species from 7 families were found to be *C. militaris* hosts. Field observations were carried out from 2007 to 2009 in three localities where the number of insects infected by *C. militaris* in the period of mass appearance of stromata ranged, depending on the location, an average from 0.5 to 1.5 specimens/m² (in some places even up to 20 specimens/m²). They also observed that with reduction in the number of caterpillars in the tree crowns also decreased the level of defoliation of trees (5-25%). Moreover, they did not find any living pupa in leaf litter on the trial sites.

Similarly, widespread epizootic caused by entomopathogenic fungi have been found in the population of the pine sawfly *Diprion pini* in Poland (Sierpińska, 1998). During outbreak of 1794 collected larvae 1826 isolates of fungi were isolated. The most frequently isolated species were *I. farinosa* (603 isolates), *C. militaris* (542), *B. bassiana* (311), and *L. lecanii* (179). These species have caused 30-80% mortality of overwintering larvae.

There are also examples (though very rare) epizootic caused by fungi of not fully proven insecticidal properties. Marcelino et al. (2009) discovered in the population of *Fiorinia externa* epizootic caused by *Colleotrichum* sp. - a fungus widely known as phytopathogen. Poprawski & Yule (1991) also determined in natural populations of *Phylophaga* spp. such phytopathogenic fungi as *Aspergillus*, *Penicillium* and *Fusarium*, but only occasionally, not as in the case of *Colleotrichum* on a large scale. Some authors consider that phytopathogenic fungi, under some circumstances, may infect and kill insects (Teetor-Barsch & Roberts, 1983; Poprawski & Yule, 1991).

Relatively few studies have investigated the influence of insects' behaviour on fungal infection and development of epizootics. For example, Hajek (2001) studied the unusual behaviour of late stage larvae of *L. dispar* to the risk of infection by *E. maimaiga*. Late instars larvae of *L. dispar* move down from the tree canopy and wander during daylight hours under leaf litter or in cryptic locations on tree boles. Hajek (2001) has found in field experiment that larvae caged over the soil were at much greater risk of *E. maimaiga* infection compared with larvae caged in the understory vegetation or on tree trunks. She also found that infections occurred with even brief exposures to the soil. The author assumes that infection of all larvae caged on the ground where initiated by germ conidia from germinating resting spores. The density of resting spores at the base of the tree is usually the highest. During epizootics, cadavers of late instars fall from tree trunks and *E. maimaiga* resting spores are leached into the soil. Therefore, such behaviour of larvae, rather uncommon in other Lepidoptera, to remain near tree base exposes them to the areas with highest titers of *E. maimaiga* resting spores.

Outbreak pest	Tree host or type of forest	Country	Epizootics/limiting factors	References
Cinara pinea	Pine forests	China	*Erynia canadensis*	Li et al., 1989
Dendrolimus pini	*Pinus silvestris*	Poland	*Isaria farinosa* *Cordyceps militaris* *Beauveria bassiana* *Lecanicillium lecanii*	Sierpińska, 1998
Dendrolimus punctatus	Masson pine	China	*Beauveria bassiana*	Ge et al., 2009,
Fiorinia externa	*Tsuga canadensis*	USA	*Colletotrichum* sp. *Lecanicillium lecanii*, *Beauveria bassiana*	Marcelino et al., 2009
Hyblaea puera	*Tectona grandis*	India	parasites, Buculovirus HpNPV and *Beauveria bassiana*	Gowda & Naik, 2008
Leucoma salicis	*Populus nigra*	Poland	nuclear polyhedrosis virus *Beauveria bassiana*	Ziemnicka, 2008a, 2008b Ziemnicka & Sosnowska, 1996
Lymantria dispar	*Larix leptolepis*	Japan	*Entomophthora aulicae* mixed infection with *Paecilomyces canadensis*	Aoki, 1974
" – "	Mixed hardwood forest	USA	*Entomophaga maimaiga* and nuclear polyhedrosis virus	Andreadis & Weseloh, 1990 Elkinton et al., 1991 Hajek, 1997
" – "	*Larix leptolepis* and *Alnus japonica*	Japan	nuclear polyhedrosis virus and mixed infection with *Entomophaga maimaiga*, and *Isaria javanica*	Shimazu & Takatsuka, 2010
Malacosoma disstria		New York and Maryland, USA	*Furia gastropachae* (=*Furia crustosa*)	Filotas et al., 2003
Multispecies communities of lepidopterans and sawflies	Different coniferous and deciduous forests	Siberia, Russia	*Cordyceps militaris* *Cordyceps sp.*	Kryukov et al., 2011
Orgyia leucostigma	*Abies balsamea*	Nova Scotia, Canada	*Entomophaga aulicae*, nucleopolyhedrovirus	van Frankenhuyzen et al., 2002
Panolis flammea	*Pinus contorta*	Scotland	*Isaria farinosa*, nuclear polyhedrosis virus *Entomophaga aulicae*, *Nomuraea rileyi*, *Beauveria bassiana*	Hicks et al., 2001 Hicks & Watt, 2000 Watt & Leather, 1988
Syntypistis (=*Quadricalcarifera*) *punctatella*	*Fagus crenata*	Japan	*Cordyceps militaris* *Isaria farinosa* *Beauveria bassiana*,	Kamata, 1998, 2000

Table 1. Examples of outbreaks of forest pests and epizootics caused by entomopathogenic fungi

Many insects spend at least part of his life in the soil. Soil is also a natural environment for entomopathogenic fungi. Therefore, such natural behaviours of insects related to their biology, such as accumulation in the soil or leaf litter to wintering or pupation, conducive to fungal infections and natural reduction of many insect pests. Tkaczuk and Miętkiewski (1998) determined the natural reduction of *Dendrolimus pini* population caused by entomopathogenic fungi during hibernation period. Microbiological analysis of the pine sawfly cocoons showed that entomopathogenic fungi were responsible for the mortality of overwintering larvae, depending on location from 12 to 52%. Epizootics described by Kryukov et al. (2011) and Sierpińska (1998) also occurred during the overwintering of insects in the form of pupae and larvae in leaf litter.

On the other hand, many forest pests occupy cryptic habitats, where they are protected from direct contact with fungi, and where thermal and moisture conditions are not conducive to the development of fungal infection. For example, such group of insects are bark beetles (Curculionidae: *Scolitynae*). In the available literature we did not find studies that describe extensive epizootics in this group of insects, although there are numerous reports of entomopathogenic fungi isolated from the bark beetles (Bałazy, 1968; Glare et al., 2008; Brownbridge et. al., 2010; Draganova et al., 2010; Tanyeli et al., 2010) and tested in the context of biological control this group of insects (Kreutz et al., 2004a; Draganova et al., 2007; Sevim et al., 2010; Tanyeli et al., 2010, Zhang et al., 2011)

4.2 Insect resistance to fungal infections

In the population of a particular host, individuals are not equally susceptible to infection. Different species of hosts are also not equally susceptible to infection from a particular species or strain of the pathogen. On the one hand, different species or even strains of the fungus may display different levels of virulence and parasitic specialization against a specific host. But on the other hand, susceptibility of the host may change with its development (larvae are usually more sensitive than adults) or may depend on its behaviour and individual resistance to infection. Insects have relatively primitive immunological system, although they can react to the entrance of the fungal pathogen inside their body. Cellular and humoral defence mechanisms, such as antimicrobial proteins, phagocitosis and multihaemocytic encapsulation of fungal structures, have been observed. Encapsulation is always associated with production of melanin. After infection, the fungal propagules are encapsulated within melanin (melanization process). Melanins act antagonistic to fungi, inhibiting their growth (Butt, 1987; Hajek & St. Leger, 1994; Boguś et al., 2007).

They are also known examples of non-specific: morphological (Smith & Grula, 1982; Saito & Aoki, 1983) behavioural (Viliani et al., 1994; Myles, 2002) and physiological (Serebrov et al., 2006; Rohlfs & Churchill, 2011) defence mechanisms to avoid fungal pathogens. Viliani et al., (1994) found in laboratory experiments that the application of mycelial particles in soil affected the behaviour of both larval and adult Japanese beetles, *Popilia japonica*. Grubs avoided soil that contained high concentrations of *Metarhizium anisopliae* mycelium for up to 20 days after application. Some insects have the ability to detect and alert the presence of an infected individual in the population. Very interesting behaviour in this regard have social insects such as termites. The presence of conidia-dusted termites in colony caused the alarm manifested by rapid bursts of longitudinal oscillatory movement by workers. The intensity of alarm peaked about 15 minutes after introduction of the conidia-dusted termites, at which time 80% of the termites were aggregated near the treated individual. Alarm and aggregation significantly subsided after 24 minutes and were then followed by grooming, biting, defecation, and burial of the infected termite (Myles, 2002). In defence against infection, insect may produce on the surface of its body fungistatic compounds that inhibit spore germination and growth. It was found that certain fatty acids on the surface of *Heliothis zea* larvae inhibit the germination of spores of

Beauveria bassiana and *Isaria fumosorosea* (formerly *Paecilomyces fumosoroseus*). Non-pathogenic (saprophytic) fungi and bacteria occurring naturally on the insect cuticle could inhibit germination of spores of entomopathogenic fungi (Smith & Grula, 1982; Saito & Aoki, 1983).

5. Entomopathogenic fungi in biological control of insects

5.1 Strategies of biological control

Chemical insecticides are commonly used in plant protection. The consequence of this is to increase the resistance of insects to various chemical substances contained in plant protection products. Over 500 arthropod species now show resistance to one or more types of chemicals (Mota-Sanchez et al., 2002). Other serious problem is invasive species that are accidentally introduced to a new country or continent and which escape their coevolved natural pathogens or predators. This forces to seek new, alternative and more environmentally safe, methods of reducing outbreaks of pests. In recent years more attention paid to the possibility of using natural enemies, including entomopathogenic fungi, in control of insect pests. Aside from playing a crucial role in natural ecosystems, entomopathogenic fungi are being developed as environmentally friendly alternatives in agriculture and forestry. They can be increasingly exploited for forest pest management as biological control agents and in the attempts to improve the sustainability of forest ecosystem.

Biological control is defined as the use of living organisms to suppress the population density, or impact of a specific pest organism, making it less abundant or less damaging than it would otherwise be (Eilenberg et al., 2001). Thus, the aim of biological control is not a complete elimination of target species, but reducing its population below the economic threshold of harmfulness.

There are four strategies for biological control: classical, inoculation, inundation and conservation biological control. However, in forestry, only classical and inundation biological control strategies are widely used. Classical biological control is the intentional introduction of an exotic, usually co-evolved, biological control agent for permanent establishment and long-term pest control (Eilenberg et al., 2001). In the case of microorganisms, widely distributed in nature, the term exotic means the use of a particular strain or biotype, which is not native to the area where the pest is controlled. Introduced species to induce long-term effect has to acclimate to the area under certain climatic conditions, multiply and spread. So it is important to understand of the biology of species "exotic" and target, as well as the ability to monitoring its presence in the area. Inoculation biological control is also the intentional release of a living organism as a biological control agent with the expectation that it will multiply and control the pest for an extended period, but not permanently (Eilenberg et al., 2001). Inoculation involves releasing small numbers of natural enemies at prescribed intervals throughout the pest period, starting when the density of pest is low. The natural enemies are expected to reproduce themselves to provide more long-term control. Inundation biological control is the use of living organisms to control pests when control is achieved exclusively by the released organisms themselves (Eilenberg et al., 2001). In practice this means the release of large numbers of mass-produced biological control agents (so-called biopesticides) to reduce a pest population without necessarily achieving continuing impact or establishment. Pest population is treated such a quantity of biopesticide to get immediate results (on the pattern of use of chemical insecticides). Conservation biological control is such modification of the environment or existing practices to protect and enhance specific natural enemies or other organisms to reduce the effect of pests (Eilenberg et al., 2001). Conservation techniques involve the identification and manipulation of factors that limit or enhance the abundance and effectiveness of natural enemies.

5.2 Field application of entomopathogenic fungi

Attempts for practical application of entomopathogenic fungi, in classical or inundation biological control strategy, are always preceded by laboratory tests. Laboratory tests are mainly aimed to select for highly virulent strains, determine the optimal dose of inoculum, to examine the impact of biotic and abiotic factors on the fungus used and testing different methods of application (Lingg & Donaldson, 1981; Markova, 2000; Wegensteiner, 2000; Kreutz et al., 2004a; Dubois et al., 2008; Shanley et al., 2009; Augustyniuk-Kram, 2010; Zhang et al., 2011). Laboratory tests do not always coincide later with their practical use, but provide valuable information on the activity of entomopathogenic fungi and their potential role in biological control of many dangerous pests. In laboratory conditions, Markova (2000) studying the susceptibility of *Lymantria dispar* larvae, *Ips typographus*, *Hylobius abietis* beetles and diapausing *Cephaleia abietis* larvae on various strains of *Beauveria bassiana*, *Isaria farinosa* (=*Paecilomyces farinosus*), *Metarhizium anisopliae* and *Lecanicillium lecanii* (=*Verticillium lecanii*) obtained very different results. Larvae of *L. dispar* were susceptible to only one strain of *B. bassiana*, but this strain acted too slowly. Beetles of *Ips typographus* were susceptible to all tested strains, while the beetles of *H. abietis* were only susceptible to one strain of *M. anisopliae* and one strain of *B. bassiana*. Similarly *Cephaelia abietis* larvae were susceptible to infections by one strain of *B. bassiana* and one strain of *I. farinosa* (Markova, 2000). From other studies it is known that good results in the control of *L. dispar* were obtained using the entomophthoralen fungus *Entomophaga maimaiga* that is a major natural enemy in endemic Asian gypsy moth population. Nielsen et al. (2005b) assessed virulence and fitness of the six strains of *E. maimaiga* originating from Japan, Russia, China and North America in different gypsy moth population (originating from Japan, Russia, Greece and USA). They found that all *E. maimaiga* isolates tested were pathogenic to all populations of *L. dispar*, regardless of the geographical origin of the fungal isolates, with at least 86% mortality. However, fungal isolates differed significantly in virulence (measured as time to death) and fitness (measured as fungal reproduction) (Nielsen et al., 2005b).

The use of a microorganism in practice is not easy. The biggest problem is that it is very difficult to predict the effects of biological control agents before their release. The success of field trials depends on many factors that must be taken into consideration. Quite often observed is the phenomenon of lower efficiency of biological control agents applied in the field compared with laboratory tests. Among many factors such features of entomopathogenic fungi as high virulence against target insect, harmless for beneficial organisms (non-target species), warm-blooded animals and humans, high resistance to biotic and abiotic environmental conditions are critical in achieving satisfactory results in field trials (van Lenteren et al., 2003; Jackson et al., 2010). Impact on non-target organisms

Pest	Pathogen	Country	Methods	Effect	References
Agrilus planipennis	*Beauveria bassiana*	USA	Pre-emergent trunk spray (10 or 100 x 10^{13} conidia/ha) and polyester fiber bands impregnated with a sporulating culture of *B. bassiana* (6.4 x 10^8 conidia/cm^2)	Infection rates ranged from 58.5 and 83% at two application rates, longevity of females and males was significantly reduced, females laid fewer eggs, prolonged larval development. Fiber bands method – 32% mortality adult on treated trees vs. 1% on control trees.	Liu & Bauer, 2008
Anoplophora glabripennis	*Beauveria bassiana*, *B. brongniartii*	China	Two methods were compared: fiber bands fastened around tree trunks was compared with trunk sprays	Longevity was decreased by both strains compared with controls, with females killed earlier by *B. brongniartii* than by *B. bassiana*. This decrease in longevity was independent of the application method used. Daily oviposition rate per female were also reduced by both strains.	Dubois et al., 2004
Ips typographus	*Beauveria bassiana*	Germany	Introduction of *B. bassiana*-inoculated beetles into untreated population or natural population of beetles were lured into a pheromone traps and treated there with conidia of *B. bassiana*	Significant reduction in the length of maternal galleries and the number of larvae and pupae were observed. In experiment with pheromone traps significant reduction in the number of bore holes were observed. Additionally, no larvae, pupae and juveniles were found under the bark of trunks of treated trees.	Kreutz et al., 2004b

Pest	Pathogen	Country	Methods	Effect	References
Lymantria dispar	*Entomophaga maimaiga*	USA	Resting spores were released around the bases of oaks in 41 plots (2-years trials).	In the second year of study *E. maimaiga* infections were detected in 40 of the 41 release plots. Infection levels in release plots averaged 72.4 % and were associated with declining egg mass densities. The next year after the first application *E. maimaiga* was abundant 1,000 m from release plots.	Hajek et al., 1996
" – "	" – "	Bulgaria	Cadavers of *L. dispar* containing *E. maimaiga* resting spores	Only 6.3% larvae collected in experimental plots contained spores and resting spores of *E. maimaiga*	Pilarska et al., 2000
Melolontha melolontha	*Beauveria brongniartii*	Denmark	Barley kernels colonized by fungus were placed in holes of 10 cm depth around trees or barley kernels were mixed with the soil and placed around the new trees during re-plantation	30% more trees without damages	Eilenberg et al., 2006
" – "	" – "	Denmark	10 or 30 g barley kernels per tree were thrown directly into the planting hole before the tree was inserted into the hole	Significant effect was achieved 1 and 1/2 year after the application. Almost 13% of the untreated trees had been killed, 25% showed substantial discoloration and needle trees. None of the treated trees died and only single trees showed discolouration.	Eilenberg et al., 2006

Pest	Pathogen	Country	Methods	Effect	References
" – "	" – "	Switzerland	Blastospores were applied by helicopter at the swarming places. The mean dose 2.0-3.7 x 10^{14} spores ha^{-1}	The infection rates of the treated adults ranged between 30 to 99 %. The treatment reduced the average reproduction rate. The population density 6 and 9 years decrease 50-80%	Keller et al., 1997
Strophosoma capitatum and *S. melanogrammum* larvae	*Metarhizium anisopliae*	Denmark	Conidial suspension – 1.0 x 10^{14} per ha; procedure was repeated twice, year after year	In the same year after the first application were not achieved satisfactory results. One year and two years after the first application the accumulated density of *S. melanogrammum* weevils was reduced by 60 and 40%. For *S. capitatum* only in the first year after the first application achieved a reduction of up to 38% of beetles.	Nielsen et al., 2007

Table 2. Examples of biocontrol of forest pests by using entomopathogenic fungi

must always be considered as a side effect during field applications of entomopathogenic fungi. The results of numerous works suggest little impact of entomopathogenic fungi on non-target organisms (James et al., 1995; Parker et al., 1997; Traugot et al., 2005; Nielsen et al., 2007). And finally, the use of entomopathogenic fungi on a large scale depends above all on the possibility of cheaper mass rearing on artificial media. Unfortunately, most fungal biopesticides are produced on the basis of hypocrealen fungi, among which the majority belongs to polyphagous species that is a broad spectrum of potential hosts. Entomophthoralen fungi amongst many species are more specialized (monophagous), therefore not of interest among potential producers of mycoinsecticides because of difficulties in their cultivation on artificial substrates and the multiplication of infective material on a mass scale (Pell et al., 2001).

Examples of practical application of entomopathogenic fungi can be found in Table 2. Each example briefly describes the method of application and the final effect measured in the reduction of the target insect populations, or health status of trees. In the described examples, depending on the target insect and fungus, different methods of application were used. In the case of soil-dwelling pests the most appropriate and most common method is introduction into the soil barley kernels colonized by fungus. In European countries, this method is widely used against larvae of *Melolontha melolontha* in various crops (Keller et al., 1997; Fröschle & Glas, 2000; Bajan et al., 2001; Vestergaard et al., 2002). Recently, a commercial product based on barley kernels colonised by *Beauveria brongniartii* (Melocont®-Pilzgerste) was tested under EU-funded project BIPESCO which aim was to study and develop entomopathogenic fungi for the control of subterranean insect pests like scarabs and weevils. Another fairly common method is the use of fiber bands (=fungal bands) impregnated with entomopathogenic fungi and placed around tree trunks or branches. This method was first developed as a biological control method to control adults of Japanese pine sawyer, *Monochamus alternatus*, the main vector of pine wilt disease caused by the pinewood nematode, *Bursaphelenchus xylophylus* (Shimazu, 2004). Currently, this method gives satisfactory results in biological control of invasive species *Anoplophora glabripennis* and *Agrilus planipennis* (Table 2). Despite rare reports of natural epizootics among bark beetles, attempt to biological control, particularly laboratory tests, are very promising. In laboratory studies efficacy of entomopathogenic fungi against bark beetles is very high, reaching up to 80-100% for some strains (Draganova et al., 2007; Sevim et al., 2010; Tanyeli et al., 2010). In field trials Kreutz et al. (2004b) also achieved significant reduction in number of larvae, pupae and juveniles. According to Hunt, as high mortality of bark beetles may be associated with the lack in their cuticle certain lipids that inhibit germination of fungal spores (Hunt, 1986).

5.3 Conservation biological control and Integrated Pest Management (IPM strategy)

Recently more attention is focused on the study of entomopathogenic fungi in natural environments, and hence in conservation biological control strategy. Conservation biological control is a strategy in which forest management and environmental manipulations are adopted to enhance conditions for the development of different groups of natural enemies of pests. In forestry conservation biological control requires actions on a large scale, not only in wooded areas, but also beyond, in adjacent areas. In Europe, an example of pest that control is not limited to one environment is the previously mentioned *Melolontha melolontha*. Adults are pests of deciduous trees. Eggs are laid in the soil usually in the areas adjacent to large forest complexes such as arable fields, nurseries, orchards, where the larvae feed on roots.

The use of entomopathogenic fungi, particularly in this strategy, requires a thorough knowledge of the biology and ecology of both pests and their natural enemies. It also requires recognition of factors that may interfere with their effectiveness. Conservation biological control also needs long-term and large-scale researches on multitrophic relationships between natural enemy and their hosts and their impact on natural regulation of serious insect pests (Tscharntke et al., 2008).

Natural enemies and their potential hosts always exist within a particular biocenosis, where are affected by abiotic and biotic factors. One of them are pesticides that kill natural enemies as well as their potential hosts (Miętkiewski et al., 1997; Chandler et al., 1998; Meyling & Eilenberg, 2007). For this reason, many countries prohibited the use of chemical insecticides to control forest pests (e.g. Denmark). Therefore the only alternative is biological or integrated pest control. Integrated Pest Management in forestry is defined as "a combination of prevention, observation and suppression measures that can be ecologically and economically efficient and socially acceptable, in order to maintain pest populations at a suitable level" (FAO, 2011). Biological control of pests through the use of natural enemies and other methods like mechanical control, planting proper trees in proper sites during reforestation is preferred over synthetic pesticides. Such actions favour sustainable control and efficiency of natural enemies (FAO, 2011). The previously mentioned project BIPESCO was intended to help replace or reduce the input of chemical pesticides in European agriculture, forestry and horticulture.

Numerous studies indicate that forest communities display greater diversity of entomogenous species than arable fields. Bałazy (2004) compared species richness of entomopathogenic fungi in forest areas (mainly in protected areas like national parks) with areas used for agriculture found that out of all 210 species over 60% was collected in the forest protected areas. Only 20 species were found in annual crops. Significantly more infected host cadavers were found in other habitats such as extensively utilized meadows, mid-field, shelterbelts, woodlots and swamps. Evans (1974) also found that even in forests with very high biodiversity as tropical rain forests, the diversity of entomopathogenic fungi in damaged forests is less than undisturbed, and that the incidence of natural epizootic decreased. Such natural and semi-natural habitats are often refuges for insects and entomopathogenic fungi. Thus, the protection of this group of pathogens and their natural habitats is also important question in the context of conservation of biodiversity. Improving our understanding of the ecology of entomopathogenic fungi is also essential to further develop these organisms in biological control of many serious insect pests.

6. References

Andreadis, T.G. & Weseloh, R.M. (1990). Discovery of *Entomophaga maimaiga* in North American gypsy moth, *Lymantria dispar*. Proc. Natl. Acad. Sci. *Proceedings of the National Academy of Sciences USA*, Vol. 87, No. 7, (1 April 1990), pp. 2461-2465, ISSN 1091-6490.

Aoki, J. (1974). Mixed infection of the gypsy moth, *Lymantria dispar japonica* Motschulsky (Lepidoptera: Lymantriidae), in a larch forest by *Entomophaga aulicae* (Reich.) Sorok. and *Paecilomyces canadensis* (Vuill.) Brown et Smith. *Applied Entomology and Zoology*, Vol. 9, No. 3, pp. 185-190, ISSN 1347-605X.

Augustyniuk-Kram, A. (2010). Mortality of the nut-leaf *Strophosoma melanogrammum* (Forster) and damage rate of needles after treatment with entomopathogenic fungi. *Journal of Plan Protection Research*, Vol. 50, No. 4, (December 2010), pp. 545-550, ISSN 1427-4345.

Aung, O.M.; Soytong, K. & Hyde, K.D. (2008). Diversity of entomopathogenic fungi in rainforests of Chiang Mai Province, Thailand. *Fungal Diversity*, Vol. 30, (May 2008), pp. 15-22, ISSN 1560-2745.

Bajan, C.; Augustyniuk, A.; Mierzejewska, E. & Popowska-Nowak, E. (2001). Grzyby owadobójcze jako alternatywne środki ochrony roślin [Entomopathogenic fungi as alternative means of plant protection]. *Biuletyn Naukowy*, Vol. 12, pp. 159-167, ISSN 1640-1395, (in Polish with English summary).

Baker, C.W. & Baker, G.M. (1998). Generalist Entomopathogens as Biological Indicators of Deforestation and Agricultural Land Use Impacts on Waikato Soils. *New Zealand Journal of Ecology*, Vol. 22, No. 2, pp. 189-196, ISSN 0110-6465.

Bałazy, S. (1968). Analysis of bark beetle mortality in spruce forests in Poland. *Ekologia Polska, Ser. A*, Vol. 16, 33 pp. 657–687.

Bałazy, S. (2004). Znaczenie obszarów chronionych dla zachowania zasobów grzybów entomopatogenicznych [Significance of protected areas for the preservation of entomopathogenic fungi]. *Kosmos*, Vol. 53, No. 1, pp. 5-16, ISSN 0023-4249 (in Polish with English summary).

Bidochka, M.J.; Kamp, A.M.; Lavender, T.M.; Dekoning, J. & De Croos, J.N.A. (2001). Habitat association in two genetic groups of the insect-pathogenic fungus *Metarhizium anisopliae*: uncovering cryptic species? *Applied and Environmental Microbiology*, Vol. 67, No. 3, (March 2001), pp. 1335-1342, ISSN 1098-5336.

Bidochka, M.J.; Menzies, F.V. & Kamp, A.M. (2002). Genetic groups of the insect-pathogenic fungus *Beauveria bassiana* are associated with habitat and thermal growth preferences. *Archives of Microbiology*, Vol. 178, No. 6, (December 2002), pp. 531–537, ISSN 1432-072X.

Blackwell, M. (2010). Fungal evolution and taxonomy. *BioControl*, Vol. 55, No. 1, (February 2010), pp. 7-16, ISSN 1573-8248.

Boguś, M.I. & Scheller, K. (2002). Extraction of an insecticidal protein fraction from the parasitic fungus *Conidiobolus coronatus* (Entomophthorales). *Acta Parasitologica*, Vol. 47, No. 1, (January 2002), pp. 66-72, ISSN 1896-1851.

Boguś, M.I.; Kędra, E.; Bania, J.; Szczepanik, M.; Czygier, M.; Jabłoński, P.; Pasztaleniec, A.; Samborski, J.; Mazgajska, J. & Polanowski, A. (2007). Different defence strategies of *Dendrolimus pinii*, *Galleria mellonella*, and *Calliphora vicina* against fungal infection. *Journal of Insect Physiology*, Vol. 53, No. 9, (September 2007), pp. 909-922, ISSN 0022-1910.

Bridge, P.D.; Prior, C.; Sagbohan, J.; Lomer, C.J.; Carey, M. & Buddie, A. (1997). Molecular characterization of isolates of *Metarhizium anisopliae* from locusts and grasshoppers. *Biodiversity and Conservation*, Vol. 6, No. 11, (November 1997), pp. 177-189, ISSN 1572-9710.

Bridge, P.D.; Clark, M.S. & Pearce D.A. (2005). A new species of *Paecilomyces* isolated from the Antarctic springtail *Cryptopygus antarcticus*. *Mycotaxon*, Vol. 92, (April-June 2005), pp. 213-222, ISSN 0093-4666.

Brownbridge, M.; Reay, S.D. & Cummings, N.J. (2010). Association of Entomopathogenic Fungi with Exotic Bark Beetles in New Zealand Pine Plantations. Mycopathologia, Vol. 169, No. 1, (January 2010), pp. 75–80, ISSN 1573-0832.

Burges, H.D. (1981). Strategy for the microbal control of pests in 1980 and beyond. In: *Microbial Control of Pests and Plant Diseases 1970-1980*, H.D. Burges, (Ed.), pp. 797-836, Academic Press, ISBN 0121433609, London and New York.

Butt, T.M. (1987). A fluorescence microscopy method for the rapid localization of fungal spores and penetration sites on insect cuticle. *Journal of Invertebrate Pathology*, Vol. 50, No. 1, (July 1987), pp. 72-74, ISSN 0022-2011.

Castrillo, L.A. & Brooks, W.M. (1998). Differentiation of *Beauveria bassiana* isolates from the darkling beetle, *Alphitobius diaperinus*, using isozyme and RAPD analyses. *Journal of Invertebrate Pathology*, Vol. 72, No. 3, (November 1998), pp. 190-196, ISSN 0022-2011.

Chandler, D.; Hay, D. & Reid, A.P. (1997). Sampling and occurrence of entomopathogenic fungi and nematodes in UK soils. *Applied Soil Ecology*, Vol. 5, No. 2, (May 1997), pp. 133–141, ISSN 0929-1393.

Chandler, D.; Mietkiewski, R.T.; Davidson, G.; Pell, J.K. & Smits, P.H. (1998). Impact of habitat type and pesticide application on the natural occurrence of entomopathogenic fungi in UK soils. *IOBC/WPRS Bulletin*, Vol. 21, No. 4, pp. 81–84, ISSN 1027-3115.

Cobb, B.D. & Clarkson, J.M. (1993). Detection of molecular variation in the insect pathogenic fungus *Metarhizium* using RAPD-PCR. *FEMS Microbiology Letters*, Vol. 112, No. 3, (September 1993), pp. 319-324, ISSN 1574-6968.

Couteaudier, Y. & Viaud, M. (1997). New insights in population structure of *Beauveria bassiana* with regard to vegetative compatibility groups and telomeric restriction fragment length polymorphisms. *FEMS Microbiology Ecology*, Vol. 22, No. 2, (March 1997), pp. 175–182, ISSN 1574-6941.

Cravanzola, F.; Platti, P.; Bridge, P.D. & Ozino, O.I. (1997). Detection of genetic polymorphism in strains of the entomopathogenic fungus *Beauveria brongniartii* isolated from the European cockchafer (*Melolontha spp.*). *Letters in Applied Microbiology*, Vol. 25, No. 4, (September 1997), pp. 289-294, ISSN 1472-765X.

Draganova, S.; Takov, D. & Doychev, D. (2007). Bioassays with isolates of *Beauveria bassiana* (Bals.) Vuill. and *Paecilomyces farinosus* (Holm.) Brown & Smith against *Ips sexdentatus* Boerner and *Ips acuminatus* Gyll. (Coleoptera: Scolytidae). *Plant Science (Sofia)*, Vol. 44, pp. 24-28, ISSN 0568-465X.

Draganova, S.A.; Takov, D.I. & Doychev, D.D. (2010). Naturally-Occurring Entomopathogenic Fungi on Three Bark Beetle Species (Coleoptera: Curculionidae) in Bulgaria. *Pesticides & Phytomedicine (Belgrade)*, Vol. 25, No. 1, 2010, pp. 59-63, ISSN 1820-3949.

Dubois, T.; Hayek, A.E.; Jiafu, H. & Li, Z. (2004). Evaluating the Efficiency of Entomopathogenic Fungi Against the Asian Longhorned Beetle, *Anoplophora glabripennis* (Coleoptera: Cerambycidae), by Using Cages in the Field. *Environmental Entomology*, Vol. 33, No. 1, (February 2004), pp. 62-74, ISSN 0046-225X.

Dubois, T.; Lund, J.; Bauer, L.S. & Hajek, A.E. (2008). Virulence of entomopathogenic hypocrealean fungi infecting *Anoplophora glabripennis*. *BioControl*, Vol. 53, No. 3, (June 2008), pp. 517-528, ISSN 1573-8248.

Eilenberg, J.; Hajek, A. & Lomer, C. (2001). Suggestion for unifying the terminology in biological control. *BioControl*, Vol. 46, No. 4, (December 2001), pp. 387-400, ISSN 1573-8248.

Eilenberg, J.; Schmidt, N.M.; Meyling, N. & Wolsted, C. (2007). Preliminary survey for insect pathogenic fungi in Arctic Greenland. *IOBC/WPRS Bulletin*, Vol. 30, No. 1, pp. 12, ISSN 1027-3115.

Eilenberg, J.; Nielsen, C.; Harding, S. & Vestergaard, S. (2006). Biological control of scarabs and weevils in Christmas trees and greenery plantations. In: *An ecological and societal approach to biological control*. J. Eilenberg & H.M.T. Hokkanen, (Eds.), pp. 247-255, Springer, ISBN 978-90-481-7108-8.

Elkinton, J. S.; Hajek, A.E.; Boettner, G.H. & Simons, E.E. (1991). Distribution and apparent spread of *Entomophaga maimaiga* (Zygomycetes: Entomophthorales) in gypsy moth (Lepidoptera: Lymantriidae) populations in North America. *Environmental Entomology*, Vol. 20, No. 6, (December 1991), pp. 1601–1605, ISSN 1938-2936.

Evans, H.C. (1974). Natural control of arthropods, with special reference to ants (Formicidae), by fungi in the tropical high forests og Ghana. *Journal of Applied Ecology*, Vol. 11, No. 1, (April 1974), pp. 37-50, ISSN 0021-8901.

Evans, H.C. (1982). Entomogenous fungi in tropical forest ecosystems: an appraisal. *Ecological Entomology*, Vol. 7, No. 1, (February 1982), pp. 47–60, ISSN 1365-2311.

FAO (2011). Guide to implementation of phytosanitary standards in forestry. *FAO Forestry Paper*, Vol. 164, pp. 17-19, ISSN 0258-6150.

Fernandes, É.K.K.; Moraes, Á.M.L.; Pacheco, R.S.; Rangel, D.E.N.; Miller, M.P.; Bittencourt, V.R.E.P. & Roberts, D.W. (2009). Genetic diversity among Brazilian isolates of *Beauveria bassiana*: comparisons with non-Brazilian isolates and other *Beauveria* species. *Journal of Applied Microbiology*, Vol. 107, No. 3, (September 2009), pp. 760-774, ISSN 1365-2672.

Filotas, M.J.; Hajek, A.E. & Humber, R.A. (2003). Prevalence and biology of *Furia gastropachae* (Zygomycetes: Entomophthorales) in populations of forest tent caterpillar (Lepidoptera: Lasiocampidae). *The Canadian Entomologist*, Vol. 135, No. 3, pp. 359-378, ISSN 1918-3240.

Fröschle, M. & Glas, M. (2000). The 1997 control campaign of *Melolontha melolontha* L. at the Kaiserstuhl area (Baden- Würtemberg) Field trials and practical experiences. *IOBC/WPRS Bulletin*, Vol. 23, No. 8, pp. 27-34, ISSN 1027-3115.

Fungaro, M.H.P.; Vieira, M.L.C.; Pizzirani-Kleiner, A.A. & de Azevedo, J.L. (1996). Diversity among soil and insect isolates of *Metarhizium anisopliae* var. *anisopliae* derected by RAPD. *Letters in Applied Microbiology*, Vol. 22, No. 6, (June 1996), pp. 389-392, ISSN 1472-765X.

Fuxa, J.R. & Tanada, Y. (1987). *Epizootiology of insect diseases*. Wiley-Interscience, ISBN 047187812X, New York.

Garrido-Jurado, I.; Márquez, M.; Ortiz-Urquiza, A.; Santiago-Álvarez, C.; Iturriaga, E.A.; Quesada-Moraga, E.; Monte, E. & Hermosa, R. (2011). Genetic analyses place most Spanish isolates of *Beauveria bassiana* in a molecular group with word-wide distribution. *BMC Microbiology*, Vol. 11, article no. 84, (April 2011).

Ge, J.; Wang, X.-l.; Wang, B.; Ding, D.-g., Du, Y.-b. & Li, Z.-z. (2009). Temporal-spatial dynamic a natural epizootic caused by *Beauveria bassiana* in a population of the masson-pine caterpillar, *Dendrolimus punctatus*. *Chinese Journal of Biological Control*, Vol. 25, No. 2, pp. 129-133, ISSN 1005-9261, (in Chinese with English abstract).

Ghikas, D.V.; Kouvelis, V.N. & Typas, M.A. (2010). Phylogenetic and biogeographic implications inferred by mitochondrial intergenic region analyses and ITS1-5.8S-ITS2 of the entomopathogenic fungi *Beauveria bassiana* and *B. brongniartii*. *BMC Microbiology*, Vol. 10, article no. 174, (16 June 2010).

Glare, T.R.; Reay, S.D.; Nelson, T.L. & Moore, R. (2008). *Beauveria caledonica* is a naturally occurring pathogen of forest beetles. *Mycological Research*, Vol. 112, No. 3, (March 2008), pp. 352-360, ISSN 0953-7562.

Gowda, J. & Naik, L.K. (2008). Key Mortality Factors of Teak Defoliator, *Hyblaea puera* Cramer (Hyblaeidae : Lepidoptera). *Karnataka Journal of Agricultural Sciences*, Vol. 21, No. 4, (October-December 2008), pp. 519-523, ISSN 0972-1061.

Grunde-Cimerman, N.; Zalar, P. & Jerm, S. (1998). Mycoflora of cave cricket *Troglophilus neglectus* cadavers. *Mycopathologia*, Vol. 144, No. 2, (November 1998), pp. 111–114, ISSN 0301-486X.

Hajek, A.E. (1997). Fungal and Viral Epizootics in Gypsy Moth (Lepidoptera: Lymantriidae) Populations in Central New York. *Biological Control*, Vol. 10, No. 1, (September 1997), pp. 58-68, ISSN 1049-9644.

Hajek, A.E. (2001). Larval behavior in *Lymantria dispar* increases risk of fungal infection. *Oecologia*, Vol. 126, No. 2, (January 2001), pp. 285–291, ISSN 1432-1939.

Hajek, A.E. & Shimazu, M. (1996). Types of spores produced by *Entomophaga maimaiga* infecting the gypsy moth *Lymantria dispar*. *Canadian Journal of Botany*, Vol. 74, No. 5, pp. 708-715, ISSN 1916-2804.

Hajek, A.E. & St. Leger, R.J. (1994). Interactions between fungal pathogens and insect host. *Annual Review of Entomology*, Vol. 39, pp. 293-322.

Hajek, A.E.; Humber, R.A. & Elkinton, J.S. (1995). The mysterious origin of *Entomophaga maimaiga* in North America. *American Entomologist*, Vol. 41, No. 1, pp. 31–42, ISSN 2155-9902.

Hajek, A.E.; Elkinton, J.S. & Witcosky, J.J. (1996). Introduction and Spread of the Fungal Pathogen *Entomophaga maimaiga* (Zygomycetes: Entomophthorales) Along the Leading Edge of Gypsy Moth (Lepidoptera: Lymantriidae) Spread. *Environmental Entomology*, Vol. 25, No. 5, (October 1996), pp. 1235-1247, ISSN 0046-225X.

Hibbett, D.S.; Binder, M.; Bischoff, J.F.; Blackwell, M.; Cannon, P.F.; Eriksson, O.E.; Huhndorf, S.; James, T.; Kirk, P.M.; Lücking, R.; Lumbsch, H.T.; Lutzoni, F.; Matheny, P.B.; McLaughlin, D.J.; Powell, M.J.; Redhead, S.; Schoch, C.L.; Spatafora, J.W.; Stalpers, J.A.; Vilgalys, R.; Aime, M.C.; Aptroot, A.; Bauer, R.; Begerow, D.; Benny, G.L.; Castlebury, L.A.; Crous, P.W.; Dai, Y.-C.; Gams, W.; Geiser, D.M.; Griffith, G.W.; Gueidan, C.; Hawksworth, D.L.; Hestmark, G.; Hosaka, K.; Humber, R.A.; Hyde, K.D.; Ironside, J.E.; Kõljalg, U.; Kurtzman, C.P.; Larsson, K.-H.; Lichtwardt, R.; Longcore, J.; Miądlikowska, J.; Miller, A.; Moncalvo, J.-M.; Mozley-Standridge, S.; Oberwinkler, F.; Parmasto, E.; Reeb, V.; Rogers, J.D.; Roux, C.; Ryvarden, L.; Sampaio, J.P.; Schüßler, A.; Sugiyama, J.; Thorn, R.G.; Tibell, L.; Untereiner, W.A.; Walker, C.; Wang, Z.; Weir, A.; Weiss, M.; White, M.M.; Winka, K.; Yao, Y.-J. & Zhang, N. (2007). A higher-level phylogenetic classification of the *Fungi*. *Mycological Research*, Vol. 111, No. 5, (May 2007), pp. 509-547, ISSN 1469-8102.

Hicks, B.J. & Watt, A.D. (2000). Fungal disease and parasitism in *Panolis flammea* during 1998: evidence of change in the diversity and impact of the natural enemies of a forest pest. *Forestry*, Vol. 73, No. 1, pp. 31-36, ISSN 0015-752X.

Hicks, B.J.; Barbour, D.A.; Evans, H.F.; Heritage, S.; Leather, S.R.; Milne, R. & Watt, A.D. (2001). The history and control of the pine beauty moth, *Panolis flammea* (D. & S.)

(Lepidoptera: Noctuidae), in Scotland from 1976 to 2000. *Agricultural and Forest Entomology*, Vol. 3, No. 3, (August 2001), pp. 161-168, ISSN 1461-9563.

Hughes, W.O.H. & Boomsma, J.J. (2006). Does genetic diversity hinder parasite evolution in social insect colonies? *Journal of Evolutionary Biology*, Vol. 19, No. 1, (January 2006), pp. 132-143, ISSN 1420-9101.

Hunt, D.W.A. (1986). Absence of fatty acid germination inhibitors for conidia of *Beauveria bassiana* on the integument of the bark beetle *Dendrocrotonus ponderosa* (Coleoptera: Scolytidae). *The Canadian Entomologist*, Vol. 118, No. 8, pp. 837-838, ISSN 1918-3240.

Hunter, M.D. (2002). Ecological causes of pest outbreaks. In: *Encyclopedia of Pest Management*. Pimentel, D. (Ed.), pp. 214-217, Marcel Dekker Inc., ISBN 9780824706326.

Hyden, T.P.; Bidochka, M.J. & Khachatourians, G.G. (1992). Entomopathogenicity of several fungi towards the English grain aphid (Homoptera: Aphididae) and enhancement of virulence with host passage of *Paecilomyces farinosus*. *Journal of Economic Entomology*, Vol. 85, No. 1, (February 1992), pp. 58-64, ISSN 1938-291X.

Jackson, M.A.; Dunlop, C.A. & Jaronski, S.T. (2010). Ecological consideration in producing and formulating fungal entomopathogens for use in insect biocontrol. *BioControl*, Vol. 55, No. 1, (February 2010), pp. 129-145, ISSN 1573-8248.

James, R.R.; Shaffer, B.T.; Croft, B. & Lighthart, B. (1995). Field evaluation of *Beauveria bassiana*: Its persistence and effects on the pea aphid and a non-target coccinellid in alfalfa. *Biocontrol Science and Technology*, Vol. 5, No. 4, pp. 425-437, ISSN 1360-0478.

Jensen, A.B. & Eilenberg, J. (2001). Genetic variation within the insect-pathogenic genus *Entomophthora*, focusing on the *E. muscae* complex, using PCR-RFLP of the ITS II and LSU rDNA. *Mycological Research*, Vol. 105, No. 3, (March 2001), pp. 307–312, ISSN 0953-7562.

Jensen, A.B.; Thomsen, L. & Eilenberg, J. (2001). Intraspecific Variation and Host Specificity of *Entomophthora muscae* sensu stricto Isolates Revealed by Random Amplified Polymorphic DNA, Universal Primed PCR, PCR-Restriction Fragment Length Polymorphism, and Conidial Morphology. *Journal of Invertebrate Pathology*, Vol. 78, No. 4, (November 2001), pp. 251-259, ISSN 0022-2011.

Kamata, N. 1998. Periodic Outbreaks of the Beech Caterpillar, *Quadricalcarifera punctatella*, and its Population Dynamics: the Role of Insect Pathogens. In: *Proceedings: Population Dynamics, Impacts, and Integrated Management of Forest Defoliating Insects*. McManus, M.L. & Liebhold, A.M. (Eds.), pp. 34-46, USDA Forest Service General Technical Report NE-247, Banska Štiavnica, Slovak Republic, August 18-23, 1996.

Kamata, N. (2000). Population dynamics of the beech caterpillar, *Syntypistis punctatella*, and biotic and abiotic factors. *Population Ecology*, Vol. 42, No. 3, (December 2000), pp. 267–278, ISSN 1438-390X.

Keller, S. & Zimmerman, G. (1989). Mycopathogens of soil insects. In: *Insect-Fungus Interactions*. Wilding, N.; Collins, N.M.; Hammond, P.M. & Webber, J.F. (Eds.), pp. 240–270, Academic Press, ISBN 0127518002, London.

Keller, S.; Schweizer, C.; Keller, E. & Brenner, H. (1997). Control of white grubs (*Melolontha melolontha* L.) by treating adults with the fungus *Beauveria brongniartii*. *Biocontrol Science and Technology*, Vol. 7, No. 1, pp. 105-116, ISSN 1360-0478.

Klingen, I.; Eilenberg, J. & Meadow, R. (2002). Effects of farming system, field margins and bait insect on the occurrence of insect pathogenic fungi in soils. *Agriculture, Ecosystems & Environment*, Vol. 91, No. 2-3, (November 2002), pp. 191–198, ISSN 0167-8809.

Kreutz, J.; Vaupel, O. & Zimmermann, G. (2004a). Efficacy of *Beauveria bassiana* (Bals.) Vuill. against the spruce bark beetle, *Ips typographus* L., in the laboratory under various conditions. *Journal of Applied Entomology*, Vol. 128, No. 6, (July 2004), pp. 384-389, ISSN 1439-0418.

Kreutz, J.; Zimmermann, G. & Vaupel, O. (2004b). Horizontal transmission of the entomopathogenic fungus *Beauveria bassiana* among the spruce bark beetle, *Ips typographus* (Col., Scolitidae) in the laboratory and under field conditions. *Biocontrol Science and Technology*, Vol. 14, No. 8, pp. 837-848, ISSN 1360-0478.

Kryukov, V.Yu.; Yaroslavtseva, O.N.; Lednev, G.R. & Borisov, B.A. (2011). Local epizootics caused by teleomorphic corycypitoid fungi (Ascomycota: Hypocreales) in populations of forest lepidopterans and sawflies of the summer-autumn complex in Siberia. *Microbiology*, Vol. 80, No. 2, (April 2011), pp. 286-295, ISSN 0026-2617.

Leal, S.C.M.; Bertioli, D.J.; Butt, T.M. & Peberdy, J.F. (1994). Characterization of isolates of the entomopathogenic fungus *Metarhizium anisopliae* by RAPD-PCR. *Mycological Research*, Vol. 98, No. 9, (September 1994), pp. 1077-1081, ISSN 0953-7562.

Leal, S.C.M.; Bertioli, D.J.; Butt, T.M.; Carder, J.H.; Burrows, P.R. & Peberdy, J.F. (1997). Amplification and restriction endonuclease digestion of the Pr1 gene for the detection and characterization of *Metarhizium* strains. *Mycological Research*, Vol. 101, No. 3, (March 1997), pp. 257-265, ISSN 0953-7562.

Li, Z.-Z.; Wang, J.-L. & Lu, X.-X. (1989). Two entomophthoralean fungi causing extensive epizootics in forest and tea plant pests. *Acta Mycologica Sinica*, Vol. 8, No. 2, pp. 81-85, ISSN 1672-6472, (in Chinese with English abstract).

Lingg, A.J. & Donaldson, M.D. (1981). Biotic and abiotic factors affecting stability of *Beauveria bassiana* conidia in soil. *Journal of Invertebrate Pathology*, Vol. 38, No. 2, (September 1981), pp. 191-200, ISSN 0022-2011.

Liu, H. & Bauer, L. S. (2008). Microbial control of emerald ash borer, *Agrilus planipennis* (Coleoptera: Buprestidae) with *Beauveria bassiana* strain GHA: greenhouse and field trials. *Biological Control*, Vol. 45, No. 1, (April 2008), pp. 124-132, ISSN 1049-9644.

Luangsa-ard, J.J.; Hywel-Jones, N.L.; Manoch, L. & Samson, R.A. (2004). On the relationships of Paecilomyces sect. Isarioidea species. *Mycologia*, Vol. 96, No. 4, (July/August 2004), pp. 773-780, ISSN 1557-2536.

Malakar, R.D.; Elkinton, J. S.; Hajek, A.E. & Burand, J.P. (1999). Within-host interactions of *Lymantria dispar* (Lepidoptera: Lymantriidae) nucleopolyhedrosis virus and *Entomophaga maimaiga* (Zygomycetes: Entomophthorales). *Journal of Invertebrate Pathology*, Vol. 73, No. 1, (January 1999), pp. 91–100, ISSN 0022-2011.

Marcelino, J.A.P.; Gouli, S.; Giordano, R.; Gouli, V.V.; Parker, B.L. & Skinner, M. (2009). Fungi associated with a natural epizootic in *Fiorinia externa* Ferris (Hemiptera: Diaspididae) populations. *Journal of Applied Entomology*, Vol. 133, No. 2, (March 2009), pp. 82–89, ISSN 1439-0418.

Markova, G. (2000). Pathogenicity of several entomogenous fungi to some of the most serious forest insect pest in Europe. *IOBC/WPRS Bulletin*, Vol. 23, No. 2, pp. 231-239, ISSN 1027-3115.

Maurer, P.; Couteaudier, Y.; Girard, P.A.; Bridge, P.D. & Riba, G. (1997). Genetic diversity of *Beauveria bassiana* and relatedness to host insect range. *Mycological Research*, Vol. 101, No. 2, (February 1997), pp. 159-164, ISSN 0953-7562.

Meyling, N.V. & Eilenberg, J. (2007). Ecology of the entomopathogenic fungi *Beauveria bassiana* and *Metarhizium anisopliae* in temperate agroecosystems: potential for conservation biological control. *Biological Control*, Vol. 43, No. 2, (November 2007), pp. 145–155, ISSN 1049-9644.

Miętkiewski, R.; Żurek, M.; Tkaczuk, C. & Bałazy, S. (1991). Occurrence of entomopathogenic fungi in arable soil, forest soil and litter. *Roczniki Nauk Rolniczych*, Seria E, Vol. 21, No. 1/2, pp. 61–68, ISSN 0080-3693 (in Polish with English summary).

Miętkiewski, R.T.; Pell, J.K. & Clark, S.J. (1997). Influence of pesticide use on the natural occurrence of entomopathogenic fungi in arable soils in the UK: field and laboratory comparisons. *Biocontrol Science and Technology*, Vol. 7, No. 4, pp. 565–575, ISSN 1360-0478.

Mota-Sanchez, D.; Bills, P.S. & Whalon, M.E. (2002). Arthropod resistance to pesticides: status and overview. In: *Pesticides in agriculture and the environment*. Wheeler, W.B. (Ed.), pp. 241-272, CRC Press, ISBN 9780824708092, New York.

Myles, T.G. (2002). Alarm, Aggregation, and defense by *Reticulitermes flavipes* in response to a naturally occurring isolate of *Metarhizium anisopliae*. *Sociobiology*, Vol. 40, No. 2, pp. 243-255.

Neuvéglise, C.; Brygoo, Y.; Vercambre, B. & Riba, G. (1994). Comparative analysis of molecular and biological characteristics of strains of *Beauveria brongniartii* isolated from insects. *Mycological Research*, Vol. 98, No. 3, (March 1994), pp. 322-328, ISSN 0953-7562.

Nielsen, C.; Milgroom, M.G. & Hajek, A.E. (2005a). Genetic diversity in the gypsy moth fungal pathogen *Entomophaga maimaiga* from founder populations in North America and source populations in Asia. *Mycological Research*, Vol. 109, No. 8, (August 2005), pp. 941–950, ISSN 0953-7562.

Nielsen, C.; Keena, M. & Hajek, A.E. (2005b). Virulence and fitness of the fungal pathogen *Entomophaga maimaiga* in its host *Lymantria dispar*, for pathogen and host strains originating from Asia, Europe, and North America. *Journal of Invertebrate Pathology*, Vol. 89, No. 3, (July 2005), pp. 232–242, ISSN 0022-2011.

Nielsen, C.; Vestergaard, S.; Harding, S. & Eilenberg, J. (2007). Microbial control of Strophosoma spp. Larvae (Coleoptera: Curculionidae) in Abies procera greenery plantation. *Journal of Anhui Agricultural University*, Vol. 34, No. 2, pp. 184-194, ISSN 1672-352X.

Ormond, E.L.; Thomas, A.P.M.; Pugh, P.J.A.; Pell, J.K. & Roy, H.E. (2010). A fungal pathogen in time and space: the population dynamics of *Beauveria bassiana* in a conifer forest. *FEMS Microbiology Ecology*, Vol. 74, No. 1, (October 2010), pp. 146-154, ISSN 1574-6941.

Parker, B.L.; Skinner, M.; Gouli, V. & Brownbridge, M. (1997). Impact of Soil Applications of *Beauveria bassiana* and Mariannaea sp. on Nontarget Forest Arthropods. *Biological Control*, Vol. 8, No. 3, (March 1997), pp. 203–206, ISSN 1049-9644.

Pell, J.K.; Eilenberg, J.; Hajek, A.E. & Steinkraus, D.C. (2001). Biology, ecology and pest management potential of Entomophthorales. In: *Fungi as biocontrol agents: progress, problems and potential.* T.M. Butt; C.W. Jackson & N. Magan, (Eds.), pp. 71–153, CABI International, ISBN 0851993567, Oxon.

Pilarska, D.; McManus, M.; Hajek, A.; Herard, F.; Vega, F.; Pilarski, P. & Markova, G. (2000). Introduction of the entomopathogenic fungus *Entomophaga maimaiga* Hum., Shim.& Sop. (Zygomycetes: Entomophtorales) to a *Lymantria dispar* (L.) (Lepidoptera: Lymantriidae) population in Bulgaria. *Journal of Pest Science*, Vol. 73, pp. 125–126, ISSN 1612-4758.

Poprawski, T.J. & Yule, W.N. (1991). Incidence of fungi in natural populations of *Phyllophaga* spp. and susceptibility of *Phyllophaga anxia* (LeConte) (Col., Scarabaeidae) to *Beauveria bassiana* and *Metarhizium anisopliae* (Deuteromycotina). *Journal of Applied Entomology*, Vol. 112, No. 1-5, (January/December 1991), pp. 359-365, ISSN 1439-0418.

Poprawski, T.J.; Riba, G.; Jones, W.A. & Aioun, A. (1988). Variation in isoesterase of geographical population of *Beauveria bassiana* (Deuteromycotina: Hyphomycetes) isolated from *Sitona* weevils (Coleoptera: Curculionidae). *Environmental Entomology*, Vol. 17, No. 2, (April 1988), pp. 275-279, ISSN 1938-2936.

Quesada-Moraga, E.; Navas-Cortés, J.A.; Maranhao, E.A.A.; Ortiz-Urquiza, A. & Santiago-Álvarez, C. (2007). Factors affecting the occurrence and distribution of entomopathogenic fungi in natural and cultivated soils. *Mycological Research*, Vol. 111, No. 8, (August 2007), pp. 947–966, ISSN 1469-8102.

Rath, A.C.; Koen, T.B. & Yip, H.Y. (1992). The influence of abiotic factors on the distribution and abundance of *Metarhizium anisopliae* in Tasmanian pasture soils. *Mycological Research*, Vol. 96, No. 5, (May 1992), pp. 378–384, ISSN 1469-8102.

Rehner, S.A.; Minnis, D.; Sung G.-H.; Luangsa-ard, J.J.; DeVoto, L. & Humber, R.A. (2011). Phylogeny and systematics of the anamorphic, entomopathogenic genus *Beauveria*. *Mycologia*, (April 2011), ISSN 1557-2536, [E-publication ahead of print].

Roberts, D.W. (1981). Toxins of entomopathogenic fungi. In: *Microbial control of pests and plant diseases 1970-1980.* H.D. Burges (Ed.), pp. 441-464, Academic press, ISBN 0121433609, London and New York.

Rohel, E.; Couteaudier, Y.; Papierok, B.; Cavelier, N. & Dedryver, C.A. (1997). Ribosomal internal transcribed spacer size variation correlated with RAPD-PCR pattern polymorphisms in the entomopathogenic fungus *Erynia neoaphidis* and some closely related species. *Mycological Research*, Vol. 101, No. 5, (May 1997), pp. 573-579, ISSN 0953-7562.

Rohlfs, M. & Churchill, A.C.L. (2011). Fungal secondary metabolites as modulators of interactions with insects and other arthropods. *Fungal Genetics and Biology*, Vol. 48, No. 1, (January 2011), pp. 23–34, ISSN 1087-1845.

Saito, T. & Aoki, J. (1983). Toxicity of free fatty acids on the larval surface of two lepidopterous insects towards *Beauveria bassiana* (Bals.) Vuill. and *Paecilomyces fumosoroseus* (Wize) Brown et Smith (Deuteromycetes: Moniliales). *Applied Entomology and Zoology*, Vol. 18, No. 2, pp. 225-233, ISSN 1347-605X.

Samson, R.A., Evans, H.C. & Latgé, J.P. (1988). *Atlas of Entomopathogenic Fungi*. Springer-Verlag, ISBN 0387188312, New York.

Sánchez-Peña, S.R.; San-Juan Lara, J. & Medina, R.F. (2011). Occurrence of entomopathogenic fungi from agricultural and natural ecosystems in Santillo, Mexico, and their virulence towards thrips and whiteflies. *Journal of Insect Science*, Vol. 11, Article 1, pp. 1-10, ISSN 1536-2442.

Serebrov, V.V.; Gerber, O.N.; Malyarchuk, A.A.; Martemyanov, V.V.; Alekseev, A.A. & Glupov, V.V. (2006). Effect of Entomopathogenic Fungi on Detoxification Enzyme Activity in Greater Wax Moth *Galleria mellonella* L. (Lepidoptera, Pyralidae) and Role of Detoxification Enzymes in Development of Insect Resistance to Entomopathogenic Fungi. *Biology Bulletin*, Vol. 33, No. 6, (December 2006), pp. 581-586, ISSN 1062-3590.

Sevim, A.; Demir, I.; Tanyeli, E. & Demirbag, Z. (2010). Screening of entomopathogenic fungi against the European spruce bark beetle, *Dendroctonus micans* (Coleoptera: Scolytidae). *Biocontrol Science and Technology*, Vol. 20, No. 1, pp. 3-11 ISSN 1360-0478.

Shanley, R.P.; Keena, M.; Wheeler, M.M.; Leland, J. & Hajek, A.E. (2009). Evaluating the virulence and longevity of non-woven fiber bands impregnated with *Metarhizium anisopliae* against the Asian longhorned beetle, *Anoplophora glabripennis* (Coleoptera: Cerambycidae). *Biological Control*, Vol. 50, N o. 2, (August 2009), pp. 94-102, ISSN 1049-9644.

Shimazu, M. (2004). A novel technique to inoculate conidia of entomopathogenic fungi and its application for investigation of susceptibility of the Japanese pine sawyer, *Monochamus alternatus*, to *Beauveria bassiana*. *Applied Entomology and Zoology*, Vol. 39, No. 3, pp. 485-490, ISSN 1347-605X.

Shimazu, M. & Takatsuka, J. (2010). *Isaria javanica* (anamorphic Cordycipitaceae) isolated from gypsy moth larvae, *Lymantria dispar* (Lepidoptera: Lymantriidae), in Japan. *Applied Entomology and Zoology*, Vol. 45, No. 3, (August 2010), pp. 497-504, ISSN 1347-605X.

Sierpińska, A. (1998). Towards an Integrated Management of *Dendrolimus pini* L.. In: *Proceedings: Population Dynamics, Impacts, and Integrated Management of Forest Defoliating Insects*, McManus, M.L. & Liebhold, A.M. (Eds.), pp. 129-142, General Technical Report NE-247, Radnor, PA: USDA Forestry Service, (June 1998), Banská Štiavnica, Slovak Republic, August 18-23, 1996.

Smith, R.J. & Grula, E.A. (1982). Toxic components on the larval surface of the corn earworm (*Heliothis zea*) and their effects on germination of *Beauveria bassiana*. *Journal of Invertebrate Pathology*, Vol. 39, No. 1, (January 1982), pp. 15-22, ISSN 0022-2011.

Sosnowska, D.; Bałazy, S.; Prishchepa, L. & Mikulskaya, N. (2004). Biodiversity of Arthropod Pathogens in the Białowieża Forest. *Journal of Plant Protection Research*, Vol. 44, No. 4, pp. 313-321, ISSN 1427-4345.

St. Leger, R.J.; May, B.; Allee, L.L.; Frank, D.C.; Staples, R.C. & Roberts, D.W. (1992). Genetic differences in allozymes and formation of infection structures among isolates of the entomopathogenic fungus *Metarhizium anisopliae*. *Journal of Invertebrate Pathology*, Vol. 60, No. 1, (July 1992), pp. 89-101, ISSN 0022-2011.

Sun, B.-D. & Liu X.-Z. (2008). Occurrence and diversity of insect-associated fungi in natural soils in China. *Applied Soil Ecology*, Vol. 39, No. 1, (May 2008), pp. 100-108, ISSN 1873-0272.

Sung, G.-H.; Hywel-Jones, N.L.; Sung, J.-M.; Luangsa-ard, J.J.; Shrestha, B. & Spatafora, J.W. (2007). Phylogenetic classification of *Cordyceps* and the clavicipitaceous fungi. *Studies in Mycology*, Vol. 57, pp. 5 – 59, ISSN 1872-9797.

Tanyeli, E.; Sevim, A.; Demirbag, Z.; Eroglu, M. & Demira, I. (2010). Isolation and virulence of entomopathogenic fungi against the great spruce bark beetle, *Dendroctonus micans* (Kugelann) (Coleoptera: Scolytidae). *Biocontrol Science and Technology*, Vol. 20, No. 7, pp. 695-701, ISSN 1360-0478.

Teetor-Barsch, G.H. & Roberts, D.W. (1983). Entomogenous *Fusarium* species. *Mycopathologia*, Vol. 84, No. 1, (January 1983), pp. 3-16, ISSN 1573-0832.

Tigano-Milani, M.S.; Gomes, A.C.M.M. & Sobral, B.W.S. (1995). Genetic variability among Brazilian isolates of the entomopathogenic fungus *M. anisopliae*. *Journal of Invertebrate Pathology*, Vol. 65, No. 2, (March 1995), pp. 206-210, ISSN 0022-2011.

Tkaczuk, C. & Miętkiewski, R. (1998). Mycoses of pine sawfly (Diprion pini L.) during hibernation period in relation to entomopathogenic fungi occurring in soil and litter. *Folia Forestalia Polonica serie A – Forestry*, No. 40, pp. 25-33, ISSN 0071-6677.

Traugott, M.; Weissteiner, S. & Strasser, H. (2005). Effects of the entomopathogenic fungus *Beauveria brongniartii* on the non-target predator *Poecilus versicolor* (Coleoptera: Carabidae). *Biological Control*, Vol. 33, No. 1, (April 2005), pp. 107–112, ISSN 1049-9644.

Tscharntke, T.; Bommarco, R.; Clough, Y.; Crist, T.O.; Kleijn, D.; Rand, T.A.; Tylianakis, J.M.; van Nouhuys, S. & Vidal, S. (2008). Conservation biological control and enemy diversity on a landscape scale. *Biological Control*, Vol. 45, No. 2, (May 2008), pp. 238–253, ISSN 1049-9644.

Tymon, A.M. & Pell, J.K. (2005). ISSR, ERIC and RAPD techniques to detect genetic diversity in the aphid pathogen *Pandora neoaphidis*. *Mycological Research*, Vol. 109, No. 3, (March 2005), pp. 285-293, ISSN 0953-7562.

van Frankenhuyzen, K.; Ebling, P.; Thurston, G.; Lucarotti, C.; Royama, T.; Guscott, R.; Georgeson, E. & Silver, J. (2002). Incidence and impact of *Entomophaga aulicae* (Zygomycetes: Entomophthorales) and a nucleopolyhedrovirus in an outbreak of the whitemarked tussock moth (Lepidoptera: Lymantriidae). *Canadian Entomologist*, Vol. 134, No. 6, 2002, pp. 825-845, ISSN 1918-3240.

van Lenteren, J.C.; Babendreier, D.; Bigler, F.; Burgio, G.; Hokkanen, H.M.T.; Kuske, S.; Loomans, A.J.M.; Menzler-Hokkanen, I.; van Rijn, P.C.J.; Thomas, M.B.;

Tommasini, M.G. & Zeng, Q.-Q. (2003). Environmental risk assessment of exotic natural enemies used in inundative biological control. *BioControl*, Vol. 48, No. 1, (February 2003), pp. 3-38, ISSN 1573-8248.

Vänninen, I. (1995). Distribution and occurrence of four entomopathogenic fungi in Finland: effect of geographical location, habitat type and soil type. *Mycological Research*, Vol. 100, No. 1, (January 1996), pp. 93–101, ISSN 1469-8102.

Vänninen, I.; Husberg, G.-B. & Hokkanen, H.M.T. (1989). Occurrence of entompathogenic fungi and entomopathogenic nematodes in cultivated soils in Finland. *Acta Entomologica Fennica*, Vol. 53, pp. 65–71, ISSN 0001-561X.

Vänninen, I.; Tyni-Juslin, J. & Hokkanen, H. (2000). Persistence of augmented *Metarhizium anisopliae* and *Beauveria bassiana* in Finnish agricultural soils. *BioControl*, Vol. 45, No. 2, (June 2000), pp. 201-222, ISSN 1573-8248.

Vestergaard, S.; Nielsen, C.; Harding, S. & Eilenberg, J. (2002). First field trials to control *Melolontha melolontha* with *Beauveria brongniartii* in Christmas trees in Denmark. *IOBC/WPRS Bulletin*, Vol. 25, No. 7, pp. 51-58, ISSN 1027-3115.

Viliani, M.G.; Krueger, S.R.; Schroeder, P.C.; Consolie, F.; Consolie, N.H.; Preston-Wilsey, L.M. & Roberts, D.W. (1994). Soil application effects of *Metarhizium anisopliae* on Japanese-beetle (Coleoptera, Scarabeidae) behaviour and survival in turfgrass microcosms. *Environmental Entomology*, Vol. 23, No. 2, (April 1994), pp. 502-513, ISSN 1938-2936.

Wang, S.; Miao, X.; Zhao, W.; Huang, B.; Fan, M.; Li, Z. & Huang, Y. (2005). Genetic diversity and population structure among strains of the entomopathogenic fungus, *Beauveria bassiana*, as revealed by inter-simple sequence repeats (ISSR). *Mycological Research*, Vol. 109, No. 12, (December 2005), pp. 1364-1372, ISSN 0953-7562.

Watt, A.D. & Leather, S.R. (1988). The pine beauty moth in Scotish lodgepole pine plantations. In: *Dynamics of Forest Insect population*. Berryman, A.A. (Ed.), pp. 243-266, Plenum Press, ISBN 0-306-42745-1, New York.

Wegensteiner, R. (2000). Laboratory evaluation of *Beauveria bassiana* (Bals.) Vuill. and *Beauveria brongniartii* (Sacc.) Petch against the four eyed spruce bark beetle, *Polygraphus polygraphus* (L.) (Coleoptera, Scolitydae). *IOBC/WPRS Bulletin*, Vol. 23, No. 2, pp. 161-166, ISSN 1027-3115.

Weseloh, R.M. & Andreadis, T.G. (1992). Mechanisms of Transmission of the Gypsy Moth (Lepidoptera: Lymantriidae) Fungus, *Entomophaga maimaiga* (Entomphthorales: Entomophthoraceae) and Effects of Site Conditions on Its Prevalence. *Environmental Entomology*, Vol. 21, No. 4, (August 1992), pp. 901-906, ISSN 0046-225X.

Widden, P. & Parkinson, D. (1979). Populations of fungi in a high arctic ecosystem. *Canadian Journal of Botany*, Vol. 57, No. 21, pp. 2408-2417, ISSN 0008-4026.

Zare, R. & Gams, W. (2001). A revision of *Verticillium* section Prostrata. IV. The genera *Lecanicillium* and *Simplicillium*. *Nova Hedwigia*, Vol. 73, pp. 1-50, ISSN 0029-5035.

Zhang, L.-W.; Liu Y.-J.; Yao, J.; Wang, B.; Huang, B.; Li, Z.-Z.; Fan, M.-Z. & Sun, J.H. (2011). Evaluation of *Beauveria bassiana* (Hyphomycetes) isolates as potential agents for control of *Dendroctonus valens*. *Insect Science*, Vol. 18, No. 2, (April 2011), pp. 209-216, ISSN 1744-7917.

Ziemnicka, J. (2008a). Outbreaks and natural viral epizootics of the satin moth *Leucoma salicis* L. (*Lepidoptera: Lymantriidae*). *Journal of Plant Protection Research*, Vol. 48, No. 1, pp. 23-39, ISSN 1427-4345.

Ziemnicka, J. (2008b). Effects of viral epizootic induction in population of the satin moth *Leucoma salicis* L. (*Lepidoptera: Lymantriidae*). *Journal of Plant Protection Research*, Vol. 48, No. 1, pp. 41-52, ISSN 1427-4345.

Ziemnicka, J. & Sosnowska, D. (1996). Entomopathogenic fungi in population of the satin moth *Stilpnotia salicis* L. *Journal of Plant Protection Research*, Vol. 37, No. 1/2, pp. 128-137, ISSN 1427-4345.

6

Quantitative Chemical Defense Traits, Litter Decomposition and Forest Ecosystem Functioning

Mohammed Mahabubur Rahman[1,2,*] and Rahman Md. Motiur[3]
[1]United Graduate School of Agricultural Science, Ehime University, Matsuyama, Ehime
[2]Education and Research Center for Subtropical Field Science,
Faculty of Agriculture, Kochi University, Kochi
[3]Silvacom Ltd., Edmonton
[1,2]Japan
[3]Canada

1. Introduction

In forest ecosystems, litter decomposition, which plays a critical role in nutrient cycling, is influenced by a number of biotic and abiotic factors, including quantitative chemical defense (Ross *et al.*, 2002), environmental conditions such as soil properties and climate (Badre *et al.*, 1998, Vanderbilt *et al.*, 2008), and decomposer community and its complex nature. However, Lavelle *et al.*, (1993) proposed a hierarchical model for the factors controlling litter decomposition. The levels of the hierarchal model are: climate (temperature and moisture)> physical properties of soil (clay and nutrients)>litter quality> macro and microorganisms (Lavelle *et al.*, 1993). Leaf litter quality, which is inherited from living leaves, has repeatedly been emphasized as one of the most important factors controlling the decomposition process (Swift *et al.*, 1979, Melillo *et al.*, 1982, Osono & Takeda 2005). Decomposition rate may be decreased with latitude and lignin content of litter but increased with temperature, precipitation and nutrient concentrations at the large spatial scale (Zhang *et al.*, 2008).

Bernhard-Reversa & Loumeto (2002) mentioned that litter fall serves three main functions in the ecosystem such as energy input for soil microflora and fauna, nutrient input for plant nutrition, and material input for soil organic matter building up. The first two functions are completed through decomposition and mineralization, and the third one through decomposition and humification.

Litter decomposition is a primary source of soil nutrients such as nitrogen and phosphorus, which are often limiting to plant growth in terrestrial ecosystems. The litter is broken down by insects, worms, fungi and microorganisms, organically-bound nutrients are released as free ions to the soil solution which are then available for uptake by plants. The variation in soil carbon and nutrient cycling has been clearly linked to variation in particular aspects of litter chemistry. For example, net N mineralization rates in monocultures of different grass species were correlated with root lignin content, suggesting that substrate chemistry is an important control over mineralization and/or immobilization processes (Hobby & Gough 2004).

De Santo *et al.*, (2009) revealed that litter decomposition rates at the boreal forest were significantly lower than at the temperate one and did not differ between needle litter and leaf litter. In the boreal forest mass-loss was positively correlated with the nutrient release. In this site, Mn concentration at the start of the late stage was positively correlated with lignin decay and Ca concentration was negatively correlated to litter mass loss and lignin decay. In the temperate forest neither lignin, N, Mn, and Ca concentration at the start of the late stage, nor their dynamics were related to litter decomposition rate and lignin decay (Santo *et al.*, 2009). On the other hand, litter quality had stronger effects on decomposition than the temperature in temperate forest (Rouifed *et al.*, 2010). In arid and semi-arid sites, photodegradation could be an influential factor for litter decomposition (Austin & Vivanco 2006). Powers *et al.*, (2009) used a short-term litter bag experiment to quantify the effects of litter quality, placement and mesofaunal exclusion on decomposition in 23 tropical forests in 14 countries. They concluded that decomposition in tropical forest is controlled by soil fauna and litter chemistry, which would vary with the precipitation regime.

Zhang *et al.*, (2008) stressed that the combination of total nutrient elements and C:N accounted for 70.2% of the variation in the litter decomposition rates. On the other hand, the combination of latitude, mean annual temperature, C: N and total nutrient accounted for 87.54% of the variation in the litter decomposition rates. They also indicated that litter quality is the most important direct regulator of litter decomposition at the global scale.

The management of scheme options can influence vegetation and wildlife value for delivering ecosystem services by modifying the composition of floral and faunal communities (Smith *et al.*, 2009). Most European temperate forests have been managed according to "classical" sustainable yield principles for a very long time. In France, vast areas of deciduous forests have been cultivated on short rotations for the production of fuelwood under so-called "low forest" regimes. Many of these forests are now being converted back to high forest management (FAO, 2011). Clear felling is common in harvesting operations in mature softwood stands in Northern America. Thinning is a very common and recommended management practice to manage the plantation and forest stands in general (Blanco *et al.*, 2011). Intensive forest management, which may include operations such as: site preparation; tree planting; tending; thinning; and fertilizer application is often influence the litter decomposition process. Litter decomposition in unmanaged systems is affected mainly by climatic variables (Aerts 1997, Blanco *et al.*, 2011).

However, for better understanding of litter decomposition process and forest ecosystem functioning, we need to know more detail about quantitative chemical defense and their effects on the litter decomposition process which influence the ecosystem functioning. This paper will discuss the quantitative defense traits of forest litter, their effects on litter decomposition and ecosystem functioning.

2. Quantitative chemical defense traits of litter

Quantitative chemicals are those that are present in high concentration in plants (5 – 40% dry weight). The most quantitative metabolites are digestibility reducers that make plant cell walls indigestible to animals. The effects of quantitative metabolites are dosage dependent and the higher these chemicals' proportion in the herbivore's diet, the less nutrition the herbivore can gain from ingesting plant tissues. Because they are typically large molecules, these defenses are energetically expensive to produce and maintain, and often take longer to synthesize and transport (Nina & Lerdau 2003). These secondary compounds may be secreted within the cells, for example, in vacuoles, or excreted extracellularly. They include poisonous compounds whose concentration in the cell tends to be relatively low, e.g. alkaloids, cyanogenic glycosides and cardenolides. Some of these secondary metabolites accumulate to levels high enough to reduce the plant's digestibility and palatability for herbivores (McKey 1979; Lindroth & Batzli 1984; Lambers 1993).

However, quantitative chemical defensive traits of litter may divide into three main categories: 1. Lignin 2. Total phenolics 3. Tannins (Hydrolyzable tannins and Proanthocyanidins or Condensed tannins).

2.1 Lignin

Lignin is a polymer of aromatic subunits usually derived from phenylalanine. It is an important constituent of plant secondary cell walls and comprises the largest fraction of plant litter. Lignified leaves are rigid in structure, and highly recalcitrant to decay. Its chemical assay is difficult and different methods may lead to different results. Because it constitutes a barrier preventing decomposition of cellulose, lignin content of litter has been reported to control litter decomposition rate (Sterjiades & Erikson 1993).

2.1.1 Chemistry and occurrence of lignin

Lignins are complex polymers formed by the dehydrogenative polymerization of three main monolignols, p-coumaryl, coniferyl, and sinapyl alcohols (Fig. 1).

Gymnosperm lignins are mainly formed from coniferyl alcohol, together with small proportions of p-coumaryl alcohol. Angiosperm lignins are mainly formed from coniferyl and sinapyl alcohols with small amounts of p-coumaryl alcohol (Lewis 1999). Table 1 shows the % lignin content of different leaf litter.

2.1.2 Effects of lignin on litter decomposition

Lignin is an essential component of plant litter, and is among the most recalcitrant compounds, and consequently is of major importance in soil humus building. Lignin concentration in leaves (or lignin to mineral ratios) has been widely used as an index of organic-matter quality. For instance, lignin concentrations alone, or lignin to N ratios in leaves could explain the rate of decomposition; negative correlations have been reported between lignin concentrations (or lignin to mineral ratios) and decomposition rates (Meentemeyer 1978; Melillo *et al.*, 1982; Vitousek *et al.*, 1994; Hobbie 1996; Kitayama *et al.*, 2004). The ratio of lignin and N as a factor that is more related to decomposition than lignin content (Fig. 2).

On the other hand, hemicellulose and lignin concentrations were reported to be negatively correlated with decomposition (Vivanco & Austin 2008). The initial lignin content of leaf litter influenced the rate of decomposition. The species exhibiting higher initial lignin contents showed lower rates of decomposition of leaf litter. For example, the decomposition of *Quercus dealbata* litter is slower than that of *Quercus fenestrata* (Laishram

& Yadava, 1988). However, the concentrations of the lignin fraction increased as decomposition proceeded, reaching relatively steady levels in the range of 45–51% (Berg 2000; Devi & Yadava, 2007). These increases showed partially linear relationships with accumulated mass loss (Berg *et al.*, 1984).

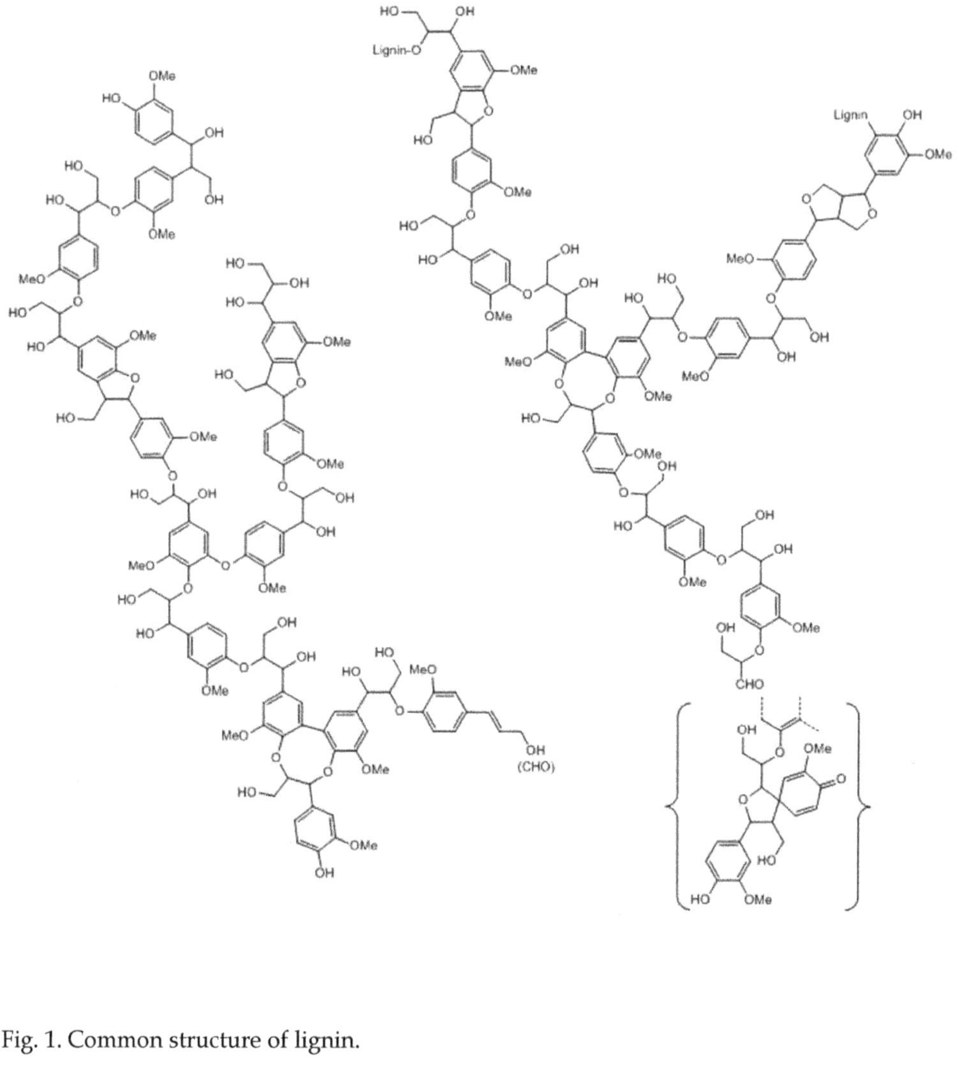

Fig. 1. Common structure of lignin.

Name of the species	Category of Plant	Lignin content of litter (%)
Acacia auriculiformis	Broadleaf tree	54.4
Acacia mangium	Broadleaf tree	43.1
Acer saccharum	Broadleaf tree	10.8
Alphitonia petriei	Broadleaf tree	40.4
Betula pubescens Ehrh	Broadleaf tree	14.0
Castanea sativa	Broadleaf tree	21.1
Dipterocarpus tuberculatus	Broadleaf tree	7.45
Eucalyptus grandis	Broadleaf tree	21.1
Gaultheria griffithiana	Shrub	5.0
Nothofagus dombeyi	Broadleaf tree	19.3
Nothofagus nervosa	Broadleaf tree	29.2
Nothofagus obliqua	Broadleaf tree	27.6
Picea orientalis	Coniferous tree	21.5
Pinus contorta Dougl	Coniferous tree	37
Pinus sylvestris L.	Coniferous tree	29.3
Populus nigra	Broadleaf tree	21.5
Quercus rubra	Broadleaf tree	23.1
Quercus dealbata	Broadleaf tree	6.0
Quercus fenestrata	Broadleaf tree	4.0
Quercus griffithii	Broadleaf tree	3.8
Rhododendron arboreum	Shrub	8.0
Tilia americana	Broadleaf tree	20.0
Broadleaves		21.57±14.23
Coniferous tree		29.26±7.75
Shrub		6.50±2.12

Table 1. The lignin content of different plant species litter (percentage of dry weight) (adapted from Rahman *et al.*, 2011).

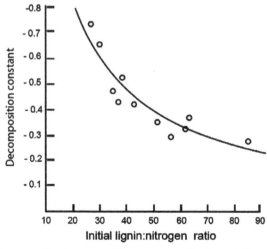

Fig. 2. Relationship between the decomposition constant (*k* = lose of dry mass, to initial lignin concentration of litter) and the lignin: nitrogen ratio of litter (adapted from Melillo *et al.*,1982).

As decomposition proceeds the litter becomes enriched with lignin and N along with other components (Devi & Yadava, 2007). Earlier works have shown that as the lignin concentration increases during litter decomposition, the decomposition rates get suppressed (Fogel & Cromack, 1977; Devi & Yadava, 2007). The suppressing effect of lignin on litter mass-loss rates can be described as a linear relationship in the later stages of decomposition, which, for pine litter, may start at ca. 20–30% mass loss (Fig. 3a). For these later stages, the slope and intercept of this negative relationship varies among sites under different climates (Berg et al., 1993). The lowest effect of lignin concentration on mass-loss rates was found near the Arctic Circle (where long-term average actual evapotranspiration was about 385–390 mm).

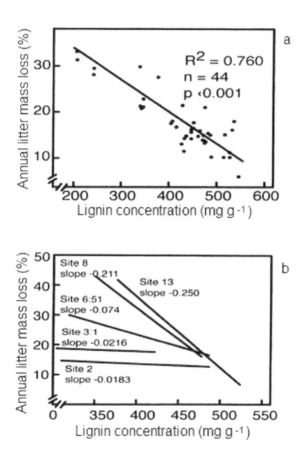

Fig. 3. Annual litter mass loss (%) as a function of initial lignin concentration at the start of each one year period. (a) The linear relationship indicated here, namely a decreasing rate with increasing lignin concentrations for one type of liter. (b) The same relationships under different climatic conditions (five climatically different sites with the AET values 385, 387, 472, 509 and 560 mm for sites 2, 3, 6:51, 8 and 13, respectively) indicate that the rate retarding effect of lignin is stronger in warmer wetter climates (adapted from Berg 2000).

In contrast, in Northern Germany and on the European continent the rate-regulating effect of lignin was found to be higher (Fig. 3 b). In a research on decomposition study from India, it is reported that lignin and fibre contents have showed a negative relation with weight loss of litter (Devi & Yadava, 2007). Many studies have reported a decline in the rate of weight loss of litter due to high initial lignin content (Singh & Gupta, 1977; Devi & Yadava, 2007). More recently, the scientists have found a highly significant, positive correlation between lignin contents and litter decay rates (Raich *et al.*, 2007).

2.2 Total phenolics

Phenolic compounds are one of the most abundant and widely distributed groups of substances in the plant kingdom with more than 8000 phenolic structures currently known (Harbone, 1980). They are products of the secondary metabolism of plants and arise biogenetically from two main primary synthetic pathways: the shikimate pathway and the acetate pathway (Paixao *et al.*, 2007).

2.2.1 Chemistry and occurrence of total phenolics

Phenolics are a heterogeneous group of natural substances characterized by an aromatic ring with one or more hydroxyl groups (Fig. 4). These substances are chemically diverse carbon-based secondary plant compounds occurring in plant tissues (Harborne 1997).

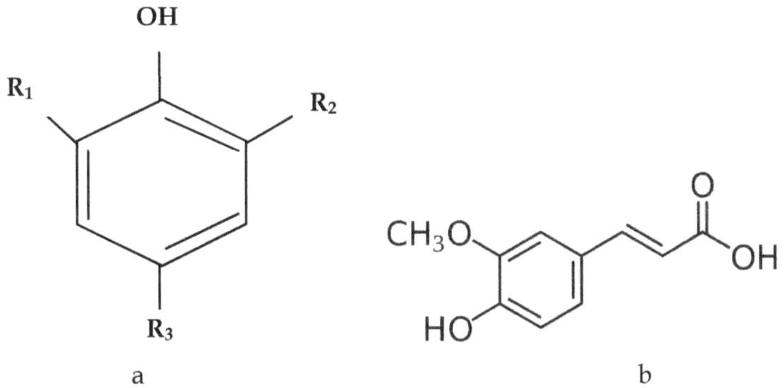

Fig. 4. Simple phenols (C_6), a. phenolic acids b. ferulic acid.

Phenols can be roughly divided into two groups: (1) low molecular weight compounds; and (2) oligomers and polymers of relatively high molecular weight (Hättenschwiler & Vitousek, 2000). Low molecular weight phenolics occur universally in higher plants, some of them are common in a variety of plant species and others are species specific. Because of the large variety of analytical methods and problems with choosing the appropriate standards, polyphenol concentrations reported in the literature vary immensely and might not be comparable with each other. Nevertheless, the two most frequently used polyphenol measurements (i.e. 'total phenolics' and Proanthocyanidins) are accepted reasonably well, and they commonly yield results in the range of about 1% to 25% of total green leaf dry mass (Hättenschwiler & Vitousek, 2000). The amount of phenolics in plant tissues varies with leaf species, age and degree of decomposition (Table 2) (Barlocher & Graca, 2005).

Name of the Species	Plant type	Phenolics (% leaf dry mass)
Acer saccharum (s)	Deciduous tree	15
		2.7
Alnus glutinosa (s)	Tree	6.6
		6.8-7.6
Carya glabra (s)	Deciduous tree	9.1
Eucalyptus globulus (s)	Evergreen tree	6.4
		9.8
Fagus sylvatica (s)	Deciduous tree	8.0
Quercus alba (s)	Tree	16.2
Sapium sebiferum (l, s)	Deciduous tree	3.0
Spartina alterniflora (y)	Perennial deciduous grass	0.4-1.5

Table 2. The phenolics concentrations (%) in selected plant tissues, including senescent leaves (s), live (l) and yellow-green to brown-dead grass leaves (y) (after Barlocher and Graca, 2005).

2.2.2 Effects of total phenolics on litter decomposition

Total phenolics are considered to be biologically active, e.g. by protecting plants against biotic (e.g. microbial pests, herbivores) or abiotic stresses (e.g. air pollution, heavy metal ions, UV-B radiation) (Hutzler *et al.*, 1998), by contributing to allelopathic reactions (Waterman & Mole 1994) and by retarding decomposition rates of organic matter (Hattenschwieler & Vitousek 2000). In particular, phenolics play a major role in the defense against herbivores and pathogens (Lill & Marquis 2001).

In addition, some phenolics may prevent leaf damage resulting from exposure to excessive light (Lee & Gould 2002). The bulk of phenolics remain present during leaf senescence and after death, these compounds may also affect microbial decomposers (Harrison 1971) and therefore delay microbial decomposition of plant litter (Salusso 2000). Canhoto & Graca (1996) observed a strong negative correlation between the phenol content of different native litter types and litter decomposition rates in a stream, whereas Canhoto & Graca (1999) showed that phenolics from *Eucalyptus* leaves decrease feeding by detritivores. Thus, effects of phenolics on detritivores may be one reason for the low decomposability of *Eucalyptus* litter. The initial concentration of total phenolics in litter is positively correlated with dry organic carbon loss (Madritch & Hunter 2004). High amount of phenolics compounds in plants tissue decrease N concentration, which impedes the litter decomposition (Xuefeng *et al.*, 2007). Barta *et al.*, (2010) confirmed that a low amount of phenolics and low phenolics/N ratio in plant litter is closely related to higher differences in microbial respiration rates and mineral N release during the four months of litter decomposition in spruce forest.

Lin *et al.*, (2006) observed a negative correlation between total phenolics and N contents for *Kandelia candel* and *Bruguiera gymnorrhiza* leaf litter at various stages of decomposition (Fig. 5). The perception of phenols as inhibitors, however, is far too simple, and the variety of phenolic compounds can have many different functions within the litter layer and the underlying soil (Hattenschwiler & Vitousek 2000). Even intraspecific variation in litter polyphenol concentrations can strongly influence soil processes and ecosystem functioning (Schweitzer *et al.*, 2004). Phenols may influence rates of decomposition as they bind to N in the leaves forming compounds resistant to decomposition (Palm & Sanchez 1990). Gorbacheva & Kikuch (2006) found that dynamics of easily oxidized phenolics may influence the litter decomposition rate in the monitored subarctic field. Some scientist mentioned that phenolics stimulate microbial activity and subsequently reduce plant available N (Madritch & Hunter 2004; Lin et al.,, 2006). These results contribute important information to the growing body of evidence, indicating that the quality of C moving from plants to soils is a critical component of plant-mediated effects on soil biogeochemistry and, possibly, competitive interactions among species. Gorbacheva & Kikuch (2006) found that the essential part of phenolics that participates in the formation of mobile forms of organic matter, leaches from the organic horizon and migrates through the soil profile.

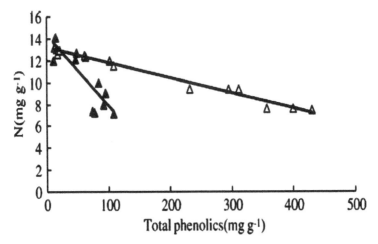

Fig. 5. Relationship between total phenolics and N contents during leaf decomposition of *Bruguiera gymnorrhiza* (Bg) and *Kandelia candel* (Kc). Symbols are: black triangle for Bg leaf; white triangle for Kc leaf (Adapted from Lin *et al.*, 2006).

2.3 Tannins

Tannin is the fourth most abundant biochemical substance in vascular plant tissue after cellulose, hemicellulose and lignin (Kraus *et al.*, 2003). Leaves and bark may contain up to 40% tannin by dry weight (Matthews *et al.*, 1997; Kraus *et al.*, 2003) and in leaves and needles tannin concentrations can exceed lignin levels (Benner *et al.*, 1990). Tannin reduce herbivore load either directly through toxicity or growth alteration or indirectly through reduction of palatability (Feeny 1970).

2.3.1 Chemistry and occurrence of Tannins

Tannins are heterogenous group of phenolics compounds derived from flavonoids and gallic acid. Bate-Smith & Swain (1962) defined tannin as water-soluble polyphenolic compounds ranging in molecular weight from 500 to 3000 Daltons that have the ability to precipitate alkaloids, gelatins and other proteins. Haslam (1998) has substituted the term "polyphenol" for "tannin", in an attempt to emphasize the multiplicity of phenolic group's characteristic of these compounds. Haslam also notes that molecular weights as high as 20,000 Daltons have been reported, and that tannins complex not only with proteins and alkaloids but also with certain polysaccharides. However, tannins found in higher plants are divided into two major classes termed Proanthocyanidins or Condensed tannins and hydrolyzable tannins (Fig. 6).

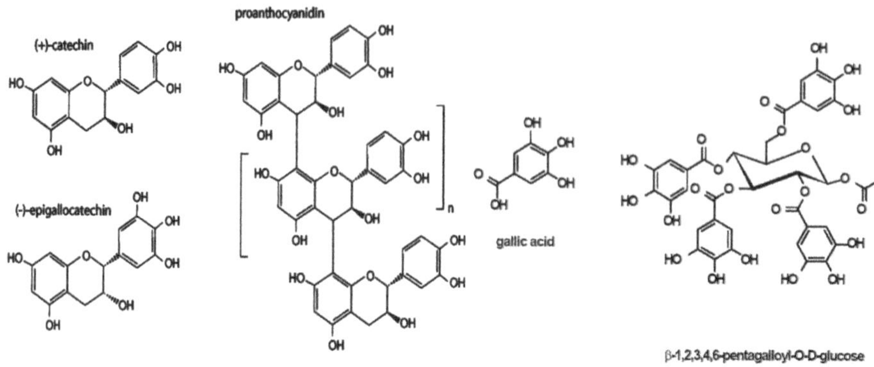

a   b

Fig. 6. Structures of tannins; a. Flavan-3-ols (+)-catechin and (-)-epigallocatechin, examples of monomeric precursors that polymerize to form macromolecular products such as linear proanthocyanidins composed of monomeric flavanoid units connected by C4-C8 linkages. b. hydrolyzable tannins β-1,2,3,4,6-pentagalloyl-O-D-glucose.

Tannins are distributed in species throughout the plant kingdom. They are commonly found in both gymnosperms as well as angiosperms. Tannins are mainly physically located in the vacuoles or surface wax of plants. Because tannins are complex and energetically costly molecules to synthesize, their widespread occurrence and abundance suggests that tannins play an important role in plant function and evolution (Zucker 1983). Tannins occur in plant leaves, roots, wood, bark, fruits and buds (Peters & Constabel, 2002; Ossipov et al., 2003). Tannin distribution in plant tissues appears to vary from species to species. In leaf tissues, tannins have been reported to occur preferentially in the epiderm, hypoderm, periderm, mesophyll, companion cells and vascular tissues, as well as throughout the leaf tissue (Grundhöfer et al., 2001). In roots, anatomical studies have identified a 'condensed tannin zone' in pine and eucalyptus located between the growing tip and the more developed cork zone (Peterson et al., 1999; Enstone et al., 2001). Hydrolyzable tannins have a more restricted occurrence than condensed tannin, being found in only 15 of the 40 orders of dicotyledons (Hättenschwiler & Vitousek 2000).

2.3.2 Effects of tannin on litter decomposition

N and lignin concentration or C: N and lignin: N ratios are often used to predict rates of litter decomposition. However, a number of studies have shown that tannin and/or polyphenol content is a better predictor of decomposition, net N mineralization and N immobilization (Palm & Sanchez, 1991; Gallardo & Merino, 1992; Driebe & Whitham, 2000; Kraus et al., 2003). Coq et al., (2010) mentioned that litter decomposition in tropical rainforest correlated well with condensed tannin concentration. They concluded that leaf litter tannins play a key role in decomposition and nutrient cycling in the tropical rainforest.

In the past decades, many studies have shown that tannins are involved in defense mechanisms of plants against attack by bacteria, fungi, and herbivores (Zucker 1983; Scalbert 1991). There is not much knowledge about the mechanisms of action of the tannin (Zucker 1983; Scalbert 1991) even though modern analytical methods have improved the analysis of these complex structures (Mole & Watermann 1987; Schofield *et al.*, 2001). Proposals on the mechanism of action include tannins forming stable complexes with plant proteins to make the tissue unattractive and difficult to digest (Schofield *et al.*, 2001) and tannins acting like a toxin through highly specific reactions with digestive enzymes or directly at the cell membranes (Zucker 1983) or through depletion of essential iron by complexation (Mila *et al.*, 1996). Leaves with high initial contents of condensed tannins, seem to decompose slowly in both terrestrial (Valachovic *et al.*, 2004) and aquatic ecosystems (Wantzen *et al.*, 2002). Condensed tannin may play an important role in aquatic leaf litter decomposition, as they may deter invertebrate shredders (Wantzen *et al.*, 2002). Condensed tannin deters herbivore feeding by acting as toxins and not as digestion inhibitors by protein precipitation. Other researchers have obtained data that suggest the toxic nature of tannins (Robbins *et al.*, 1987; Provenza *et al.*, 1990; Clausen *et al.*, 1990). Alongi (1987) noticed that if decomposers are inhibited by high contents of tannins in their food, strong effects on litter breakdown would be expected. Handayanto *et al.*, (1997) found a strong negative correlation between N mineralization rates and the protein precipitation capacity of litter material, a measure of tannin reactivity. Litter material high in tannin content is commonly associated with reduced decomposition rates (Gallardo & Merino, 1992; Kalburtji *et al.*, 1999). The convergent evolution of tannin-rich plant communities has occurred on nutrient-poor acidic soils throughout the world. Tannins were once believed to function as anti-herbivore defenses, but more and more ecologists now recognize them as important controllers of decomposition and nitrogen cycling processes. Tannins inhibit soil nitrogen accumulation and the rate of terrestrial and aquatic decomposition (Hissett & Gray 1976). Tannins make plant tissues unpalatable and indigestible for animals. Tannins impede digestion of plant tissues by blocking the action of digestive enzymes, binding to proteins being digested or interfering with protein activity in the gut wall (Howe & Westley 1988; Lambers 1993). Tannins may also reduce insect predation because they increase the leaf toughness (Haslam 1988). Kraus *et al.*, (2003) summarized that tannins may limit litter decomposition in a number of different ways: (1) by themselves being resistant to decomposition (2) by sequestering proteins in protein-tannin complexes that are resistant to decomposition (3) by coating other compounds, such as cellulose, and protecting them from microbial attack (4) by direct toxicity to microbes, and (5) by complexing or deactivating microbial exoenzymes.

The studies by Schimel *et al.*, (1998) in the Alaskan taiga provide some of the most comprehensive examinations of the diversity of phenolics and condensed tannin effects on soil processes.

Secondary succession in these forests starts with *Salix/Alnus* communities and continues to an *Alnus/Populus*, a *Populus*, and finally a *Picea alba*–dominated community. *Populus balsamifera* was found to play a key role during succession by the production of polyphenols that interfere with soil processes. Plants from strongly N-limited ecosystems are generally defended by tannins, whereas N-based and structural defenses become more abundant with increases in N supply (Gartlan *et al.*, 1980).

For example, in the savannas of southern Africa, infertile miombo woodlands and savannas on soils derived from highly weathered granites have trees whose leaves are defended by tannins, while on nearby savannas on higher-nutrient soils such as shales and young volcanic soils, plants are defended by spines. The prevalence of chemical defenses depends on ecosystem nutrient supply (Craine *et al.*, 2003). Fig. 7 represents the schematic overview of the effects of quantitative chemicals from leaf litter on various soil processes and its consequences for the nitrogen cycle and successional dynamics in terrestrial ecosystems.

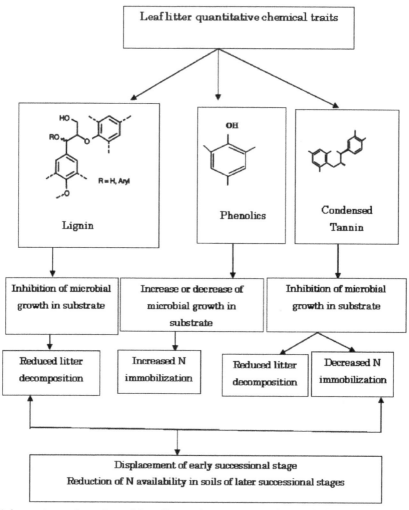

Fig. 7. Schematic reorientation of the effects of quantitative chemicals from leaf litter on various soil processes and its consequences for the nitrogen cycle and successional dynamics in terrestrial ecosystems (modified after Schimel *et al.*, 1998).

3. Litter decomposition and ecosystem functioning

Ecosystem is composed of three subsystems i.e., producer- consumer- and decomposer-subsystem. Ecosystem functioning is affected not only by the function of each subsystem but also by interactions between them. Quantitative defense is the driving force of ecosystem functioning (fig. 8).

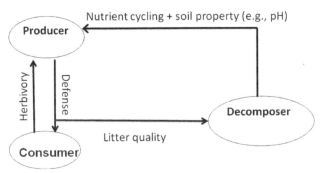

Fig. 8. Ecosystem functioning.

Wardle *et al.*, (1997) demonstrated the importance of tree species composition in determining turnover rate of organic matter and N mineralization by comparing the forests dominated by early successional, fast-growing species and those dominated by slow-growing climax species. This is a verification of the hypothesis in the boreal zone the climax of which is characterized by slow rate of decomposition of hardly decomposable spruce needle litter.

Plant litter decomposition is important to many ecosystem functions such as the formation of soil organic matter, the mineralization of organic nutrients, and the carbon balance (Austin & Ballaré 2010). It is estimated that the nutrients released during litter decomposition can account for 69-87% of the total annual requirement of essential elements for forest plants (Waring & Schlesinger 1985).

Decomposition and nutrient cycling are fundamental to ecosystem biomass production. Most natural ecosystems are nitrogen (N) limited and biomass production is closely correlated with N turnover (Vitousek & Howarth, 1991; Reich *et al.*, 1997). Typically external input of nutrients is very low and efficient recycling of nutrients maintains productivity (Likens *et al.*, 1970). Decomposition of plant litter accounts for the majority of nutrients recycled through ecosystems. Rates of plant litter decomposition are highly dependent on litter quality; high concentration of phenolic compounds, especially lignin, in plant litter has a retarding effect on litter decomposition (Hattenschwiler & Vitousek 2000). At the ecosystem level, chemical defense can influence litter decomposition and nutrient cycling rates (Hattenschwiler & Vitousek 2000; Kraus *et al.*, 2003). Ecosystems dominated by plants with low-lignin concentration often have rapid rates of decomposition and nutrient cycling (Chapin *et al.*, 2003). Simple carbon (C) containing compounds are preferentially metabolized by decomposer microorganisms which results in rapid initial rates of decomposition. More complex C compounds are decomposed more slowly and may take many years to completely breakdown. Phenols and tannins, affect nutrient cycling in soil by inhibiting organic matter degradation, mineralization rates and N availability (Kraus *et al.*, 2003).

3.1 Effects of management on litter decomposition

Litter decomposition rate may be regulated by species composition, leaf litter chemistry, management activities, or any of the combinations of them. The forest management activities had significant effects on leaf litter decomposition. They also had significant effects on leaf litter chemistry during decomposition process. Clear cut or selective thinning method for timber harvesting may change litter decomposition by altering microclimatic conditions of the forest floor, leaf litter chemistry and composition of the microbial community (Li et al., 2009; Blanco et al., 2011). In a cold climate, selective thinning can increase the temperatures of the forest floor, and thus increase decomposition rate whereas selective thinning in a warm climate may slow decomposition by reducing the moisture content of surface organic matter (Li et al., 2009; Blanco et al., 2011). Blanco et al., (2011) observed that thinning effects clearly influenced litter chemistry in Continental forest and Mediterranean forest, generally appeared sequentially, first nutrients and cellulose and then total C and lignin.

4. Concluding remarks and future direction

Leaf litter decomposition is the fundamental process of ecosystem functioning. Quantitative chemical is hypothesized to play a key role in regulating litter decomposition and to be important for the production of dissolved organic matter and CO_2. Lignin decomposition may be relatively slower in boreal forest than tropical forests. Results after three years of decomposition experiments of Canadian boreal forest, Moore et al., (1999) concluded that lignin/N ratio, and some other climatic variables were valuable parameters for predicting mass loss.

Plants growing in tropical regions have higher polyphenolics concentrations in their tissues compared to temperate zone species (Coly 1983). Tannins are abundant in tropical tree foliage and have the potential effects on litter decomposition. Wieder et al., (2009) used natural variations in species litter chemistry combined with a through fall removal experiment to understand how climate–chemistry interactions regulate tropical forest litter decomposition. Their results suggested that widely used predictors of litter decomposition based on chemical quality are still useful in tropical forests and that these wet systems also require an understanding of litter solubility to best prediction rates of decomposition.

The above discussion justifies that quantitative chemical traits can be used as a predictive tool for litter decomposability and ecosystem functioning. The decomposition of plant litter is an essential process in terrestrial ecosystems, resulting in carbon and nutrients being recycled for primary production. While a great deal of research has addressed quantitative chemical defense and their effect on decomposition and ecosystem functioning, there are many areas of quantitative chemical's biogeochemistry that are not known. There is still little information found regarding how different types of quantitative chemicals influence soil organisms, and how these chemicals' biodegradation affects the soil quality. An understanding of the role of quantitative chemicals in plant litter decomposition will allow for more accurate predictions of carbon dynamics in terrestrial ecosystems. When we can make a relationship between quantitative chemicals and decomposition then we can easily predict ecosystem functioning, which is important for conservation and restoration management of endangered ecosystems. Hence, it is imperative that future research focuses more attention on quantitative chemicals' biogeochemistry and their effects on litter decomposition, CO_2 emission and soil quality. There is a specific need to understand the role of lignin and of lignified cellulose, and their interactions during the late stages of decomposition. How do different types of lignin building blocks influence litter decomposition processes? What is the fate of quantitative chemicals after litter decomposition? Identifying and quantifying links between quantitative chemical defense traits and litter decomposability would enhance our understanding of ecosystem functioning and will provide us with a predictive tool for modeling decomposition rates under different vegetation types.

5. Acknowledgements

We are thankful to Biogeochemistry, and Ecology journal editors for giving permission to use the figure in this manuscript.

6. References

Alongi D.M. (1987) The influence of mangrove-derived tannins on interidal meiobenthos in tropical estuaries. Oecologia, 71, 537–540.

Aerts R. (1997) Climate, leaf litter chemistry and leaf litter decomposition in terrestrial ecosystems: a triangular relationship. Oikos, 79:439–449.

Austin A.T., Ballaré C.L. (2010) Dual role of lignin in plant litter decomposition in terrestrial ecosystems. Proceedings of the National Academy of Sciences USA, 107, 4618–4622.

Austin A.T., Vivanco L. (2006) Plant litter decomposition in a semi-arid ecosystem controlled by photodegradation. Nature, 442:555–558.

Badre B., Nobelis P., Tre´molie`res M. (1998) Quantitative study and modeling of the litter decomposition in a European alluvial forest. Is there an influence of overstorey tree species on the decomposition of ivy litter (Hedera helix L.)? Acta Oecologica, 19, 491–500.

Barlocher F., Graca M.A.S. (2005) Total phenolics, Methods to study litter decomposition: a practical guide. Springer, Berlin Heidelberg New York. Bärlocher F., Newell S.Y. (1994) Phenolics and protein affecting palatability of Spartina leaves to the gastropod Littoraria irrorata. P.S.Z.N.I. Marine Ecology, 15, 65–75.

Bärlocher F., Canhoto C., Graça M.A.S. (1995) Fungal colonization of alder and eucalypt leaves in two streams in Central Portugal. Archiv für Hydrobiologie, 133, 457–470.

Barta J., Applova M., Vanek D., Kristufkova M., Santruckova H. (2010) Effect of available P and phenolics on mineral N release in acidified spruce forest: connection with lignin-degrading enzymes and bacterial and fungal communities. Biogeochemistry, 97, 71–87.

Bate-Smith E.C., Swain T. (1962) Flavonoid compounds, Comparative Biochemistry, Vol. 3A. Academic Press, New York.

Berg B. (2000) Litter decomposition and organic matter turnover in northern forest soils. Forest Ecology and Management, 133, 13–22.

Berg B., McClaugherty C. (2003) Plant litter. Decomposition. Humus Formation. Carbon Sequestration. Springer-Verlag Heidelberg, Berlin, Germany.

Berg B., Ekbohm G., McClaugherty C. (1984) Lignin and holocellulose relations during long term decomposition of some forest litters: Long-term decomposition in a Scots-pine forest. IV. Canadian Journal of Botany, 62, 2540–2550.

Berg B., McClaugherty C., Johansson M. (1993) Litter mass-loss rates in late stages of decomposition at some climatically and nutritionally different pine sites. Long-term decomposition in a Scots pine forest VIII. Canadian Journal of Botany, 71, 680–692.

Bernhard-Reversat F., Loumeto, J.J. (2002) The litter system in African forest tree plantations. In: Reddy, M.V. (Ed.), Management of Tropical Plantation–Forests and their Soil Litter System: Litter, Biota and Soil–Nutrient Dynamics. Science Publishers, Inc., Plymouth, UK, pp. 11–39.

Blanco J.A., Imbert J.B., Castillo F.J. (2011) Thinning affects *Pinus sylvestris* needle decomposition rates and chemistry differently depending on site conditions, Biogeochemistry, DOI 10.1007/s10533-010-9518-2.

Cameron G.N., LaPoint T.W. (1978) Effects of tannins on the decomposition of Chinese tallow leaves by terrestrial and aquatic invertebrates. Oecologia, 32, 349–366.

Canhoto C.M., Graca M.A.S. (1996) Decomposition of *Eucalyptus globulus* leaves and three native leaf species (*Alnus glutinosa, Castanea sativa, Quercus faginea*) in a Portuguese low order stream. Hydrobiologia, 333, 79–85.

Canhoto C.M., Graca M.A.S. (1999) Leaf barriers to fungal colonization and shredders (Tipula lateralis) consumption of decomposing *Eucalyptus globulus*. Microbial Ecology, 37, 163--172.

Chapin F.S.III., Matson P.A., Mooney H.A. (2003) Principles of terrestrial ecosystem ecology. Springer-Verlag, New York, USA.

Clausen T.P., Provenza., F.D., Burritt E.A., Reichardt P.B., Bryant J.P. (1990) Ecological implications of condensed tannin structure: a case study. Journal of Chemical Ecology, 16, 2381–2392.

Coley P.D. (1983) Herbivory and defensive characteristics of tree species in a lowland tropical forest .Ecological Monograph, 53, 209–233.

Coq S., Jean-Marc S., Meudec E., Cheynier V., Hättenschwiler S. (2010) Interspecific variation in leaf litter tannins drives decomposition in a humid tropical forest in French Guiana. Ecology, 91, 2080–2091.

Craine J., Bond W., Piter L., Reich B., Ollinger. S. (2003) The resource economics of chemical and structural defenses across nitrogen supply gradients. Oecologia, 442,547–556.

De Santo AV., De Marco A., Fierro A., Berg B., Rutigliano F.A. 2009. Factors regulating litter mass loss and lignin degradation in late decomposition stages, Plant Soil, 318,217–228.

Devi A.S., Yadava P.S. (2007) Wood and leaf litter decomposition of *Dipterocarpus tuberculatus* Roxb. in a tropical deciduous forest of Manipur, Northeast India. Current Science, 93, 243–246.

Driebe E. M. Whitham T.G. (2000) Cottonwood hybridization affects tannin and nitrogen content of leaf litter and alters decomposition. Oecologia, 123:99–107.

Enstone D.E., Peterson C.A., Hallgren S.W. (2001) Anatomy of seedling tap roots of loblolly pine (*Pinus taeda* L.). Trees, 15, 98–111.

FAO. (2011) Forest management, harvesting, and silviculture, http://www.fao.org/DOCREP/003/X4109E/X4109E03.htm.

Feeny P. (1970) Seasonal changes in oak leaf tannins and nutrients as a cause of spring feeding by winter moth caterpillars. Ecology, 51, 565–581.

Fogel R., Cromack K.Jr. (1997) Effect of habitat and substrate quality on Douglas-fir litter decomposition in western Oregon. Canadian Journal of Botany, 55, 1632–1640.

Gallardo A., Merino J. (1992) Nitrogen immobilization in leaf litter at two Mediterranean ecosystems of SW Spain. Biogeochemistry, 15, 213–228.

Gartlan J.S., Waterman P.G., McKey D.B., Mbi C.N., Struhsaker T.T. (1980) A comparative study of the phytochemistry of two African rainforests. Biochemical Systematics and Ecology, 8, 401–422.

Gessner M.O. (1991). Differences in processing dynamics of fresh and dried leaf litter in a stream ecosystem. Freshwater Biology, 26, 387–398.

Gorbacheva T.T., Kikuchi R., (2006) Plant-to-soil pathways in the subarctic – qualitative and quantitative changes of different vegetative fluxes, Environmental Biotechnology, 2: 26–30.

Graca M.A.S., Barlocher, F. (1998). Proteolytic gut enzymes in Tipula caloptera – Interaction with phenolics. Aquatic Insects, 21, 11–18.

Graça M.A.S., Newell S.Y., Kneib, R.T. (2000) Grazing rates of organic living fungal biomass of decaying Spartina alterniflora by three species of salt-marsh invertebrates. Marine Biology, 136, 281–289.

Grundhöfer P., Niemetz R., Schilling B., Gross G.G. (2001) Biosynthesis and subcellular distribution of hydrolyzable tannins. Phytochemistry, 57, 915–927.

Handayanto E., Giller K. E., Cadisch G. (1997) Regulating N release from legume tree prunings by mixing residues of different quality. Soil Biology and Biochemistry, 29, 1417–1429.

Harbone J.B. (1980) Plant phenolics. In: E.A. Bell & B.V. Charlwood, (Eds). Encyclopedia of plant physiology, secondary plant products Vol. 8, Springer-Verlag, Berlin Heidelberg, New York:329–395.

Harborne J.B. (1997) Role of phenolic secondary metabolites in plants and their degradation in the nature. In: G. Cadisch & K.E. Giller (Eds). Driven by nature: Plant litter quality and decomposition. CAB International, Wallingford: 67–74.

Harrison A. F. (1971) The inhibitory effect of oak leaf litter tannins on the growth of fungi, in relation to litter decomposition. Soil Biology and Biochemistry, 3: 167–172.

Haslam E. (1998) Practical polyphenolics: from structure to molecular recognition and physiological function, Cambridge University Press: Cambridge, UK.

Hättenschwiler S., Vitousek P.M. (2000) The role of polyphenols in terrestrial ecosystem nutrient cycling. Tree, 15, 238–243.

Hissett R., Gray T.R.G. (1976) Microsites and time changes in soil microbe ecology. In: J.M. Anderson & A. Macfadyen (Eds.) The role of terrestrial and aquatic organisms in decomposition processes. Blackwell, London: 23–40.

Hobbie S.E. (1996) Temperature and plant species control over litter decomposition in Alaskan tundra. Ecological Monograph, 66,503–522.

Hobbie S.E., Gough L. 2004. Litter decomposition in moist acidic and non-acidic tundra with different glacial histories, Oecologia, 140, 113–124.

Howe H.F., Westley L.C. (1988) Ecological relationships of Plants and Animals. Oxford University Press, New York.

Hutzler P., Fischbach R.J., Heller W., Jungblut T.P., Reuber S., Schmitz R., Veit M., Weissenböck G., Schnitzler J.P. (1998) Tissue localization of phenolic compounds in plants by confocal laser scanning microscopy. Journal of Experimental Botany, 49, 953–965.

Kalburtji K.L., Mosjidis J.A., Mamolos A.P. (1999) Litter dynamics of low and high tannin Sericea lespedesa plants under field conditions. Plant Soil, 208, 217–281.

Kitayama K., Suzuki S., Hori M., Takyu M., Aiba S.H., Lee N.M., Kikuzawa K. (2004). On the relationships between leaf-litter lignin and net primary productivity in tropical rain forests. Oecologia, 140, 335–339.

Kraus T. E.C., Yu Z., Preston C.M., Dahlgren R.A., Zasoski R.J. (2003) Linking chemical reactivity and protein precipitation to structural characteristics of foliar tannins. Journal of Chemical Ecology, 29, 703–73.

Laishram I.D., Yadava P.S. (1988) Lignin and nitrogen in the decomposition of leaf litter in a subtropical forest ecosystem at Shiroi hills in north-eastern India. Plant Soil, 106, 59-64.

Lambers H. (1993) Rising CO_2, secondary plant metabolism, plant-herbivore interactions and litter decomposition. Vegetatio, 104/105, 263–271.

Lambers H., Poorter H. (1992) Inherent variation in growth rate between higher plants: A search for physiological causes and ecological consequences. Advances in Ecological Research, 23, 187-261.

Lavelle E., Blanchart E., Martin A., and Martin S. (1993) A hierarchical model for decomposition in terrestrial ecosystems: application to soils of the humid tropics, Biotropica, 25, 130-150.

Lee D.W., Gould K.S. (2002) Anthocyanins in leaves and other vegetative organs: An introduction. Advances in Ecological Research, 37, 1-16.

Lewis N.G. (1999) A 20th century roller coaster ride: a short account of lignifications. Current Opinion in Plant Biology, 2, 153-162.

Lewis N.G., Yamamoto E. (1990) Lignin: occurrence, biogenesis and biodegradation. Annual Review of Plant Physiology and Plant Molecular Biology, 41, 455-496.

Li Q, Moorhead DL., DeForest J.L., Henderson R., Chen J., Jensen R. (2009) Mixed litter decomposition in a managed Missouri Ozark forest ecosystem. Forest Ecology and Management, 257, 688-694.

Likens G.E., Bormann F.H., Johnson N.M., Fisher D.W., Pierce R.S. (1970) Effects of forest cutting and herbicide treatment on nutrient budgets in the Hubbard Brook watershed-ecosystem. Ecological Monographs, 40, 23-47.

Lill J.T., Marquis R.J. (2001) The effects of leaf quality on herbivore performance and attack from natural enemies. Oecologia, 126, 418-428.

Lin Y.M., Liu J.W., Xiang P., Lin P., Ye G.F., Sternberg L., da S.L. (2006) Tannin dynamics of propagules and leaves of Kandelia candel and Bruguiera gymnorrhiza in the Jiulong River Estuary, Fujian, China. Biogeochemistry, 78, 343-359.

Lindroth R.L., Batzli G.O. (1984) Plant phenolics as chemical defenses: effects of natural phenolics on survival and growth of prairie voles (Microtus ochrogaster). Journal of Chemical Ecology, 10, 229-244.

Matthews S., Mila I., Scalbert A., Donnelly D.M.X. (1997) Extractable and non-extractable proanthocyanidins in barks. Phytochemistry, 45, 405-410.

McKey D. 1979. The distribution of secondary compounds within plants. In: G.A. Rosenthal & D.H. Janzen, (Eds.) Herbivores, their Interaction with SA Secondary Plant Metabolites. Academic Press, New York : 56-134.

Meentemeyer V. (1978) Macroclimate and lignin control of litter decomposition rates. Ecology, 56, 465-472.

Melillo J.M., Aber J.D., Muratore J.F. (1982) Nitrogen and lignin control of hardwood leaf litter decomposition dynamics. Ecology, 63, 621-626.

Mila I., Scalbert A., Expert D. (1996) Iron withholding by plant polyphenols and resistance to pathogens and rots. Phytochemistry, 42, 1551-1555.

Mole S., Watermann P.G. (1987) A critical analysis of techniques for measuring tannins in ecological studies. Oecologia, 72, 137-147.

Moore T.R., Trofymow J.A., Taylor B., Prescott C., Camire´ C., Duschene L., Fyles J., Kozak L., Kranabetter M., Morrison I., Siltanen M., Smith S., Titus B., Visser S., Wein R., Zoltai S. (1999) Litter decomposition rates in Canadian forests. Global Change Biology, 5, 75-82.

Nina S., Lerdau M. (2003) The evolution of function in plant secondary metabolites. International Journal of Plant Science, 164, 93-102.

Osono T., Takeda T. (2005) Limit values for decomposition and convergence process of lignocellulose fraction in decomposing leaf litter of 14 tree species in a cool temperate forest. Ecological Research, 20, 51-58.

Ossipov V., Salminen J-P., Ossipova V., Haukioja E., Pihlaja K. (2003) Gallic acid and hydrolyzsable tannins are formed in birch leaves from an intermediate compound of the shikimate pathway. Biochemical Systematics and Ecology, 31, 3-16.

Paixao, N; Perestrelo, R; Marques, J.C., Camara J.C. (2007) Relationship between antioxidant capacity and total phenolic content of red rose and white wines, Food Chemistry, 105, 204–214.

Palm C.A., Sanchez P.A. (1991) Nitrogen release from the leaves of some tropical legumes as affected by their lignin and polyphenolic contents. Soil Biology & Biochemistry, 23, 83–88.

Palm C.A., Sanchez P.A. (1990) Decomposition and nutrient release patterns of the leaves of three tropical legumes. Biotropica, 22, 330–338.

Pereira A.P., Graça, M.A.S., Molles M. (1998) Leaf decomposition in relation to litter physicochemical properties, fungal biomass, arthopod colonization, and geographical origin of plant species. Pedobiologia, 42, 316–327.

Peters D.J., Constabel C.P. (2002) Molecular analysis of herbivore induced condensed tannin synthesis: cloning and expression of dihydroflavonol reductase from trembling aspen (Populus tremuloides). Plant Journal, 32, 701–712.

Peterson C.A., Enstone D.E., Taylor J.H. (1999) Pine root structure and its potential significance for root function. Plant Soil, 217, 205–213.

Powers J.S, Montgomery R.A, Adair E.C, Brearley F.Q, DeWalt S.J, Castanho C.T, Chave J, Deinert E, Ganzhorn J.U, Gilbert M.E et al., (2009) Decomposition in tropical forests: a pan-tropical study of the effects of litter type, litter placement and mesofaunal exclusion across a precipitation gradient. Journal of Ecology, 97, 801–811.

Provenza F.D., Burritt E.A., Clausen T.P., Bryant J.P., Reichardt P.B., Distel R.A. (1990) Conditioned taste aversions: A mechanism for goats to avoid condensed tannins in blackbrush. American Naturalist, 136, 810-828.

Rahman M.M., Motiur M.R., Yoneyama, A. (2011) Lignin chemistry and its effects on litter decomposition in forest ecosystem. Chemistry & Biodiversity, DOI: 10.1002/cbdv.201100198.

Raich J.W., Russell A.E., Ricardo B.A. (2007) Lignin and enhanced litter turnover in tree plantations of lowland Costa Rica. Forest Ecology and Management, 239, 128–135.

Reich P.B., Grigal D.F., Aber J.D., Gower S.T. (1997) Nitrogen mineralization and productivity in 50 hardwood and conifer stands on diverse soils. Ecology, 78, 335–347.

Robbins C.T., Hagerman A.E., Austin P.J., McArthur C., Hanley T.A. (1991) Variation in mammalian physiological responses to a condensed tannin and its ecological implications. Journal of Mammology, 72, 480–486.

Rouifed S., Handa I.T., David J.F., Ha¨ttenschwiler S. (2010) The importance of biotic factors in predicting global change effects on decomposition of temperate forest leaf litter. Oecologia, 163, 247–256.

Salusso M.M. (2000) Biodegradation of subtropical forest woods from north-west Argentina by Pleurotus laciniatocrenatus. New Zealand Journal of Botany, 38, 721–724.

Scalbert A. (1991) Antimicrobial properties of tannins. Phytochemistry, 30, 3875–3883.

Schimel J.P., Cates R.G., Ruess R. (1998) The role of balsam poplar secondary chemicals in controlling soil nutrient dynamics through succession in the Alaskan taiga. Biogeochemistry, 42, 221–34.

Schofield P., Mbugua D.M., Pell A.N. (2001) Analysis of condensed tannins: a review. Animal Feed Science and Technology, 91, 21–40.

Schweitzer J.A., Bailey J.K., Rehill B.J., Hart S.C., Lindroth R.L, Keim P., Whitham T.G. (2004) Genetically based trait in dominant tree affects ecosystem processes. Ecology Letters, 7, 127–34.

Singh J.S., Gupta S.R. (1977) Plant decomposition and soil respiration in terrestrial ecosystems. Botanical Review, 43, 449-528.

Smith J., Potts S.G., Woodcock B.A. Eggleton P. (2009) The impact of two arable field margin management schemes on litter decomposition. Applied Soil Ecology, 41, 90–97.

Sterjiades R., Erikson K.E.L. (1993) Biodegradation of lignins. In: Scalbert A. (Eds.) Polyphenolic Phenomena, INRA Editions, Paris: 115–126.

Suberkropp K., Godshalk G.L., Klug M.J. (1976) Changes in the chemical composition of leaves during processing in a woodland stream. Ecology, 57, 720–727.

Swift M.J., Heal O.W., Anderson J.M. (1979) Decomposition in Terrestrial Ecosystems, Studies in Ecology 5. Blackwell, Oxford.

Valachovic Y.S., Caldwell B.A., Cromack K., Griffiths R.P. (2004) Leaf litter chemistry controls on decomposition of Pacific Northwest trees and woody shrubs. Canadian Journal of Forest Research, 34, 2131–214.

Vitousek P.M., Turner D.R., Parton W.J., Sanford R.L.Jr. (1994) Litter decomposition on the Mauna Loa environmental matrix Hawaii: patterns, mechanisms, and models. Ecology, 75, 418–429.

Vanderbilt K.L., White C.S., Hopkins O., Craig J.A. (2008) Aboveground decomposition in arid environments: results of a long-term study in central New Mexico. Journal of Arid Environments, 72, 696–709.

Vitousek P.M., Howarth R.W. (1991) Nitrogen limitation on land and in the sea: how can it occur? Biogeochemistry, 13, 87–115.

Vivanco L., Austin A. (2006) Intrinsic effects of species on leaf litter and root decomposition: a comparison of temperate grasses from North and South America. Oecologia, 150, 97–107.

Wantzen K. M., Wagner R., Suetfeld R., Junk W.J. (2002) How do plant herbivore interactions of trees influence coarse detritus processing by shredders in aquatic ecosystems of different latitudes? Verhandlungen Internationale Vereinigung für Theoretische und Angewandte Limnologie, 28, 815–821.

Wardle D.A., Zackrisson O., Hornberg G., Gallet C. (1997) The influence of island area on ecosystem properties. Science, 277, 1296–1299.

Waring R.H., Schlesinger W.H. (1985) Forest ecosystems: concepts and management. Orlando, FL, Academic Press.

Waterman P.G., Mole S. (1994) Analysis of phenolic metabolites. London, UK.

Wieder W.R., Cleveland C.C., Townsend A.R. (2009) Controls over leaf litter decomposition in wet tropical forests. Ecology, 90, 3333–3341.

Xuefeng L., Shijie H., Yan H. (2007) Indirect effects of precipitation variation on the decomposition process of Mongolian oak (*Quercus mongolica*) leaf litter. Frontiers of Forestry in China, 2: 417–423.

Zhang D., Hui D., Luo Y., Zhou G. (2008) Rates of litter decomposition in terrestrial ecosystems: global patterns and controlling factors. Journal of Plant Ecology, 1, 1–9.

Zucker W.V. (1983) Tannins: does structure determine function? An ecological perspective. American Naturalist, 121, 335–365.

7

Genetic Sustainability of Fragmented Conifer Populations from Stressed Areas in Northern Ontario (Canada): Application of Molecular Markers

K.K. Nkongolo[1,2,*], R. Narendrula[1], M. Mehes-Smith[1,2],
S. Dobrzeniecka[1], K. Vandeligt[1], M. Ranger[1] and P. Beckett[1]
*[1]Department of Biology and
[2]Biomolecular Science Program,
Laurentian University, Sudbury
Canada*

1. Introduction

The Earth's genes, species, and ecosystems are the product of over 3 billion years of evolution and the basis for the survival of our own species. Biological diversity, the measure of the variation in genes, species and ecosystems is valuable because future practical uses and values are unpredictable and our understanding of ecosystems is insufficient to be certain of the impact of removing any component. Genetic diversity is an indicator of ecosystem condition and sustainability. It is a fundamental component of biodiversity and it encompasses all of the genetically determined differences that occur between individuals of a species. The loss of biodiversity is due above all to economic factors, especially the low values given to biodiversity and ecological functions such as watershed protection, nutrient cycling, pollution control, soil formation and photosynthesis. Biodiversity is very much a cross-sectoral issue, and virtually all sectors have an interest in its conservation and the sustainable use of its components. Biological resources are renewable and with proper management can support human needs indefinitely. These resources, and the diversity of the systems which support them, are therefore the essential foundation of sustainable development.

The past two decades have been a time of great change in the management of natural resources in Ontario and around the world. Ontario's forest policy has shifted to a more balanced ecological approach as the forest is now viewed as part of a larger ecosystem OMNR, 2001). All forest policies and associated management practices in Ontario conform to the Policy Framework for Sustainable Forests (OMNR, 2001). Many Ontario communities especially in the North depend on forests. There are some 60.9 million hectares of forested land in the province, representing approximately 57% of the 106.8

million hectare provincial land base (including water). Crown forest accounts for approximately 91% of this forested land located mostly in the Northern Ontario region. It represents the boreal and the Great Lake St. Lawrence Forests that are composed mainly of conifer species (Natural resources of Canada, 2003; OMNR, 2001). The Forest Resource Inventory is the primary survey for sustainable forest management. Information generated by forest inventories has contributed greatly to our knowledge of one of Ontario's renewable resources and continues to serve as the basis for major forest resource planning and policy decisions in the OMNR. But this information is not sufficient to ensure the sustainable management of the forest resource. To achieve this goal, information on genetic diversity of tree populations is essential.

In addition, forest management practices must change to keep pace with climatically induced changes in forest ecosystems. The sustainability, biodiversity, health, and economic benefits of forests will be affected to varying degrees by climate change. A detailed analysis of the level of genetic variability in species and populations is essential in developing climate change models (Colombo et al., 1998).

Evolutionary adaptation to new climate conditions can only occur where sufficient genetic variation exists to allow selective forces to discriminate between adaptive and maladaptive traits. Adaptation may occur more rapidly in species with shorter life cycles, as long as conditions are favourable for reproduction, than in long-lived species such as trees which will undergo a time lag response to changing conditions (Colombo et al., 1998). Forest tree species generally have high levels of genetic variability and gene dispersal rates.

On the other hand, genetic structure of Northern Ontario forests has been seriously affected by past forest management, mining, and forest fire activities. In an effort to maintain the long term viability of the forest landbase in Northern Ontario, forest companies and local government organizations have concentrated on artificial regeneration of conifer seedlings as a primary means of reforestation. To date, over nine millions of forest trees mostly conifers have been planted within the Greater Sudbury Region and surrounding areas.

The Sudbury region in Ontario, Canada has a history over the past 100 years of logging, mining, and sulphide ore smelting, releasing more than 100 million tonnes of SO_2 and tens of thousands of tonnes of cobalt, copper, nickel, and iron ores into the atmosphere from the open roast beds (1888-1929) and smelters (1888-present) (Freedman and Hutchison, 1980). These factors have caused acidification, severe metal contamination of the soils and water at sites within approximately 30 km of the smelters in the Sudbury region. Sudbury area is one of the most ecologically disturbed regions in Canada. There have been numerous studies documenting the effects of SO_2 in the Sudbury region (Cox and Hutchinson, 1980; Amiro and Courtin, 1981; Gratton *et al.*, 2000). In general, information on landscape degradation, soil toxicity, acidification, plant metal accumulation and forest composition in Northern Ontario is readily available but knowledge of genetic variation within and among forest tree populations is lacking. This genetic diversity information is crucial to ensure sustainability of the forest resource. The impoverished plant communities that are currently found in the Greater Sudbury Region (GSR) are not only structurally and floristically different from plant communities found in uncontaminated areas in the basin, but they appear to have a different genetic make-up.

Many studies have used morphological markers to assess genetic variability within and among species and populations. Those markers are not usually reliable since phenotypic variation is often related to environmental factors. Molecular markers are an important and very powerful tool for genetic analyses of plant species. Molecular markers such as Random Amplification of Polymorphic DNA (RAPD), Inter-Simple Sequence Repeats (ISSR), Simple Sequence Repeats (SSR), and Amplified Fragment Length Polymorphism (AFLP) have been successfully used to assess the genetic diversity in many plant species (Semagn et al., 2006; Sharma *et al.*, 2008). Each one of these marker systems offers a unique combination of advantages and disadvantages (Sharma *et al.*, 2008). They differ in the type of sequence polymorphism detected (insertion/deletions vs. point mutation), information content, the dominance relationships between alleles (dominant vs. codominant markers), amount of DNA required, the need for DNA sequence information in the species under analysis, development costs, the ease of use, and the extent to which they can be automated.

The overall objective of this chapter is to provide current stage of knowledge from several studies on genetic variability in planted and natural fragmented conifer populations from Northern Ontario using ISSR and microsatellite (SSR) markers.

2. Materials and methods

2.1 Genetic material

Needles from White pine (*Pinus strobus*), jack pine (*Pinus banksiana*), red pine (*Pinus resinosa*), white spruce (*Picea glauca*), and black spruce (*Picea mariana*) individual trees were sampled from natural (Na) and planted (P) populations. The locations of some sampling sites are illustrated in Figure 1. Additional samples were from the nursery used for the Sudbury land reclamation program and were considered as introductions. For each site, needles and seed samples from first and second generations trees were collected separately. In general 10% to 20 % of each population was analyzed. For each tree, 15 grams of needles were weighed in duplicates, frozen in liquid nitrogen and stored at -80 °C until DNA extraction.

2.2 Soil characterization

Soil samples were analyzed in collaboration with TESTMARK Laboratories Ltd. Sudbury, Ontario, Canada. The laboratory is ISO/IEC 17025 certified, a member of the Canadian Council of Independent Laboratory (CCIL) and the Canadian Association of Environmental Analytical Laboratories (CAEAL), and is accredited by the Standards Council of Canada (SSC). The laboratory employs standard QA/QC procedures, involving blank and replicate analyses and with recovery rate of 98 ± 5% in analyses of spiked samples depending on element selected, in their inductively coupled plasma mass spectrometry (ICPMS) analyses reported here. The minimum detection limits (MDL) following microwave digestion of plant tissue Aqua Regia for elements reported here, were: Aluminum 0.05 µg/g (0.5 µg/g), Arsenic 0.05 µg/g (0.5 µg/g), Cadmium 0.05 µg/g (0.5 µg/g), Cobalt 0.05 µg/g (0.5 µg/g), Copper 0.05 µg/g (0.5 µg/g), Iron 1.0 µg/g (10 µg/g), Lead 0.05 µg/g (0.5 µg/g),

Magnesium 0.2 µg/g (2.0 µg/g), Manganese 0.05 µg/g (0.5 µg/g), Nickel 0.05 µg/g (0.5 µg/g) and Zinc 0.05 µg/g (0.5 µg/g). These MDLs reflect actual sample weights and dilutions; instrument detection limits were lower.

The data for the metal levels in soil samples were analyzed using SPSS 7.5 for Windows. All the data were transformed using a $\log_{10}$ transformation to achieve a normal distribution. Kruskal-Wallis test the non-parametric analog of a one-way ANOVA was used to compare independent samples, and tests the hypothesis that several populations have the same continuous distribution. ANOVA followed by Tukey's HSD multiple comparison analysis were performed to determine significant differences ($p < 0.05$) among the sites.

2.3 DNA extraction

The total cellular DNA from individual samples was extracted from seedling tissue using the method described by Nkongolo (1999), with some modifications. The modification involved addition of PVP (polyvinylpyrrolidone) and β-mercaptoethanol to the CTAB extraction buffer. The DNA concentration was determined using the fluorochrome Hoechst 33258 (bisbensimide) fluorescent DNA quantitation kit from Bio-Rad (cat. # 170-2480) and the purity was determined using a spectrophotometer (Varian Cary 100 UV-VIS spectrophotometer).

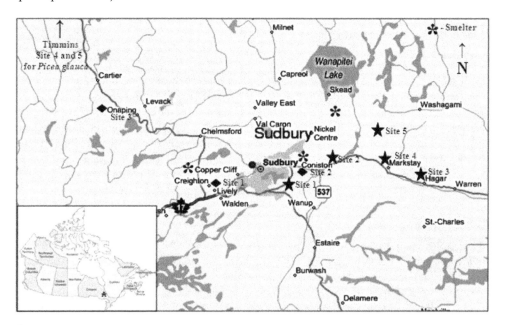

◆*Picea glauca* ; Site 1: Lively; Site 2: Coniston; Site 3: ≈ 40 km from Sudbury HW144 North towards Timmins; Site 4: ≈ 16 km from site 3 HW144 North; Site 5 (control): ≈ 35 km from Site 4 HW144 North..
★*Pinus strobus:* Site 1: Daisy Lake (HW17 Bypass); Site 2: Coniston (HW17); Site 3: Hagar ≈ 60km from Sudbury; Site 4: Markstay ≈ 38km from Sudbury; Site 5: Kukagami Road (≈ 9 km from HW17).

Fig. 1. Sudbury (Ontario) map showing locations of some sampling sites.

2.4 ISSR analysis

The ISSR amplification was carried out in accordance with the method described by Nagaoka and Ogihara (1997), with some modifications described by Mehes et al. (2007). All DNA samples were primed with each of the ten primers used (Table 1). All PCR products were loaded into 2% agarose gel in 1X Tris-Borate-EDTA (TBE) buffer. Gels were pre-stained with 4 µl of ethidium bromide and run at 3.14V/cm for approximately 120 minutes. These agarose gels were visualized under UV light source, documented with the Bio-Rad ChemiDoc XRS system and analyzed for band presence or absence with the Discovery Series Quantity One 1D Analysis Software.

The resulting data matrix of the ISSR phenotype was analyzed using POPGENE software (version 1.32) to estimate genetic diversity parameters (Yeh and Boyle, 1997a, 1997b). POPGENE is computer software used for the analysis of genetic variation among and within populations using co-dominant and dominant markers and quantitative traits. The program was used to determine the intra and inter-population genetic diversity parameters such as percentage of polymorphic loci (P%), Nei's gene diversity (h), Shannon's information index (I), observed number of alleles (Na) and effective number of alleles (Ne). The genetic structure was investigated using Nei's gene diversity statistics, including the within population diversity (H_s) and total genetic diversity (H_t) (Nei, 1973) calculated within the species using the same software. The mean and the total gene diversities, the variation among populations and gene flow were also calculated. The genetic distances were calculated using Jaccard's similarity coefficient estimated with the RAPDistance program version 1.04 (Armstrong *et al.*, 1994).

Primer identification	Nucleotide sequence (5'→3')	G + C content (%)
ISSR Primers		
Echt 5	AGAC AGAC GC	60.00
HB 13	GAG GAG GAG GC	72.70
HB 15	GTG GTG GTG GC	72.70
ISSR 1	AG AG AG AG AG AG AG AG RG	50.00
ISSR 5	ACG ACG ACG ACG AC	64.28
ISSR 9	GATC GATC GATC GC	57.14
UBC 825	AC AC AC AC AC AC AC AC T	88.88
UBC 841	GA AG GA GA GA GA GA GA YC	45.00
17899A	CA CA CA CA CA CA AG	50.00
17898B	CA CA CA CA CA CA GT	50.00

Table 1. The nucleotide sequences of ISSR primers used to screen DNA samples of *Picea glauca* and *Pinus strobus*.

2.5 Microsatellite analysis

The microsatellite analysis involved three species (*Pinus banksiana, Pinus resinosa,* and *Picea mariana*). Ten microsatellite primers, synthesized by Invitrogen, were chosen for amplification of DNA from *Pinus banksiana* and *P. resinosa* populations. These primers described in Vandeligt et al. (2011) include PtTX 3013, PtTX 3030, PtTX 3098, PtTX 309, PtTX 2123, PtTX 3088, RPS 2, RPS 20, RPS 25b, and RPS 84. For *Picea mariana,* the primers used are described in Dobrzeniecka et al. 2009. DNA amplification was performed following the procedure described by Mehes et al. 2009. The Popgene software, version 1.32 (Yeh and Boyle, 1997) was used to assess the intra- and interpopulation genetic diversity parameters such as the mean number of alleles (N_A) across loci, the total number of alleles (N_T) per locus and Shannon's information index (i)(Yeh and Boyle, 1997). The observed and expected heterozygosities (H_O and H_E respectively) were calculated using the Genepop software, version 3.4 (Raymond and Rousset, 1995). The probability test was computed using the Markov chain method (1000 iterations) in order to determine populations in Hardy-Weinberg Equilibrium (Genepop). Hardy-Weinberg equilibrium deviations were tested using alternative hypotheses, deficiency and excess of heterozygotes, for each locus and across loci and populations using Fisher's method. A test for null allele was also done using the EM algorithm of Dempster et al. (1977). The average effective number of migrants exchanged between populations in each generation, or gene flow (N_M) is estimated from F_{ST} (subdivision among populations).

3. Results

3.1 Soil analysis

Recovery and precision for all elements in reference soil samples were within acceptable range. The estimated levels of metal content in different sites from the Greater Sudbury Region in Canada are illustrated in Table 2. The levels of the metals measured were low in the control sites. Overall, the results indicated that nickel and copper continue to be the main contaminants of top soil (Table 2) in sites near the smelters (site 1 and 2). The values ranged from 30.9 to 1600.0 mg kg^{-1} and from 52.3 to 1330.3 mg kg^{-1} for nickel and copper respectively (Table 1). Arsenic concentration exceeded the OMEE (Ontario Ministry of Environment and Energy) guidelines in site 1 and manganese level exceeded the guideline in site 2. Their concentration ranged from 2.2 to 46.0 mg kg^{-1} and 163.6 to 6610.3 mg kg^{-1} for arsenic and manganese, respectively (Table 2).

Aluminum, iron and magnesium concentrations were significantly higher in sites 1 to 4 (top layer, Table 2) compared to the control site 5. The values ranged from 1673.3 to 9193.3 mg kg^{-1}, 2193.3 to 31433.3 mg kg^{-1} and 349.6 to 6866.6 mg kg^{-1} for aluminum, iron and magnesium, respectively (Table 2). Cadmium, cobalt, lead and zinc levels were within the OMEE guideline. The values for these metals ranged from 0.3 to 2.1 mg kg^{-1}, 1.6 to 37.9 mg kg^{-1}, 18.2 to 176.0 mg kg^{-1} and 52.0 to 86.8 mg kg^{-1} (Table 2). The control site 5 was always among the least contaminated for the metals analysis. All the metal concentrations obtained from the bottom layer (5 – 20 cm) were within the OMEE guideline (data not shown). Surprisingly, the data from tailings were similar or significantly lower than other contaminated sites. The pH for all the sites including the controls were low (acidic).

Sampling sites	Elements											pH
	Aluminum	Arsenic	Cadmium	Cobalt	Copper	Iron	Lead	Magnesium	Manganese	Nickel	Zinc	
Site 1	7360b	46b	2b	29b	1330bc	31433bc	176bc	2970b	163a	1600b	86bc	4.6
Site 2	9193b	10a	1b	37b	373ac	12933ac	46ac	6866b	6610c	318bc	52ab	4.5
Site 3	7670b	10a	0.9b	22a	170a	6476a	39ac	2706b	539b	120bd	60ac	-
Site 4	7090b	2a	0.3a	6a	52a	10233ac	18a	2996b	165a	54acd	32a	4.0
Site 5 (control)	1673a	2a	0.7b	2a	93a	2913a	88ab	349a	206a	30acd	55a	3.8
Tailings	-	-	0.3a	34b	300ac	-	46ac	-	-	292bc	59ac	-

Means in columns with a common subscript are not significantly different based on Tukey multiple comparison test (P ≥ 0.05). Site 1: Lively; Site 2: Coniston (close to HW17); Site 3: ≈ 40 km from Sudbury HW144 towards Timmins; Site 4: ≈ 16km from site 3 HW144 towards Timmins; Site 5 (control): ≈ 35km from Site 4 HW144 towards Timmins. Tailings data represent means of two tailings located near smelters.

Table 2. Metal concentrations in top layer (0 – 5 cm) of soil from the Sudbury region sites, concentrations are in mg kg^{-1}, dry weight*.

3.2 Analysis of populations using ISSR markers

Ten ISSR oligonucleotides (Table 1) were used for the amplification of spruce and pine populations. For each population in each species, the levels of polymorphism for the two generations analyzed were similar. Thus, the data were compiled and analyzed per population.

3.2.1 *Pinus strobus* (white pine)

The percentage of polymorphic loci within each population varied between 22% observed in the site 5 (control) to 36% in site 4 (Table 2). The level of genetic variation was similar between natural and planted populations in site 1. For site 2, the polymorphic loci were significantly higher in planted populations compared to the natural population. Data for the Nei' gene diversity (h) ranged from 0.05 (S5) to 0.14 (S4) with a mean of 0.19. A similar pattern was observed for the Shannon's information index (I), with the high value of 0.20 observed in S4 and a low value of 0.08 observed in S5. The observed number of alleles (Na) and the effective number of alleles (Ne) ranged from 1.22 to 1.36 and 1.08 to 1.25 respectively. The genotype diversity among population (H_t) was 0.15 and the within population diversity (H_s) was 0.09. Mean coefficient of gene differentiation (G_{st}) was 0.366 indicating that 63.4% of the genetic diversity resided within the population. The observed structure of genetic variability shows that there is a low level of differentiation among the *Pinus strobus* populations. The overall rate of gene flow (N_m) among population was 0.87.

Population*	P (%)	Na	Ne	h	I
S1P	30	1.30	1.14	0.08	0.13
S1Na	30	1.30	1.18	0.10	0.15
S2P	34	1.34	1.21	0.12	0.18
S2Na	22	1.22	1.12	0.07	0.11
S3M	32	1.32	1.19	0.11	0.17
S4M	36	1.36	1.25	0.14	0.20
S5M	22	1.22	1.08	0.05	0.08
Mean	29	1.29	1.17	0.10	0.15

*Population: P represents Plantation and Na represents Natural populations. M represents mixed populations including natural and planted trees.

Table 2. Genetic diversity parameters of *Pinus strobus* based on ISSR data.

3.2.2 *Pinus banksiana* (jack pine)

For Jack pine, a low to moderate levels of genetic variation was revealed within each population. The percentage of polymorphic loci (P %) ranged from 14.6 % to 45.8 % with a mean of 31.6 %. The mean level of polymorphism for the eight populations from the greater Sudbury area was 27.6% while this value was higher for populations from the nurseries with an average of 42.4% detected polymorphic loci. The levels of genetic variation detected in populations from metal-contaminated areas were similar to those found in control sites. The Nei's gene diversity (h) for all jack pine populations analyzed varied from 0.046 to 0.169 with an average of 0.100, and Shannon's index (I) ranged from 0.070 to 0.250 with an average of 0.153. The mean observed number of alleles (Na) ranged from 1.146 to 1.458, while the mean effective number of alleles (Ne) varied from 1.107 to 1.31 (Table 3).

Populations	P (%)	h	I	Ne	Na
Nursery 1 (Introduction 1)	39.58	0.0961	0.1535	1.1579	1.3958
Nursery 2 (Introduction 2)	41.67	0.1380	0.2106	1.2248	1.4167
Nursery 3 (Introduction 3)	45.83	0.1687	0.2501	1.2946	1.4583
Inco 1	31.25	0.1120	0.1653	1.2035	1.3125
Inco 2	31.25	0.1171	0.1727	1.2061	1.3125
Falconbridge 1	14.58	0.0456	0.0701	1.0756	1.1458
Falconbridge 2	27.08	0.0995	0.1467	1.1758	1.2708
Falconbridge 3	20.83	0.0630	0.0982	1.1004	1.2083
Inco Tailing	35.42	0.0977	0.1552	1.1514	1.3542
Temagami (control)	29.17	0.0818	0.1284	1.1310	1.2917
Low Water Lake (control)	31.25	0.0812	0.1297	1.1256	1.3125
Mean	31.63	0.1001	0.1528	1.1679	1.3163

P represents percentage of polymorphic loci; h, Nei's gene diversity; I, Shannon's information index; Ne, effective number of alleles; Na, observed number of alleles.

Table 3. Genetic variability parameters of *Pinus banksiana* populations growing in the Sudbury area based on ISSR data.

3.2.3 *Pinus resinosa* (red pine)

The level of genetic variation was much lower in the red pine populations. For this species, the level of polymorphic loci varied from 4.55 % to 27.27 % (Table 4). The mean level of polymorphic loci for populations from the greater Sudbury region excluding the population from the nursery was only 8.3%. Like in jack pine populations, the polymorphism detected in contaminated populations was similar to that found in non contaminated site used as a control. Overall, the mean for Nei's gene diversity and Shannon's information index, were 0.034 and 0.053, respectively for all the red pine populations analyzed. The mean observed number of alleles (Na) ranged from 1.045 to 1.27 while the mean effective alleles (Ne) varied from 1.00 to 1.17 (Table 4). The highest genetic diversity values were observed in the populations used for the Sudbury reforestation program. High levels of metal content did not affect the level variation for both species.

Population	P (%)	h	I	Ne	Na
Near Falconbridge	4.55	0.0044	0.0092	1.0049	1.0455
Very near Falconbridge	13.64	0.0411	0.0638	1.0672	1.1364
Falconbridge	4.55	0.0226	0.0314	1.0450	1.0455
Coniston	9.09	0.0180	0.0309	1.0244	1.0909
Daisy Lake	9.09	0.0272	0.0433	1.0389	1.0909
Verner (control)	9.09	0.0267	0.0423	1.0398	1.0909
Introduction 1 (control)	27.27	0.0988	0.1465	1.1710	1.2727
Mean	11.04	0.0341	0.0525	1.0559	1.1104

P represents percentage of polymorphic loci; h, Nei's gene diversity; I, Shannon's information index; Ne, effective number of alleles; and Na, observed number of alleles.

Table 4. Genetic variability parameters of *Pinus resinosa* populations growing in the Sudbury area based on ISSR data.

3.2.4 *Picea glauca* (white spruce)

All the selected primers amplified 11 to 21 fragments across the six populations studied. The amplified fragment size ranged from 170 bp to 2,240 bp. The percentage of polymorphic loci within each population varied between 50% observed in the natural site 5 Na (control) to 61% in site 1, P (Table 5). Nei' gene diversity (h) ranged from 0.17 (site 1, P) to 0.21 (site 5, Na; control) with a mean of 0.19. A similar pattern was observed for the Shannon's information index (I), with the highest value of 0.32 observed in the planted population of site 1P and the lowest value of 0.26 observed in site 5Na (control). The observed number of alleles (Na) and the effective number of alleles (Ne) ranged from 1.50 to 1.61 and 1.29 to 1.37 respectively. The genotype diversity among population (H_t) was 0.19 and the within population diversity (H_s) was 0.23. The mean coefficient of gene differentiation (G_{st}) was 0.168 indicating that 83.2% of the genetic diversity resided within the population. The observed structure of genetic variability shows that there is a low level of differentiation among the *Picea glauca* populations in the target regions even when the populations located as far as 100 km from the Sudbury were included. The overall rate of gene flow (N_m) among population was 2.47.

3.2.5 *Picea mariana* (black spruce)

The genetic diversity within each population was high. For each population, the percentage of polymorphic loci was the same for the parental and the offspring generations analyzed. Thus, the data from the two generations were combined. The percentage of polymorphic loci (P%) ranged from 65% to 90 % with a mean of 75% . Nei's gene diversity (h) varied from 0.264 to 0.359 with an average of 0.310, and Shannon's index (I) ranged from 0.381 to 0.524 with an average of 0.449 (Table 6). The mean observed number of alleles (Na) ranged from 1.650 to 1.900, while the mean effective number of alleles (Ne) varied from 1.168 to 1.632 (Table 6). Among the nine populations investigated, the highest genetic diversity was observed in population 9 from lowland in Timmins while the lowest level of diversity was detected in population 4 from upland in Chelmsford. Overall, the average level of polymorphic loci was much higher in lowlands (85%) than in uplands (68%). There was no difference between metal contaminated and uncontaminated sites for genetic variation.

Population*	P (%)	Na	Ne	h	I
Site 1 (P)	61	1.61	1.37	0.22	0.32
Site 2 (P)	53	1.53	1.33	0.19	0.29
Site 3 (Na)	55	1.55	1.32	0.19	0.29
Site 4 (Na)	53	1.53	1.33	0.19	0.28
Site 5 (control) (Na)	50	1.50	1.30	0.18	0.26
Nursery	57	1.57	1.35	0.20	0.30
Mean	55	1.55	1.33	0.19	0.29

*Population: P represents Plantation and Na represents Natural populations

Table 5. Genetic diversity parameters of *Picea glauca* based on ISSR data.

Populations	P (%)	h	I	Ne	Na
Site 1	80	0.328	0.473	1.603	1.800
Site 2	85	0.350	0.508	1.630	1.850
Site 3	70	0.269	0.396	1.473	1.700
Site 4	65	0.264	0.381	1.490	1.650
Site 5	75	0.317	0.456	1.582	1.750
Site 6	70	0.274	0.402	1.482	1.700
Site 7	70	0.308	0.441	1.567	1.700
Site 8	70	0.325	0.459	1.168	1.700
Site 9	90	0.359	0.524	1.632	1.900
Mean	75	0.310	0.449	1.514	1.750

P represents percentage of polymorphic loci; h, Nei's gene diversity; I, Shannon's information index; Na, observed number of alleles; and Ne, effective number of alleles.

Table 6. Genetic variability parameters of black spruce (*Picea mariana)* populations growing in the Sudbury area based on ISSR data.

3.2.6 Genetic differentiation among populations

For *Pinus banksiana*, the mean gene diversity within populations (Hs) and the total gene diversity (Ht) were 0.100 and 0.1438, respectively. The variation among populations (Gst)

was 0.304 indicating that 30.4 % of total genetic diversity were attributed to the differences among populations. The observed structure of genetic variability shows that there is a sensitive level of differentiation among the jack pine populations in the target regions. The overall rate of gene flow (Nm) among populations was 1.144. For *Pinus resinosa* the Hs and HT values were 0.0341 and 0.0437, respectively. About 22% of the total genetic diversity in *Pinus resinosa* was attributed to differences among populations. For *Pinus strobus*, the genotype diversity among population (H_t) was 0.15 and the within population diversity (H_s) was 0.09. Mean coefficient of gene differentiation (G_{st}) was 0.366 indicating that 63.4% of the genetic diversity resided within the population. The observed structure of genetic variability shows that there is a low level of differentiation among the *Pinus strobus* populations. The overall rate of gene flow (N_m) among population was 0.87.

For *Picea glauca* Ht and Hs were 0.19 and 0.23, respectively. The mean coefficient of gene differentiation (Gst) was 0.168 indicating that 83.2% of the genetic diversity resided within the population. The observed structure of genetic variability shows that there is a low level of differentiation among the *Picea glauca* populations in the target regions even when the populations located as far as 100 km from the Sudbury were included. The overall rate of gene flow (N_m) among population was 2.47.

For *P. mariana,* the mean gene diversity within populations (Hs) and the total gene diversity (Ht) were 0.310 and 0.385, respectively. The variation among populations (Gst) was 0.19. This indicates that 19.3% of total genetic diversity was attributed to the differences among populations. Like in *P. glauca,* the observed structure of genetic variability shows that there is a low level of differentiation among the *P. mariana* populations. The overall rate of gene flow (Nm) among populations was 2.088.

3.3 Genetic relationships among conifer populations based on ISSR analysis

3.3.1 *Pinus banksiana*, *Pinus strobus*, and *Pinus resinosa*

Because of limited genetic variation in *Pinus resinosa* samples analyzed, the genetic relatedness was analyzed only for *Pinus banksiana* and *Pinus strobus* populations The Jaccard similarity coefficients and genetic distance were calculated using ISSR data. The genetic distance scale runs from 0 (identical) to 1 (different for all criteria). In general, the genetic distance values were low as they ranged from 0.06 to 0.21 for *Pinus strobus* and from 0.037 to 0.365 (Table 7) for *Pinus banksiana*. Overall the genetic distance values revealed that all the eleven *P. banksiana* and *P. strobus* populations were genetically closely related (Table 7). For *P. banksiana*, the two populations from control site (uncontaminated), Low Water Lake and Temagami were the most closely related. The largest genetic distance was observed between population 5 from INCO 2 and the new population used in 2006 for reclamation (called introduction 2 in the present study). The dendrogram constructed, based on ISSR data revealed a particular clustering (Fig. 2). All the populations from the greater Sudbury that we analyzed clustered together while the three newly introduced populations from nurseries were grouped in a separate cluster (Fig. 2). For *Pinus strobus* the genetic distance values ranged from 0.06 (S1P and S2P) to 0.21 (S2P and S2Na) (Table 8). Dendrogram was not constructed considering the low levels of genetic distances. For *Pinus resinosa*, the level of genetic variation was too low to calculate genetic distance among populations or to construct a dendrogram.

	1	2	3	4	5	6	7	8	9	10	11
1	0	0.132	0.229	0.243	0.321	0.321	0.333	0.250	0.259	0.247	0.280
2		0	0.186	0.273	0.365	0.345	0.356	0.277	0.286	0.274	0.306
3			0	0.219	0.341	0.321	0.333	0.250	0.280	0.268	0.280
4				0	0.225	0.250	0.198	0.175	0.185	0.195	0.207
5					0	0.134	0.146	0.190	0.221	0.188	0.179
6						0	0.085	0.085	0.096	0.084	0.073
7							0	0.120	0.108	0.096	0.108
8								0	0.038	0.049	0.038
9									0	0.037	0.049
10										0	0.037
11											0

1 represents introduction 1; 2, Introduction 2; 3, Introduction 3; 4, Inco 1 site; 5, Inco 2 site ; 6, Falconbridge 1 site; 7, Falconbridge 2 site; 8, Falconbridge 3 site; 9, Inco Tailing; 10, Temagami site; and 11, Low Water Lake site.

Table 7. Distance matrix generated using bulk sample analysis from various populations of *Pinus banksiana* ISSR data (RAPDistance version 1.04).

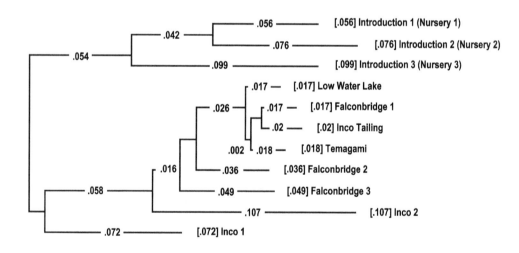

Fig. 2. Dendrogram of the genetic relationships among *Pinus banksiana* populations based on Jaccard similarity matrix using ISSR data. The values above the branches indicate the patristic distances based on the neighbor-joining (NJ) analysis.

	S1P	S1Na	S2P	S2Na	S3M	S4M	S5M
ISSR							
S1P	0.0000	0.1489	0.0667	0.1956	0.1304	0.1739	0.2128
S1Na		0.0000	0.1702	0.1778	0.1111	0.1556	0.1956
S2P			0.0000	0.2174	0.1522	0.1556	001956
S2Na				0.0000	0.0714	0.1191	0.2045
S3M					0.0000	0.0930	0.1778
S4M						0.0000	0.1818
S5M							0.0000

Population: P represents Plantation and Na represents Natural populations. M represents mixed populations including natural and planted trees. Site 1: Daisy Lake (HW17 Bypass); Site 2: Coniston (HW17 Hydro Dam ≈ 10 km from Bypass); Site 3: Hagar ≈ 60 km from Sudbury; Site 4: Markstay ≈ 38km from Sudbury; Site 5: Kukagami Road (≈ 9 km from HW17).

Table 8. Distance matrix generated from ISSR data using the Jaccard similarity coefficient analysis for *Pinus strobus* populations (Free Tree Program).

3.3.2 *Picea glauca* and *Picea mariana*

The genetic distance values were close to 0 as they varied between 0.02 (site 3, Na and site 4, Na) and 0.07 (site 2, P and site 5, Na) (Table 9) for *Picea glauca*. For *Picea mariana*, the genetic values ranged from 0.171 to 0.351 (Table 10). Overall the genetic distance values revealed that all the populations were genetically closely related (Table 10) for each of the Picea species. For *P. mariana*, the dendrogram constructed, based on ISSR data revealed a particular clustering between upland (dry) and lowlands (wet lands) (Figure 3). With the exception of site 7, no upland (dry land) population clusters with a population from a lowland (wet land). For example, the low – land (wetland) population 1 from Falconbridge clusters with the lowland (wetland) population 9 from Timmins; the upland (dry land) population 4 from Chelmsford clusters with the upland (dry land) population 8 from Timmins; the lowland (wetland) population 2 from Falconbridge clusters with the lowlands (wetland) population 5 from Cartier; and the up-land (dry land) population 3 from Capreol clusters with the upland (dry land) population 6 from Cartier.

	Site 1	Site 2	Site 3	Site 4	Site 5 (control)	Nursery
ISSR						
Site 1 (P)	0.0000	0.0520	0.0417	0.0417	0.0626	0.0209
Site 2 (P)		0.0000	0.0729	0.0729	0.0737	0.0316
Site 3 (Na)			0.0000	0.0213	0.0632	0.0417
Site 4 (Na)				0.0000	0.0632	0.0417
Site 5 (Na)					0.0000	0.0625
Nursery						0.0000

Population: P represents Plantation and Na represents Natural populations

Table 9. Distance matrix generated from ISSR data using the Jaccard similarity coefficient analysis for *Picea glauca* populations.

	Site1	Site 2	Site 3	Site 4	Site 5	Site 6	Site 7	Site 8	Site 9
Site 1	0	0.236	0.235	0.171	0.282	0.270	0.307	0.222	0.212
Site 2		0	0.324	0.263	0.230	0.307	0.256	0.350	0.351
Site 3			0	0.235	0.297	0.181	0.324	0.235	0.225
Site 4				0	0.325	0.270	0.263	0.171	0.314
Site 5					0	0.236	0.230	0.325	0.324
Site 6						0	0.263	0.222	0.264
Site 7							0	0.307	0.351
Site 8								0	0.314
Site 9									0

Table 10. Distance matrix generated using the neighbour-joining analysis from *Picea mariana* ISSR data (RAPDistance version 1.04).

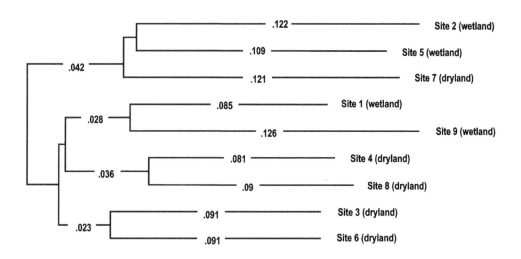

Fig. 3. Dendrogram of the genetic relationships among *Pinus banksiana* populations based on Jaccard similarity matrix using ISSR data. The values above the branches indicate the patristic distances based on the neighbor-joining (NJ) analysis.

3.4 Analysis of populations using microsatellites

3.4.1 Genetic diversity

The microsatellite loci analyzed in *Pinus banksiana* and *P. resinosa* populations are summarized in Table 11. For *P. banksiana*, the mean number of alleles per locus was 9 and the mean effective number of alleles was 3.5 (Table 11). The mean number of alleles across loci per population ranged from 3.00 to 4.67 with the samples from the INCO 3 (site 4) and the introduction 3 (population or site 11) having the highest allelic diversity. The lowest allelic diversity was observed in samples from the INCO tailing (site 7) populations (Table 11).

For *Pinus resinosa* the mean number of alleles per locus was six and the mean effective number of alleles was 2.50. The mean number of alleles across loci per population ranged from 2.33 to 3.00 for the *P. resinosa* populations (Table 11). The highest allelic diversity was observed the samples from site 2 near Falconbridge and site 7 located in Verner. The lowest allelic diversity was found in the samples from site 4 in Coniston/Wahnipitae and the newly introduced population (introduction or site 2) from nursery 2 identified as population or site 8 (Table 11).

Species/Population	N_A	N_{Ap}	HO	HE	i
Pinus banksiana					
Val Caron (site 1)	4.3333	3.3604	0.6667	0.6983	1.2967
Introduction 1 (site 2)	4.0000	2.7951	0.4667	0.6096	1.1051
Introduction 2 (site 3)	4.0000	2.6056	0.4000	0.5850	1.0650
Inco 3 (site 4)	4.6667	2.9764	0.7333	0.6133	1.1734
Inco 1 (site 5)	4.0000	2.9054	0.5000	0.6500	1.1727
Inco 2 (site 6)	3.6667	2.7763	0.6333	0.5367	0.9716
Inco tailing (site 7)	3.0000	2.1221	0.3333	0.5129	0.8551
Falconbridge (site 8)	3.6667	2.1973	0.5852	0.5421	0.9443
Temagami (site 9)	4.0000	2.6277	0.6333	0.6050	1.1054
Low Water Lake(site 10)	3.6667	2.7987	0.2593	0.6235	1.1098
Introduction 3 (site 11)	4.6667	3.7420	0.4000	0.7283	1.3936
Introduction 4 (site 12)	3.6667	2.0854	0.2583	0.4554	0.8578
Mean	4.0000	2.7422	0.4912	0.7194	1.5155
Standard error	±0.4678	±0.1723	±0.3679	±0.0133	±0.0939
Pinus resinosa					
Introduction 1 (site 1)	2.6667	2.0994	0.1000	0.4150	0.7118
Falconbridge (site 2)	3.0000	2.1847	0.5506	0.4258	0.7809
Falconbridge (site 3)	2.6667	1.8039	0.1667	0.3267	0.5926
Falconbridge (site 4)	2.6667	1.9750	0.1333	0.3748	0.6653
Coniston(site 5)	2.3333	1.7365	0.0667	0.3431	0.5497
Daisy Lake (site 6)	2.6667	1.7188	0.0667	0.3346	0.5914
Verner (site 7)	3.0000	1.5541	0.1000	0.3017	0.5459
Introduction 2 (site 8)	2.3333	1.7060	0.0667	0.2783	0.5041
Mean	2.6667	1.8473	0.0892	0.4606	0.9477
Standard error	±0.2520	±1.3076	±0.0941	±0.3992	±0.828

N_A = mean allele number per locus; N_{Ap} = mean number of polymorphic alleles per locus; HO = observed heterozygosity; HE = expected heterozygosity; I = Shannon's information index;

Table 11. Genetic diversity estimates for 12 *Pinus banksiana* and 8 *Pinus resinosa* populations from the Sudbury, Ontario region using microsatellite primers.

The observed heterozygosity (HO) at the population level ranged from 0.26 to 0.67 and the expected heterozygosity (H_E) varied from 0.46 to 0.72 for *Pinus banksiana* populations. The samples from INCO 3 (site 4) produced the highest HO values and the samples from Low Water Lake (site 10) used as control showing the lowest observed heterozygosity (Table 11). The degree of population differentiation (FST) was 17 % for *P. banksiana*. For *Pinus resinosa*, the observed heterozygosity (HO) at the population level, ranged from 0.07 to 0.55. Samples from site 2 located near Falconbridge produced the highest heterozygosity and the samples from nursery 2 (introduction 2) called population or site 8 showing the lowest values (Table 11). HE values ranged from 0.28 to 0.43. The degree of population differentiation (FST) was 23.9 % for *P. resinosa*.

For *Picea mariana*, the microsatellite analysis confirmed the high level of genetic diversity within each population but revealed no significant difference between wetland and upland populations for all the genetic parameters analyzed. Overall, 11% of the total genetic diversity was attributed to differences among populations. The mean number of alleles and effective number of alleles per locus were 10.3 and 5.6, respectively. The observed and expected heterozygosity values ranged from 0.425 to 0.732 and 0.584 to 0.768, respectively (Table 12).

Population	N_A	N_{Ap}	H_O	H_E	I
Site 1 (wetland)	5.67	5.33	0.482	0.619	1.333
Site 2 (wetland)	5.33	5.33	0.587	0.752	1.512
Site 3 (dryland)	5.67	5.67	0.652	0.768	1.572
Site 4 (dry land)	5.33	5.33	0.641	0.740	1.503
Site 5 (wetland)	6.33	6.33	0.648	0.729	1.532
Site 6 (dry land)	3.67	3.33	0.577	0.584	1.040
Site 7 (dry land)	6.00	6.00	0.559	0.744	1.541
Site 8 (dryland)	6.00	6.00	0.732	0.740	1.540
Site 9 (wet land)	5.67	5.67	0.425	0.772	1.585
Mean	5.52	5.44	0.589	0.717	1.462
Standard dev.	±0.765	±0.867	±0.094	±0.067	±0.175

N_A = mean allele number per locus; N_{Ap} = mean number of polymorphic alleles per locus; Ho = observed heterozygosity; H_E = expected heterozygosity (Nei 1973); I = Shannon's information index; F_{IS} = measure of heterozygote deficiency or excess (Wright 1978).

Table 12. Genetic diversity estimates for black spruce (*Picea mariana*) populations using microsatellite primers.

After the correction for null alleles, exact test for Hardy-Weinberg Equilibrium revealed, the majority of the populations deviated significantly from the Hardy Weinberg Equilibrium. The results revealed that the null allele frequency estimates were negligible for all populations (data not shown). The HWE deviation for these populations might be the result of other factors than null alleles. The global tests revealed significant heterozygote deficiency for most populations. Overall, the present study indicates that the long-term exposure of *P. mariana* populations to metal (more than 30 years) is not associated with the level of genetic diversity.

3.4.2 Gene flow

The gene flow estimates were considered low for both species, Nm = 1.21 for *Pinus banksiana* and Nm = 0.79 for *P. resinosa* based on Slatkin (1985). There was also no significant difference in the inbreeding coefficients among the stands within the same species. The mean inbreeding coefficients were considered high for *P. resinosa* and low *for P. banksiana*.

4. Discussion

Loss of rare alleles, lower heterozygoty and directional selection have been concerns of plant populations (Slatkin, 1985; Bergmann and Scholz, 1989). Most of the forest ecosystems within the Sudbury area have improved considerably during the last 30 years (Dudka et al., 1995; Gratton et al., 2000). Vascular and nonvascular plants such as conifers, birches and lichens have re-invaded semi-barren landscapes. More than nine millions trees mostly conifers have been planted in the Greater Sudbury Region. Genetic diversity is the foundation for forest sustainability and ecosystem stability. Bench marking genetic diversity in forest tree populations can provide resource managers with an indicator of long-term forest sustainability and ecosystem health (Mosseler and Rajora, 1998; Rajora and Mosseler, 2001a, 2001b).

For *Pinus banksiana* and *Pinus glauca*, the levels of genetic variation were low to moderate. In fact, genetic variation in *Pinus strobus* (White pine) studied varied from 24 to 40%. The newly planted populations of *Pinus banksiana* and *Pinus glauca* revealed a higher level of genetic variation compared to natural populations. The genetic distances among the pine populations growing in the Greater Sudbury area revealed that all the populations analyzed were genetically close to each other. The highest genetic diversity values were observed in new plantations being developed by the Sudbury reforestation program (Ranger et al., 2007). The level of genetic variation was low (less than 10%) for *P. resinosa*. This was attributed to other events that took place during the history of this species in North America (Mosseler et al., 1992).

Genetic variation and genetic structure of *P. mariana* (black spruce) populations growing in wet and dry lands with different levels of metal contaminations was high in all the populations analyzed with the percentage of polymorphic loci (P %) ranging from 65% to 90 %. For *Picea glauca* populations polymorphism levels ranged from 50% to 61% for ISSR markers and from 70% to 80% for RAPD markers. The level of variation in newly introduced populations of *P. mariana* and *P. glauca* from the Sudbury Reclamation program was also high. Variation within populations accounts for most of total genetic variation. Moreover, genetic tests with species-specific molecular markers revealed that all the trees from *P. mariana* and *P. glauca* planted and natural populations were pure genotypes with no introgression of other species.

In all the conifer species, metal content in soil was not associated with the level of diversity in populations analyzed. Within each species, the different populations studied were genetically closely related. Overall, the results of the present study indicate that the conifer populations from the Greater Sudbury region and other surrounding areas meet most genetic criteria of sustainability. Moreover, the levels of genetic variation observed in the targeted species were similar to data reported for other fragmented populations across Canada for the same species (Mehes et al., 2007).

Elevated accumulations of metal accumulations in soils and vegetation have been documented within short distances of the smelters in Sudbury compared to control sites (Freedman and Hutchinson, 1980; Gratton et al., 2000; Nkongolo et al., 2008). Among the sites analyzed in the present study, the highest level of metal content in soil and plant tissues were detected in samples from populations 1 and 2 located near Falconbridge Smelters in Sudbury (Gratton et al., 2000; Nkongolo et al. 2008). These populations showed the highest level of genetic variability for *Picea mariana* for example along with the control population 9 from Timmins. The same level of genetic variation was observed in parents and progenies within the same populations. This clearly indicated that the exposure to metals for more than 30 years has no effect on genetic structure and diversity of black spruce populations in Northern Ontario. This lack of association between the level of genetic variation and metal content can be attributed to the long life span of conifer species. In fact, the populations analyzed were only the first and second generations of progenies from parents exposed to metal contamination

This is in contrast to data observed in herbaceous species such as *Deschampsia cespitosa* where the level of metal accumulation reduced significantly the level of genetic variation (Nkongolo et al. 2008). Metals impose severe stress on plants, especially in the rooting zone, which has led to the evolution of metal-resistant ecotypes in several herbaceous species like *D. cespitosa* (Cox and Hutchinson 1980). Evidence of loss of genetic variation based on enzymatic analysis at the population level caused by pollution has been demonstrated in some species (Lopes et al. 2004; Prus-Glowacki et al. 2006; van Straalen and Timmermans 2002). But, plants possess homeostatic cellular mechanisms to regulate the concentration of metal ions inside the cell to minimize the potential damage that could result from the exposure to nonessential metal ions. These mechanisms serve to control the uptake, accumulation and detoxification of metals (Foy et al. 1978). This might be the case in black spruce trees exposed to certain levels of metals.

Genetic variation is the foundation for ecosystem stability and population sustainability. In tree populations this information is an indicator of long term population sustainability and health. For example, environmental stressors, such as anthropogenic factors, can affect the genetic frequencies by increasing mutation or selection. This further leads to differences among populations and increase uniformity within a population, thus increasing homozygosity and inbreeding (Dimsoski and Toth, 2001). Studies of genetic variation of impacted and unimpacted populations have defined a positive relationship between the exposure to the stressor and diversity.

Using various types of markers, several authors have reported differences in genetic structure of plants growing in contaminated areas (Muller-stark, 1985; Scholz and Bergmann, 1984). Enzymatic studies of *Picea abies* (Norway spruce) revealed genetic differences between groups of sensitive trees in polluted areas (Scholz and Bergmann, 1984). Higher heterozygosity was reported in tolerant plants of European beech in *Pinus sylvestris* (Scots pine) in Germany and Great Britain (Muller- starck, 1985; Geburek *et al.*, 1987). Berrang *et al.*, (1986) also reported a high heterozygosity in *Populus tremuloides* (Trembling aspen) and *Acer rubrum* (Red maple) populations in the USA.

No significant differences were observed among natural and planted *Picea glauca* populations. All the populations revealed high levels of polymorphic loci for the ISSR markers. This suggests that the *Picea glauca* populations are likely sustainable in long term. For *Pinus strobus*, the levels of genetic variations were in general low to moderate. The newly introduced populations revealed higher levels of polymorphic loci compared to natural populations. This confirms that the land reclamation by planting *Pinus strobus* trees and other pine species is increasing the sustainability of pine populations in the Sudbury region.

Genetic distance values were calculated according to the Jaccard similarity coefficient. In general, the genetic distance values revealed that the different *Pinus spp.* and *Picea spp.* populations were genetically closely related. Overall, the genetic distance analysis showed a high level of homogeneity among populations which could be due to the species characteristics. In fact, the relative small genetic distance values reported in the present analysis are consistent with other studies on *Picea glauca* populations in various provinces that used various molecular markers and allozymes (Rajora *et al.*, 2005; Tremblay and Simon, 1989; Alden and Loopstra, 1987). In general, the genetic similarity among the populations suggests that these populations could have originated from a common source. In addition, *Picea glauca* is an anemophilous species and its pollen is transported over great distances. The fact that *Picea glauca* populations are fairly distributed should promote the exchange of genes among populations. Hence, it is rare to find alleles that are unique to a given populations, and the frequencies of the main alleles are generally similar from one population to another (Rajora *et al.*, 2005; Tremblay and Simon, 1989; Alden and Loopstra, 1987).

In Sudbury (Canada) during the last 25 years, production of nickel, copper and other metals has been maintained at high levels while industrial sulphur dioxide (SO_2) emissions have been reduced by approximately 90% through combination of industrial technological developments and legislated controls. This has allowed for some degree of recovery to occur such as improved air quality and natural recovery of damaged ecosystems during this period of reduced emissions at Sudbury. The recovery has been further done through the reforestation program by planting over 9 million trees such as conifers in the Sudbury region. On the other hands, the African Copper belt, on the border between Zambian and DR-Congo, are among the ten most polluted areas worldwide (The Blacksmith institute, 2008; Banza et al., 2009). Like in many other regions producing heavy metals, such as Senegal, Tanzania, China, Russia, Romania, India, Philippines, Thailand, Indonesia etc.,. there are virtually no controls on the discharge of pollutants from mining and smelters. There are no land reclamation programs and environment degradations from past mining activities have not been addressed. Studies on the effect of metal contamination on genetic diversity of plant populations in those regions are limited. Prus-Glowack et al. (2006) demonstrated in a small scale study in Poland that the stress resulting from gaseous pollution and contamination of the soil with heavy metals exerts a significant effect on phenotype of individuals and on genetic structure of *Pinus sylvestris* L populations. Such data needs to be validated at larger scale using molecular markers.

5. Conclusion

The present study indicates that *Pinus spp* and *Picea spp.* populations from the Sudbury region, Ontario, are genetically variable. Metal contamination levels were not associated with genetic variation in *Picea glauca* populations. Overall, the results indicate that the conifer populations from the Greater Sudbury region and other surrounding areas meet most genetic criteria of sustainability. This conclusion was confirmed by molecular analysis using ISSR, SSR markers, and cytological studies. The effects of metals, if any, may require several generations to be detected. The reclamation of Sudbury forest lands with new populations increases the sustainability specifically for *Pinus* (Pine) species. Since Sudbury is not among the ten most polluted areas in world, a replication of this study in areas with higher soil metal content is recommended to validate the effects of metal populations in tree populations.

6. Acknowledgement

We express our appreciation to the Natural Sciences and Engineering Research Council of Canada (NSERC), Vale (Sudbury), and Xstrata nickel Limited for financial support.

7. References

Alden, J. and C. Loopstra, 1987. Genetic diversity and population structure of *Picea glauca* on an altitudinal gradient in interior Alaska. Can. J. Forest Res. 17: 1519–1526.

Amiro, B.D. and G.M. Courtin, 1981. Patterns of vegetation in the vicinity of an industrially disturbed ecosystem, Sudbury, Ontario. Can J Bot 59: 1623-1639.

Armstrong, J.S., A.J. Gibbs, R. Peakall, G. Weiller, 1994. The RAPDistance Package. 2005, May/20, 30.

Banza, C.L.N., Nawrot, T.S., Haufroid, V., Decree, S., De Putter, T., Smolders E., Kabyla, B.I., Luboya, O.N., Ilunga, A.N., Mutombo, A. Mwanza, Nemery, B. 2009. Environ Res, 1009: 745-752.

Bergmann, F. and F. Scholz, 1989. Selection effects of air pollution in Norway spruce (*Picea abies*) populations. In: Sholz, H., Gregorius, R., and Rudin, D. (Editors). Genetic effects of air pollutants in forest tree populations. Springer-Verlag, Berlin. pp. 143-160.

Berrang, P., D.F. Karnosky., R.A. Mickler and J.P. Bennett, 1986. Natural selection for ozone tolerance in *Populus tremuloides*. Can. J. For. Res. 16: 1214-1216.

The Blacksmith Institute. 2008. World's Most Polluted Places 2007. www.blacksmithinstitute.org. (accessed 2008).

Colombo, S.J., Cherry, M., Graham, C., Greifenhagen, S., McAlpine, R.S. et al. 1998. The Impacts of climate change on Ontario's forests. Forest Research Information Paper No 143. Ontario Forest Research Institute. Ontario Ministry of natural resources. 53 Pp.

Cox, R.M. and T.C. Hutchinson, 1980. Multiple metal tolerances in the grass *Deschampsia cespitosa* (L.) Beauv. From the Sudbury smelting area. New Phytol. 84: 631 -647.

Dempster, A.P., N.M. Laird, D.B. Rubin, 1977. Maximum likelihood from incomplete data via the EM algorithm. J. R. Stat. Soc. B. 39, pp. 1-38

Dimsoski, P. and G.P. Toth , 2001. Development of DNA-based microsatellite marker technology for studies of genetic diversity in stressor impacted populations. Ecotoxicology 10: 229-232.

Dobrzeniecka, S., Nkongolo, K.K., Michael, P., Wyss, S. and Mehes, M. 2009. Development and characterization of microsatellite markers useful for population analysis in *Picea mariana*. Silvae Genet. 58 (4): 168-172.

Dudka, S., R. Ponce-hernandez and T.C. Hutchinson, 1995. Current levels of total element concentrations in the surface layers of Sudbury's soils. Sci. Total Environ. 162: 161-171.

Foy, C.D., R.L. Chaney, and M.C. White, 1978. The physiology of metal toxicology in plants. Annu. Rev. Plant Physiol. 29: 11-566.

Freedman, B. and T.C. Hutchinson, 1980. Pollutant inputs from the atmosphere and accumulations in soils and vegetation near a nickel-copper smelter at Sudbury, Ontario, Canada. Can. J. Bot. 58: 108-132.

Geburek, T., F. Scholz, W. Knabe, and A. Vornweg, 1987. Genetic studies by isozyme gene loci on tolerance and sensitivity in air polluted *Pinus sylvestris* field trial. Silvae Genet. 36: 49-53.

Gratton, W.S., K.K. Nkongolo, and G.A. Spiers, 2000. Heavy metal accumulation in Soil and jack pine (*Pinus banksiana*) needles in Sudbury, Ontario, Canada. Bull Environ Contam Toxicol 64: 550 –557.

Lopes, I, D.J. Baird, R. Ribeiro, 2004. Genetic determination of tolerance to lethal and sublethal copper concentrations in field populations of *Daphnia longispina*. Arch. Environ. Contam. Toxicol. 46: 43-51.

Mehes, M.S., K.K. Nkongolo and P. Michael, 2007. Genetic analysis of *Pinus strobus* and *Pinus monticola* populations from Canada using ISSR and RAPD markers: development of genome-specific SCAR markers. Plant Syst. Evol. 267: 47-63.

Mehes M. and Nkongolo K.K. 2009. Assessing Genetic Diversity and Structure of Fragmented Populations of Eastern White Pine (*Pinus strobus*) and Western White Pine (*P. monticola*) for conservation management. J. Plant Ecol. 2: 143-151.

Mosseler, A. and O.P. Rajora, 1998. Monitoring population viability in declining tree species using indicators of genetic diversity and reproductive success. In: Sassa, K. (ed.) Environmental forest science. Kluwer Academic Publishers, Dordrecht, The Netherlands, pp. 333-344.

Mosseler, A., K.N. Egger, and G.A. Hugues, 1992. Low levels of genetic diversity in red pine confirmed by random amplified polymorphic DNA markers. Can. J. For. Res. 22: 1332-1337.

Muller-starck, G. 1985. Genetic differences between tolerant and sensitive beeches (*Fagus sylvatica* L.) in an environmentally stressed adult forest stand. Silvae Genet. 34: 241-246.

Nagaoka, T. and Y. Ogihara, 1997. Applicability of inter-simple sequence repeat polymorphisms in wheat for use as DNA markers in comparison to RFLP and RAPD markers. Theor. Appl. Genet. 94: 597-602.

Natural Resources Canada 2003. The State of Canada's Forests (2002-2003): Looking ahead. Canadian Forest Services. 95 pp.

Nei, M. 1973. Analysis of gene diversity in subdivided populations. Proc. Natl. Acad. Sci. USA 70: 3321-3323.

Nkongolo KK, 1999. RAPD variations among pure and hybrid populations of *Picea mariana, P. rubens,* and *P. glauca (Pinaceae),* and cytogenetic stability of *Picea* hybrids: identification of species – specific RAPD markers. Plant Syst Evol 215: 229-239.

Nkongolo, K.K., A. Vaillancourt, S. Dobrzeniecka, M. Mehes, P. Beckett, 2008.

Metal Content in Soil and Black Spruce (*Picea mariana*) Trees in the Sudbury Region (Ontario, Canada): low concentration of Arsenic, cadmnium, and nickel detected near smelter sources. Bull. Environ. Contam. Toxicol. 80: 107 - 111.

Ontario Ministry of Natural Resources (OMNR), 2001. Critical review of historical and current tree planting progras on private lands in Ontario. pp. 42. (2001).

Prus-Glowacki, W., E. Chudzinska, A. Wojnicka-Poltorak, L. Kozacki, K. Fagiewicz, 2006. Effects of heavy metal pollution on genetic variation and cytological disturbances in the *Pinus sylvestris* L. population. J. Appl. Genet. 47 (2): 99-108.

Rajora, O.P., I.K. Mann and Y-Z. Shi, 2005. Genetic diversity and population structure of boreal white spruce (*Picea glauca*) in pristine conifer-dominated and mixedwood forest stands. Can. J. Bot. 83: 1096-1105.

Rajora, O.P., and A. Mosseler, 2001a. Molecular markers in sustanaible management, conservation, and restoration of forest genetic resources. In: Muller-Starck G., Schubert R. (eds) Genetic response of forest systems to changing environmental conditions. Kluwer Academic Publishers, Volume 70, pp. 187-201.

Rajora, O.P. and A. Mosseler, 2001b. Challenges and opportunities for conservation of forest genetic resources. Euphytica 118: 197-212.

Ranger, M., K.K. Nkongolo., P. Michael and P. Beckett, 2007. Genetic differentiation of jack pine (*Pinus banksiana*) and red pine (*P. resinosa*) populations from metal contaminated areas in Northern Ontario (Canada) using ISSR markers. Silvae Genet. 57: 333-340.

Raymond M. and F. Rousset, 1995. *GENEPOP, (Version 1.2): population genetics software for exact tests and ecumenicsm,* J. Hered. 86: 248-249

Saitou, N., Nei, M. 1987. The neighbor-joining method: A new method for reconstructing phylogenetic trees. Mol Biol Evol: 406-425.

Scholz, F. and F. Bergmann, 1984. Selection pressure by air pollution as studied by isozyme gene-systems in Norway spruce exposed to sulphur dioxide. Silvae Genet. 33: 238-241.

Semagn, K., A. Bjornstad, and M.N. Ndjiondjop, 2006. An overview of molecular markers methods for plants. *African Journal of Biotechnology* 5 (25): 2540-2568.

Sharma, A., A.G. Namdeo, and K.R. Mahadik. 2008. Molecular markers: New prospects in plant genome analysis. *Pharmacognosy Reviews* 2 (3): 23-34.

Slatkin, M. 1985. Rare alleles as indicators of gene flow. Evolution 39: 53-65.

Tremblay, M. and J.P. Simon, 1989. Genetic structure of marginal populations of white spruce (*Picea glauca*) at its northern limit of distribution in nouveau-Quebec. Can. J. For. Res. 19: 1371-1379.

Vandeligt, K., K.K. Nkongolo, K., and M. Mehes-Smith, 2011. Genetic analysis of *Pinus banksiana* and *P. resinosa* populations from stressed sites in Ontario. Chemistry & Ecology. DOI: 10.1080/02757540.20.561790

Van Straalen, N.M. and M.J.T.N. Timmermans, 2002. Genetic variation in toxicant-stressed populations: An evaluation of the "genetic erosion" hypothesis. Human Ecology and Risk Assessment 8: 983-1002.

Yeh F., R. Yang, and T. Boyle, 1997a. Popgene, version 1.32 edition, Software Microsoft Window-Based Freeware for Population Genetic Analysis. University of Alberta, Edmonton, Canada.

Yeh, F.C, and T.J.B. Boyle, 1997b. Population genetic analysis of co-dominant and dominant markers and quantitative traits. Belgian J. Bot. 129: 157.

Part 2

Ecosystem-Level Forest Management

8

Restoration of Forest Ecosystems on Disturbed Lands on the Northern Forest Distribution Border (North-East of European Russia)

Irina Likhanova and Inna Archegova
Institute of Biology Komi SC UrD RAS
Russia

1. Introduction

Since the second half of the 20th century, due to active growth of minerals extraction in the European North of Russia, the area of disturbed forest ecosystems is steadily enlarges. It is known that disturbance of forest ecosystems leads to environmental unsteadiness and biodiversity decrease. It is vital to mention, that forest destruction negatively affects the traditional way of life of local people who depend on forest resources.

Natural forest destruction is a global problem. According to Losev K.S. et al. (2005) it is noticed that intensive growth of cultivated areas in the world is generally connected with forest destruction. About 63% of land on the planet is developed, what leads to infringement of balance stability if elements biogens (biological circulation of substances) and infraction of biosphere steadiness – habitat of human beings. It is noticed that in present time it is essential "to stop the destruction of this the most important ecological resource and then start natural restoration of forest ecosystems" (p. 78). It is vital for northern conditions, where forest ecosystems are particularly vulnerable to technogenic impact. Taiga forest ecosystems are easily disturbed and slowly self-recover after technogenic impact. This phenomenon was thoroughly surveyed by V.V. Ponomareva (1970, 1980). She wrote that forests are adjusted to strongly expressed eluvial conditions, "…forests minimize leaching of biophil elements from soil by accumulation them not in soil but in their huge always-living phytomass; (forests – ed.) exist from the autonomous above-soil circulation of elements between living organisms and their dying remnants which concentrate on soil surface" (Ponomareva, Plotnikova, 1980; P. 188). Thus, soils under taiga forests are characterized with high moisture content and so biological circulation in such ecosystems is nearer to an autonomous and close type. Huge perennial phytomass of tree plants holds organic elements it has assimilated from the upper earth's crust layer and partly gives them back with tree waste; then dead leaves and branches on soil surface are decomposed and provide for a new portion of nutrient for roots in soil litter. In this connection, forest soils have a very thin organic (productive) layer underlain by almost non-productive mineral layers with low absorbing capacity and containing practically none of plant nutrients. Technogenic interference easily destroys a thin organic soil layer and bares biologically inert mineral soil horizons which are not appreciated for plants. Restoration of disturbed forests on poor and strongly moist substratum is kind of difficult. In consequence, formerly forested areas undergo quick erosion and so their restoration further slows down.

The above-said information evidences a necessity for man-induced maintenance of ecosystem restoration on disturbed lands. In this article the results of investigation method are presented in comparative way. Method that accelerates the process of taiga forest ecosystem restoration is compared to traditional ways of forest cultivation.

2. Objects and investigation methods

We have studied different restoration modes of forest ecosystems in the Usinsk region of the Komi Republic (Russia). This particular region is characterized by severe climatic conditions. Annual air temperature is -3.2 °C. The coldest month is January with mean temperature -18.4 °C. Snow cover holds 200 days and is 48 cm high. The period with mean daily temperatures over +5°C is 110 days. Mean air temperature in July, the warmest month, is +13.8 °C. Annual precipitation is 474 mm, among them 159 mm precipitate during vegetation period (June-August) (Scientific-applied reference book…, 1989).

Principal vegetation of the study region, which is located in far north taiga subzone, is forests that intermixed with large marshes; 10% of the area is covered by tundra vegetation (Yudin, 1954). Dominant are forest stands sparse spruce and spruce-birch forests with crown density of 0.3-0.5, height of trees 8-15 m, quality (bonitet) classes of tree stands V-Va. Forest composition mainly includes *Pinus sylvestris* and *Larix sibirica*. Most popular are long-moss forests, less represented are green-moss and sphagnum forest types.

The typical soil types of the region are boggy-podzolic, gley-podzolic, tundra-boggy, and boggy peaty soils (Podzolic soils…, 1981). Soil-forming rocks are moraine loams and sandy deposits formed in the glacier period. Sandy rocks are overlain by differently-moist soils as illuvial-humus-iron podzols, weakly peaty-podzolic-gley illuvial-humus soils; fine-textured loams are overlain by gley-podzolic soils and weakly peaty-podzolic-weakly gley soils.

The base of region's economy is oil-gas extracting and processing industries. Expansion of those industries leads to enlargement of the area of disturbed lands, including forest ecosystems. Sand pits are the most common technogenic objects within the Usinsk region. Sandy material excavated from pits is used for building roads, making bore sites and bore drills, etc. Restoration of vegetation cover in severe climatic conditions on sandy technogenic substrata, poor in nutrients, proceeds extremely slowly. Consequently, the restoration method of forest ecosystems on such lands is very much required for development.

Experimental plots were used in order to study the efficiency rate of forest ecosystems' restoration. Particularly, vegetation cover was studied using common geo-botanical methods (Field geo-botany, 1964), tree species that are planted on experimental plots were monitored by common study methods of forest cultures (Ogievskiy, Khirov, 1964). Soil type's description and soil samples' analysis (soil pH, organic C, exchangeable Ca, Mg, hydrolysable N, P_2O_5, K_2O) were done by the general methods (Agrochemical methods…, 1975; Theory and practice of soil chemical investigation, 2006). Humus composition was evaluated by the Tyurin method in Ponomareva-and-Plotnikova modification (1975).

3. Results and its discussion

Investigations of new ways of forest ecosystem were done in a way of comparative analysis with traditional methods of forest recultivation.

Traditional restoration methods of forest ecosystems in the North The main methodological position of traditional restoration methods of disturbed forest ecosystems is the resource approach aiming at planting forest cultures of the principal forest-forming species as spruce and pine. In other words, the task of traditional technology is to create forest plantations and not to restore earlier destroyed forest ecosystem of previous quality (Losev et al., 2005). Coniferous cultures are planted at the age of 1-3 years and have open roots; no ground treatment is meant.

First forest cultures in the Usinsk region were planted in 1958 by the personnel of the Usinsk leskhoz. From 1991 to 2007 forest cultures were planted on area of 1020.8 ha. Among this figure, the share of *Pinus sylvestris* made 53.6%, that of *Picea obovata* 27.6%, and that of *Pinus sibirica* 1.5%. Willow young trees were also planted for ground fixing with portion of 17.3%. Most forest cultures (44.8%) were planted on pits.

We have observed the sites being reforested by common restoration methods. The sites are located on the most usual sample of technogenic disturbance, in our case on 8 b technogenic pit (N 66°16′, E 57°16′).

Sandy material on 8 b pit is characterized by low content of clay (sum of particles <0.01 is less than 6%) (Table 1). This is responsible for low absorption and moisture content values; ground can be easily water- or wind-eroded.

Sampling depth (cm)	Hygroscopic moisture, %	HCl ignition losses	Number of particles (%) with diameter of:						Sum of particles >0.01	Sum of particles <0.01
			1.0-0.25	0.25-0.05	0.05-0.01	0.01-0.005	0.005-0.001	<0.001		
0-20	0.39	0.00	11.35	81.88	0.66	1.47	0.30	4.34	93.89	6.11
20-40	0.43	0.37	11.94	80.59	2.11	0.04	0.64	4.67	94.64	5.36

Table 1. Texture composition of technogenic material on 8 b pit.

The 8 "b" pit was partly planted with 3-year-old pine trees in 2001 without previous ground treatment.

On the second year after planting, the content of nitrogen, an important biogenic element, made 0.2 mg / 100 g a.d.s. which corresponded with low organic carbon content (Table 2). On the eighth year, no significant quantitative changes in composition of nutrients and absorbed bases were observed; content of organic carbon resisted low.

Seven years after planting the survival rate of plantings made 50%, tree height 53 cm, and crown diameter 43 cm. Above presented data is general. Pine plantings were underdeveloped because of poor concentration of nutrients in substratum and so were susceptible to the (snow) Schütte disease which stroke 60% of remaining pines. This disease additionally inhibits the growth of pines and often causes their depth. Soil cover was underdeveloped with total projection cover under 1%. 7 plant pioneers were identified (*Festuca ovina, Chamaenerion angustifolium, Hieracium umbellatum, Equisetum arvense, Carex arctisibirica, Rumex acetosell)*, also in microdepressions mosses of the genus *Polytrichum* and *Ceratodon purpureus*, lichens of the *Stereocaulon* genus.

Year	Sampling depth, cm	pH$_{water}$	C$_{org.}$, %	N$_{hygr.}$	P$_2$O$_5$	K$_2$0	Ca^{2+}	Mg^{2+}
				mg /100 g a.d.s.			mM/100 g a.d.s.	
2002	0-10	5.9	0.1	0.2	8.4	2.5	0.3	0.2
	10-20	5.9	0.3	0.3	8.0	2.8	0.2	0.2
2008	0-10	6.0	0.3	0.1	9.4	2.3	0.3	0.2
	10-20	5.9	0.4	0.4	7.9	3.1	0.2	0.2

Table 2. Agrochemical indices of substratum planted with 3-year-old pine cultures.

Thus, unfavorable properties of ground material did not provide for the active self-restoring process.

The other common restoration method of disturbed area is planting willow. Willow cultures were planted on the above-mentioned 8 "b" pit under the leadership of the ecologist of the OSC "Northern Oil" V.I. Parfenyuk in 1991 without previous substratum treatment. The distance between plantings in a row was 25 cm and between rows 2 m. On the twelfth planting year (2002) only 20% of planted trees remained alive and were about 1 m high. Single herbaceous plants *Festuca ovina, Chamaenerion angustifolium, Hieracium umbellatum, Equisetum arvense* were observed between rows. Those plant species are typical of the initial stage of self-restoring succession (Table 3). Herbaceous layer projective cover made less than 1%.

Species	Availability, %			Projective cover, %			Height, cm
Herbaceous plants:	2002	2006	2011	2002	2006	2011	
Carex arctisibirica (Jurtz.) Czer.	-	7	5	-	<1	<1	15
Chamaenerion angustifolium (L.) Scop	16	29	20	<1	<1	<1	20-30
Equisetum arvense L.	48	56	50	1	<1	<1	10
Festuca ovina L.	25	53	60	<1	<1	<1	10-25
Hieracium umbellatum L.	24	29	20	<1	<1	<1	20-30
Leucanthemum vulgare Lam.	4	-	-	<1	-	-	10
Rumex acetosella L.	36	14	5	<1	<1	<1	15-20
Solidago virgaurea L.	-	17	10	-	<1	<1	10-25
Tripleurospermum perforatum (Merat.) M.Lainz	8	7	-	<1	<1	-	10
Avenella flexuosa L.	-	-	5	-	-	<1	25
Mosses:							
Bryum sp.	-	7	5	-	<1	<1	1
Ceratodon purpureus (Hedw.) Brid.	-	7	10	-	<1	1	1
Polytrichum piliferum Hedw.	-	7	60	-	<1	5	1
Lichens:							
Stereocaulon paschale (L.) Hoffm.	-	7	15	-	<1	<1	1-2
Cladonia sp.	-	-	7	-	-	<1	1-2

Note: «-» - not found.

Table 3. Species composition of soil cover at site planted with willow trees.

20 years after planting (2011), the health status of shrubby layer did not practically change; soil cover remained thin (Table 3). The surface of ground material was partly covered with algae film and protonema of mosses. Microdepressions hosted mosses of the *Polytrichum* genus and *Ceratodon purpureus*. Totally, 8 species of herbaceous plants and 3 mosses were identified at the area.

Chemical analysis of ground material samples on the twelfth willow planting year revealed a low content of biogenic elements and organic matter. On the sixteenth planting year the agrochemical parameters did not practically change (Table 4). As the herbaceous layer was very thin, sandy ground was susceptible to erosion. On the twelfth planting year we observed washed-out erosion hollow 2.5 m wide, 1 m deep, 5 m long. On the sixteenth planting year it increased in size with a depth of 1.5 m and a length over 10m.

Year	Sampling depth, cm	pH_{water}	$C_{org.}$, %	$N_{hygr.}$	P_2O_5	K_2O	Ca^{2+}	Mg^{2+}
				mg / 100 g a.d.s.			mM / 100 g a.d.s.	
2002	0-10	5.9	0.2	0.1	9.4	2.3	0.3	0.2
	10-20	5.9	0.2	0.4	7.8	3.1	0.2	0.2
2005	0-2*	5.9	1.2	1.7	9.7	0.7	1.4	0.2
2006	0-5	5.8	0.2	0.1	7.9	6.5	0.8	0.6

Note: * - crust of algae and protonema in a small depression.

Table 4. Agrochemical parameters of ground material planted with willow trees.

Consequently, there was almost no positive effect from willow planting for post-technogenic substratum restoration.

The above-sited data allow for the following conclusion. While using the traditional restoration methods of disturbed forested areas, soil and vegetation cover formation is slow, what hampers restoration of the forest ecosystem as whole. To speed up restoration there is a need to apply complex methods aiming at development plant biogeocenosis, i.e. maintenance plant matter biological cycle, ensuring conditions for intensive plant cover formation on ground surface and organic matter accumulation in substratum. So, restoration the technogenically disturbed forest ecosystem in the North cannot be efficient without development the basic system components, first of all plant community including not only tree layer but also ground plant cover.

4. Main principles of the "nature restoration" conception and the complex of methods aimed at accelerated restoration of forest ecosystems on technogenically disturbed lands

The "nature restoration" conception was developed at the Institute of Biology Komi SC UrD RAS under the leadership of Dr. I.B. Archegova. This conception means restoration of forest ecosystems including their initial structure and "functions" which, finally, ensures the integrity of the biosphere (Archegova, 1998). The "nature restoration" methodological conception operates forest as a system and ecosystem self-restoration as a succession process. From this point of view, ecosystem presents a system of the three main components as plant community, fauna-microbe complex that processes plant remnants and soil that is a productive (biogenic-accumulative) layer. These three components are integrated into ecosystem by means of organic (plant) matter biological cycle. In practice, the "nature restoration" system aims at restoration the ecosystem as a whole, not its single components, tree layer in particular. The "nature restoration" methods should correspond with the regional climatic conditions, also taking into consideration regional economy, traditional regional nature management.

Northern ecosystems poorly resist technogenic impacts and slowly self-restore because of not only severe climatic conditions together with the presence of permafrost rocks but also because of a thin productive organic-accumulative layer (soil) which hosts the majority of nutrition elements, plant roots, and active microbiota. Mineral layer becomes visual after organic-accumulative layer disturbance. Mineral layer is unfavorable for biota self-restoration and so hampers the process of nature self-restoration, first of all plant community restoration. Consequently, any organic layer technogenic disturbance always has total nature ecosystem destruction as an aftereffect. The absence of vegetation cover for a long period of time speeds up erosion processes that only aggravate self-restoration of plant-soil cover. This situation requires development an active and up-to-date approach to restoration of disturbed forest ecosystems.

Based on the "nature restoration" conception we have developed the two-stage system of rapid (managed) "nature restoration" practical methods (Fig. 1). At the first "intensive" stage, we form herbaceous ecosystem and corresponding biogenic-accumulative layer in a short period of time, namely in 3-5 years, using complex agrotechnical treatments as applying organic and mineral fertilizers and sowing local perennial herbs. In the other words, this way reduces the usually long (up to 30-40 years) initial self-restoration succession stage. At the second "assimilation" stage, no agrotechnical treatments are used. The previously formed herbaceous community is gradually self-replaced by a zonal type of plant community or generally by forest ecosystem.

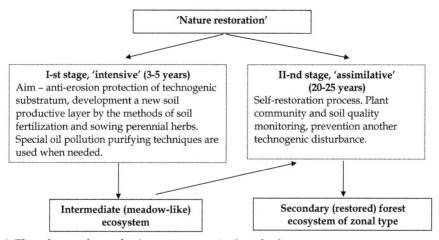

Fig. 1. The scheme of complex 'nature restoration' methods.

According to this sketch in 1991 we initiated the experiment on the 8 "b" pit located near the above discussed plots being restored by traditional methods. Intensive "nature restoration" methods included soil surface fertilization with peat in a 15-cm layer and with mineral fertilizers ($N_{60}P_{60}K_{60}$), sowing the herbaceous mixture of *Alopecurus pratensis* and *Poa pratensis* in a dose of 20 kg seeds/ha with proportion of seeds 1:1. Initial sandy substratum contained 0.2% organic carbon, 0.1 hydrolysable nitrogen, 6.7 phosphorus and 2.0 mg/100 g a.d.s. potassium oxides. Sawn grasses were annually given a complex mineral fertilizer in a rate of 30 kg each mineral element/ha.

Three years later the plot grew into an intermediate herbaceous community (TPC 100%) of meadow type with corresponding meadow-like soil constituted of typical soddy horizon and humus horizon. Later there was no any kind of treatment. It was the beginning of the second restoration stage thereby tree plants inhabited the plot with gradual replacement of herbaceous ecosystem to forest ecosystem.

After intensive agrotechnical treatments, already the first ten years saw formation of woody-shrubby layer with a crown density of 0.1 of *Betula pubescens, Larix sibirica, Salix* species 1.5-2.5 m high (Table 5). Ground cover (TPC 100%) 10 years after restoration start was considerably composed of sawn *Alopecurus pratensis* (PC 44%) and less by *Poa pratensis*. At the same time, 19 new non-sown herbaceous species were observed. Among them, *Festuca ovina, Chamaenerion angustifolium, Erigeron acris,* and *Solidago virgaurea* had highest projective cover figures (Table 8). Ground cover was largely composed of synanthropic species (*Rumex acetosella, Chamaenerion angustifolium, Crepis tectorum, Tripleurospermum perforatum, Tussilago farfara, Equisetum arvense*), characteristic of initial restoration stages of disturbed lands. The plot was inhabited by mosses and single *Cladonia* and *Peltigera* lichens; 5 mosses were totally identified dominated by polytrichum mosses (16%) (Table 8). Consequently, the first ten years after the experiment start there is an active replacement of intermediate herbaceous ecosystem by forest ecosystem with formation of woody layer of quickly-growing tree species. Ground cover was identified for numerous non-sown vascular plant species and still numerous sawn grasses and mosses.

Species	Quantity, inds./100 m²		Height, m	
	2002	2011	2002	2011
Developing tree story:				
1st layer:				
Betula pubescens Ehrh.	2	13	2.5-3	4-6
Larix sibirica Ledeb.	9	15	1.3-1.5	4-5,5
Salix caprea L.	2	2	2.5	6
Salix dasyclados Wimm.	2	1	3	4
2nd layer:				
Betula pubescens Ehrh.	-	16	-	2-3
Larix sibirica Ledeb.	-	6	-	2-3
3rd layer:				
Betula pubescens Ehrh.	-	13	-	0.5-1.5
Larix sibirica Ledeb.	-	3	-	0.5-1.5
Picea obovata Ledeb.	-	1	-	0,5
Developing tree understory:				
Betula nana L.	1	1	0.7	1-1.5
Ribes rubrum L.	1	1	0.7	1
Salix phylicifolia L.	11	6	1.5	1-3
Salix hastata L.	1	-	0.9	-
Salix lapponum L.	1	-	1.3	-

Table 5. Species composition and structure of tree story and tree understory on the experimental plot.

Species	Abundance, %			Projective cover, %			Height, cm
	2002	2006	2011	2002	2006	2011	
Shrubs:							
Arctostaphylos uva-ursi (L.) Spreng.	5	7	5	<1	<1	<1	10
Empetrum hermaphroditum (Lange) Hagerup	5	7	10	<1	<1	<1	10
Vaccinium uliginosum L.	-	-	5	-	-	<1	10
Vaccinium myrtillus L.	-	-	5	-	-	<1	10
Herbaceous plants:							
Agrostis tenuis Sibth.	-	7	35	-	<1	5	20-50
Avenella flexuosa (L.) Drey.	-	-	70	-	-	25	20-50
Alopecurus pratensis L.	100	80	75	44	46	25	80
Antennaria dioica (L.) Gaertn.	-	-	5	-	-	<1	15
Calamagrostis epigeios (L.) Roth	10	7	-	<1	<1	-	60-80
Calamagrostis lapponica (Wahl.) Hartm.	-	-	15	-	-	1	50-60
Carex brunnescens (Pers.) Poir	5	-	5	<1	-	<1	25
Chamaenerion angustifolium (L.) Scop.	100	67	65	9	5	7	40-80
Crepis tectorum L.	5	7		<1	<1		20-25
Deschampsia cespitosa (L.) Beauv.	30	7	25	2	1	7	40-80
Epilobium palustre L.	10	7	-	<1	<1	-	20
Equisetum arvense L.	10	7	15	<1	<1	<1	20
Equisetum sylvaticum L.	5	-		<1	-	-	25
Euphrasia frigida Pugsl.	-	-	5	-	-	<1	15
Erigeron acris L.	70	40	40	4	<1	1	35-45
Festuca ovina L.	100	67	40	23	18	16	15-30
Festuca rubra L.	-	7	15	-	<1	1	40-45
Hieracium umbellatum L.	40	33	25	1	1	1	40
Hieracium vulgatum L.	10	20	10	<1	1	<1	40
Omalotheca sylvatica (L.) Sch.Bip.	60	33	70	2	1	2	10-30
Phalaroides arundinacea (L.) Rausch.	-	-	5	-	-	<1	50
Poa pratensis L.	30	20	30	1	4	5	60
Rumex acetosella L.	10	7	25	<1	<1	<1	20-25
Solidago virgaurea L.	80	60	80	3	3	10	10-45
Taraxacum officinale Wigg.	40	20	5	1	<1	<1	15-20
Tripleurospermum perforatum (Merat.) M.Lainz	5	7	-	<1	<1	-	15
Tussilago farfara L.	20	7	5	<1	<1	<1	10-15
Trientalis europaea L.	-	-	5	-	-	<1	5-10
Lycopodium annotinum L.	-	-	5	-	-	<1	10
Lathyrus pratensis L.	-	-	5	-	-	<1	30-40
Orthilia secunda (L.) House	-	-	5	-	-	<1	15
Mosses:							
Brachythecium campestre (Bruch) B. S. G.	-	-	20	-	-	5	3-5
Brachythecium reflexum (Starke) Schimp.	-	-	10	-	-	5	3-5
Brachythecium salebrasum (Wed et Mohr) Bryol.	-	-	20	-	-	5	3-5

Species	Abundance, %			Projective cover, %			Height, cm
	2002	2006	2011	2002	2006	2011	
Brachythecium sp.	20	20	-	1	15	-	3-5
Ceratodon purpureus (Hedw.) Brid.	20	20	5	<1	<1	<1	2-3
Dicranum polysetum Sw.	-	-	10	-	-	1	3-4
Plagiothecium denticulatum (Hedw.) B. S. G.	-	-	10	-	-	1	1-2
Pleurozium schreberi (Brid.) Mitt.	-	20	80	-	1	10	3-5
Polytrichum commune Hedw.	20	20	20	2	1	2	5-7
Polytrichum juniperinum Hedw.	60	40	90	11	11	30	5-7
Polytrichum piliferum Hedw.	30	33	15	5	5	3	3-5
Sciurohypnum oedipodium (Mitt.) Ignatov et Huttunen.	-	-	10	-	-	1	3
Sciuro-hypnum starkei (Brid.) Ignatov et Huttunen (Brachythecium starkei(Brid.) B.S.G.)	-		65	-	-	30	3-5
Lichens:							
Cladonia anomaea (Ach.) Ahti & P.James	-	-	4	-	-	<1	1-3
Cladonia arbuscula (Wallr.) Flot.	-	7	10	-	<1	<1	1-3
Cladonia borealis Stenroos	-	-	4	-	-	<1	1-2
Cladonia botrytes (Hag.) Willd.	-	-	4	-	-	<1	1
Cladonia carneola (Fr.) Fr.	-	7	-	-	<1	-	1-2
Cladonia cervicornis (Ach.) Flot. ssp. verticillata (Hoffm.)	-	-	4	-	-	<1	1-2
Cladonia chlorophaea (Florke ex Sommerf.) Spreng.	-	-	4	-	-	<1	1-2
Cladonia cornuta (L.) Hoffm.	-	7	20	-	<1	<1	1-4
Cladonia crispata (Ach.) Flot.	-	-	4	-	-	<1	1-2
Cladonia deformis (L.) Hoffm.	-	7	-	-	<1	<1	1-3
Cladonia fimbriata (L.) Fr.	-	7	10	-	<1	<1	1-3
Cladonia gracilis (L.) Willd.	-	7	20	-	<1	<1	1-3
Cladonia phylophora Hoffm.	-	-	20	-	-	<1	1-2
Cladonia pleurota (Floerke) Schaer.	-	-	4	-	-	<1	1-2
Cladonia rangiferina (L.) Web.	-	7	20	-	<1	<1	1-3
Cladonia subulata Weber.	-	-	4	-	-	<1	1-2
Cladonia sp.	10	7	-	<1	<1	-	1
Peltigera didactyla (With.) Laundon	-	-	10	-	-	<1	1-2
Peltigera leucophlebia (Nyl.) Gyeln.	-	-	4	-	-	<1	1-2
Peltigera rufescens (Weis.) Humb.	-	-	4	-	-	<1	1-2
Peltigera sp.	10	7	-	<1	-	<1	1-2

Table 6. Composition and structure of ground plant cover on the experimental plot.

The newly-formed 10-year-old soil had the following morphological structure. Loose layer of weakly-decomposed plant waste (dead grass) was penetrated with rare moss stems.

AOA1 0-8(12) cm	Well-decomposed plant waste with inclusions of mineral particles, moist, abundant roots.
A1 layer 8(12)-21 cm	Sandy, dark-grey (humus color), structureless, loose, moist, many roots. Transition to the next horizon is abrupt by color.
III-rd layer 21-29 cm	Sandy, light-yellowish with whitish and dark-ochre spots, loose, moist, rare roots.
IV-th layer 29-45 cm	Sandy, grey-yellowish, lighter than the previous horizon, loose, moist, without roots.

In the first ten years there is a soil profile formation with organic horizons with features of soddy layer, typical for meadow ecosystems. The upper biogenic-accumulative layers (A0A1, A1) had weakly-acid medium reaction and accumulated the maximum of nitrogen, humus, and exchangeable bases (Table 7).

Year	Horizon, sampling depth, cm		pH_{water}	C, %	N_{hydr}.	P_2O_5	K_2O	Ca^{2+}	Mg^{2+}
					mg/100 g a.d.s.			mM/100 g a.d.s.	
11th year (2002)	A0A1	0-8(12)	5.4	5.0	2.9	8.5	7.9	7.1	1.1
	A1	8(12)-21	5.7	3.3	1.6	10.1	4.3	5.9	0.8
	III	21-29	5.7	1.6	0.8	9.9	3.8	1.7	0.5
	IV	29-45	5.6	0.2	0.8	10.0	2.5	1.2	0.4
20th year (2011)	In group of trees								
	A0	0-5	5.4	6.7	11.5	10.0	29.0	17.7	3.6
	A1	5-15	5.2	4.9	9.0	9.9	9.0	10.0	1.6
	A0buried *	15-21	5.0	7.2*	5.9	8.3	8.9	14.8	2.1
	BC	21-35	5.3	0.2	7.8	14.0	8.6	2.3	0.6
	Open area								
	A0	0-2(4)	5.3	5.7	9.5	13.3	36.0	10.6	2.2
	A_1	2(4)-13,5(14)	4.9	5.4	5.9	11.1	11.4	10.1	1.6
	A0 buried 13,5(14)-21		4.9	14.8*	5.5	5.3	14.9	17.5	2.2
	BC 21-24		4.8	0.3	2.4	12.8	5.5	2.5	0.6
	BC/ 24-35		5.1	0.2	1.1	13.2	3.8	2.4	0.6

Note: * – buried organic residues (peat).

Table 7. Agrochemical indices of newly-formed soil on the experimental plot.

The forest community was structurally formed on the 20th year after restoration start. The first tree story consisted of *Betula pubescens* and *Larix sibirica* with few tree-like willow species (*Salix dasyclados, Salix caprea*); tree height was 4-6 m, stem diameter 4-6(9) cm (Table 5). Crown density increased to 0.4. Since 2002, tree re-growth (with a height of less than 50 cm) was clearly dominated by *Betula pubescens*, 40 individuals per 100 m², for contrast only 5 individuals of *Larix sibirica*. *Betula pubescens* was also a dominant species in the second and third tree stories which were formed on the 20th restoration year. The plant waste accumulation dynamics of woody plants evidenced the active development of tree layer (Tables 8, 9). Thus, the community on its 20th restoration year was at the stage of quickly-growing woody plants, typical of self-restoration succession in the taiga zone (Shennikov,

1964). Low-height young growth of *Picea obovata* and *Pinus sylvestris* appeared in amount of 1-3 individuals / 100 m².

Plant waste sampling period	Plant waste weight
June 2002 – September 2002	8.5±1.7
October 2002 – May 2003	27.5±2.0
Year total:	36.1
June 2007 – September 2007	15.0±6.5
October 2007 – May 2008	52.2±8.6
Year total:	67.2
June 2008 – September 2008	13.6±5.4
October 2008 – May 2009	53.05±10.2
Year total:	66.65
June 2009 – September 2009	11.9±2.8
October 2009 – May 2010	73.3±15.3
Year total:	85.2
June 2010 – September 2010	14.7±3.2
October 2010 – May 2011	106.01±31.2
Year total:	110.71

Table 8. Plant waste weight by years on the experiment plot (air-dried weight, g/m²).

Fraction	October 2008 – May 2009		June 2009 – September 2009		October 2009 – May 2010		June 2010 – September 2010		October 2010 – May 2011	
	Weight g/m²	Share %	Weight g/m²	Share %	Weight g/m²	Share %	Weight g/m²	Share, %	Weight g/m²	Share, %
Branches	0.45	1	0.498	4	0.604	1	0.756	5	1.96	2
Betula pubescens leaves	30.68	58	2.263	19	35.88	49	1.028	7	68.59	65
Herbs	2.17	4	0.169	1	3.664	5	0.412	3	2.80	3
Salix leaves	2.48	5	4.421	37	1.972	3	0.780	5	2.95	3
Bark	0.43	1			0.14	0	0.132	1	0.37	0
Larix sibirica needles	8.46	16	1.046	9	21.352	29	9.792	67	20.50	19
Dust of rotten wood	8.38	16	3.54	30	9.736	13	1.848	13	8.72	8
Inflorescences	-	-	-	-	-	-	-	-	0.11	0

Note: «-» - not found.

Table 9. Plant waste fraction composition on the experiment plot (air-dried weight, g/m²).

At the end of the second decade of experimental years the TPC of herbaceous-dwarfshrub layer comprised 85%. This retreat in TPC was related to woody plants' shadowing. Forest dwarfshrubs increased in species number, among them *Arctostaphylos uva-ursi, Empetrum hermaphroditum, Vaccinium uliginosum, Vaccinium myrtillus* (Table 6). Among 24 herbaceous species found on the experimental plot, the forest species *Avenella flexuosa* and *Solidago virgaurea* had essential PC, 25 and 10%, correspondingly. The sawn meadow grass *Alopecurus pratensis* significantly reduced its PC (25%). Mosses counted 11 species on the 19th experimental year. The highest shares in PC belonged to *Sciuro-hypnum starkei* (30%), *Polytrichum juniperinum* (30%) and the common forest species *Pleurozium schreberi* (8%). Lichens were highly diverse with 17 species, mainly from the *Cladonia* genus. Thus, the end of the second experimental decade saw formation the forest community where sown plants of the first "intensive" restoration stage were normally replaced by forest species together with mosses and lichens.

On the 20th experiment year, soil pits on open area and in group of trees were excavated.

Soil pit №1 was dug in a group of trees (Betula pubescens, Larix sibirica). Ground cover (TPC 75%) was found for the herbs (Avenella flexuosa, Solidago virgaurea, Alopecurus pratensis, Chamaenerion angustifolium, Orthilia secunda) and the mosses (Sciuro-hypnum starkei, Pleurozium schreberi). Moss cover was well developed (PC 60%) with practically full-formed mossy litter.

A0 0-5 cm	Loose layer of mossy litter, upper part contains weakly-decomposed and lower part stronger decomposed plant remnants with inclusions of sand, abundant roots.
A1 5-21 cm	Sandy, grey-black, loose, inclusions of weakly- to well-decomposed plant remnants from outside peat (brought at the 1st restoration stage), many roots, transitional boundary is abrupt by color.
BC 21-35 cm	Sand, grey-yellowish, with whitish spots, structureless, few roots, moist.

Soil pit №2 was made on open area. Herbaceous cover (TPC 100%) was dominated by *Alopecurus pratensis, Solidago virgaurea, Omalotheca sylvatica.* There was a 2-cm-thick layer of dead grass on surface. Dead grass was the development base for the *Brachythecium* mosses and *Sciuro-hypnum starkei.*

A0 0-2(4) cm	Loose layer of weakly-mean-decomposed plant remnants, dark-grey, sand inclusions in lower part, abundant roots.
A1 2(4)-21 cm	Sandy, dark-grey to black, structureless, moist, abundant roots, inclusions of decomposed peat remnants brought at the 1st restoration stage, many rain worms, transitional boundary is abrupt by color.
BC 21-36 cm	Sandy, grey-yellowish, with whitish spots, moist, upper part with single roots.

Agrochemical parameters of the studied soils (Table 7) provide evidence that the biogenic-accumulative layers (litter and humus horizons) have been formed on the 20th restoration year. Those horizons were marked through high content of organic carbon, nitrogen, exchangeable bases, and other biogenic elements. Humus of the biogenic-accumulative layer was dominated by humic substances (Table 10).

Horizon, depth, cm	$C_{org.}$ total, %	Humic acids				Fulvic acids					Non-soluble residue	C_{HA} / C_{FA}
		1	2	3	$\sum$	1a	1	2	3	$\sum$		
A1 3-13	4.2	20.94	11.76	20.7	53.4	3.18	14.7	6.24	9.88	34.0	12.6	1.57
A1 13-23	2.4	10.98	14.22	14.4	39.6	4.88	12.6	6.5	6.1	30.08	30.32	1.32
23-28	0.2	9.52	1.91	4.76	16.19	28.6	9.5	17.17	7.14	62.41	21.4	0.26
28-45	0.1	7.7	0.76	4.61	13.07	15.38	2.32	22.29	1.51	41.5	45.43	0.31

Table 10. Fraction-group humus composition of organic-accumulative layer of the newly-formed soil on the 18th experiment year (% of total content).

Consequently, the biological cycle of organic (plant) matter started at the "intensive" stage resulted in forest ecosystem formed to the 20th year as the integrity of two components, plant community (or biotic complex) and soil. Organic (plant) matter biological cycle restoration initiated active soil restoration visualized by formation of the biogenic-organic-accumulation layer. This layer's structure depends on plant community type. It determines the significance of soil as a system structure, capable of holding and accumulating plant nutrition elements and ensuring stable conditions for self-restoration of ecosystem. These properties are formed during the transformation processes of plant waste called humus formation, the main soil formation process (Ponomareva, Plotnikova, 1980).

Restoration of nature medium components is a complete process that functionally unifies biota with its habitat. Soil can be formed when technogenic substratum reaches some "critical" mass of plant material to start the biological cycle, including humus accumulation.

In the North, the process of self-restoring succession is a long-term process. To speed up (manage) the self-restoration process on post-technogenic bare areas, is to apply a complex of agrotechnical methods, i.e. fertilization, sowing perennial grasses, that is called an "intensive" restoration period. Accumulation of organic matter (plant remnants of perennial herbs etc.) in substratum, its transformation (humus formation) with help of zoo-microbe complex, accumulation and consequent assimilation of biogenic elements by plants provide favorable conditions for the next stage, forest ecosystem development.

The conducted study has evidenced the efficiency of agromethods ("intensive" stage) for speeding up the restoration process of forest ecosystem. It was demonstrated that the first experimental decade was already indicated by the most advanced restoration succession stage, i.e. formation the herbaceous community and its transition to forest community of quickly-growing woody species under whose canopy conifers started growth. Transformation of herbaceous community and corresponding soil type during self-restoring succession in taiga zone into quickly-growing woody species stand is a normal process (Shennikov, 1964).

Acceleration in forest ecosystem formation becomes more prominent when comparing the study plot with the near self-restoring plot. On its 28th restoration year the TPC figure remained under 1 % without woody plants and with active erosion signs.

5. Optimization the "nature restoration" methods

As said above, the preliminary "intensive" stage ensures favorable substratum conditions for acceleration of woody layer self-restoration, replacement of herbaceous ecosystem by forest ecosystem. However, restoration of conifers proceeds slowly and under the canopy of quickly-growing deciduous (birch, asp) species. To further accelerate restoration of forest ecosystem on the second restoration stage, complex of methods was developed. These methods are to optimize restoration of conifers in woody layer of forest ecosystems in north taiga zone and consist in planting conifers simultaneously with agrotechnical treatments on the first ("intensive") restoration stage.

Another experiment on the territory of 8b sand-pit was started, where *Pinus sylvestris* two-year-old trees, traditionally used for restoration purposes, with open root system were planted with a planting density of 5000 individuals/ha. Herb mixture composition being sown on "intensive" stage included *Poa pratensis, Festuca rubra, Festuca pratensis, Bromopsis inermis, Phleum pratense* in proportion 1:1:1:1:1. Annual additional fertilization with complex mineral fertilizer was done during 4 years. By our data, only 30-40% of pine plantings remained alive on the second year and resisted few for the whole study period. The two-year-old *Pinus sylvestris* plantings did not develop well on the "intensive" restoration stage with a height of 12-17 cm at the fifth year. Herb stand was already 90 cm high at that period of time with 80-90% TPC. So, the study has identified two-year-old *Pinus sylvestris* plantings with open root system not a promising material to be used on the "intensive" restoration stage of the "nature restoration" experiment. Herbaceous layer developed quicker than *Pinus sylvestris* plantings. Low growth rates did not allow the plantings to overgrow herb layer in a short period of time which was particularly responsible for their future underdevelopment.

Absolutely other results were obtained on usage the high-growth material, *Pinus sylvestris* wildlings about 50 cm high with a ground clot 30x30 cm. Planting density was 2500 individuals/ha. The same herb species as sown in the trial with two-year-old plantings were used. Additional fertilizing with complex mineral fertilizer (N45P45K45) was done every spring for 4 years.

By the observation results, the planted wildlings remained alive by almost 100% 5 years afterwards (Table 11). High surviving rate of the plants was related to their sufficient height, planting with ground clot, and caring for 5 years.

Year(s) after planting	Survival rate, %	Height, cm	Stem diameter, cm	Crown diameter, cm
1	100	59.1±2.4	1.1±0.1	32.9±1.4
2	100	60.9±2.8	1.5±0.2	37.3±1.6
3	100	68.8±3.1	1.7±0.1	46.5±2.1
4	96	79.5±3.1	1.8±0.1	48.7±3.1
5	96	100.3±4.3	2.1±0.2	55.4±4.5

Table 11. Biometric parameters of *Pinus sylvestris* plants in optimization experiment (autumn observations).

Beginning from the third planting year, *Pinus sylvestris* steadily increased in height and was over 20 cm high on the fifth year (Fig. 2).

On the fifth year the mean height of planted trees was about 1 m, consequently, the trees showed high survival rate and well development.

Herbaceous cover on the experimental plot actively developed. On the third year TPC of herb layer was 30% and already 70-75% to the fourth-fifth year (Table 12). Herb layer practically lost such sown herbs as *Festuca pratensis*, soil moisture-dependent, and *Trifolium pratense*. The rhizome grasses *Bromopsisinermis, Poa pratensis* and the rhizome loose-bunch *Festuca rubra* remained. The latter species as least dependent of soil richness and moisture had the highest projective cover among sown herbs. New non-sown herbs appeared and were prevailed by *Festuca ovina* that normally grows in lichen pine forests. There were species typical of anthropogenically disturbed areas as *Chamaenerion angustifolium, Equisetum arvense*. The species *Solidago virgaurea, Deschampsia cespitosa, Avenella flexuosa,* usual for forest and meadow were fixed but were few in number. Forest sub-shrubs (*Empetrum hermaphroditum, Vaccinium vitis-idaea*) transported there within ground clot were identified. Moss cover started formation and included pioneer species (Table 12). The majority of newly-appeared species were single in number. The vivid species composition on the study plots characterized the plant community as young and unstable.

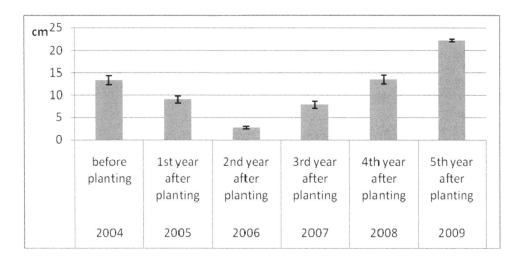

Fig. 2. Increment dynamics of *Pinus sylvestris* wildlings in optimization experiment.

Species	Number of years after experimental start			
	2	3	4	5
Sown herbs:				
Bromopsis inermis (Leyss.) Holub	2	6	10	7
Festuca pratensis Huds.	1	3	<1	2
Festuca rubra L.	2	3	7	26
Phleum pratense L.	4	7	5	5
Poa pratensis L.	2	5	8	6
Trifolium pratense L.	1	1	<1	1
Invasive species:				
Avenella flexuosa (L.) Drey.	-	-	-	1
Agrostis tenuis Sibth.	<1	<1	-	-
Carex arctisibirica (Jurtz.) Czer.	-	<1	<1	<1
Chamaenerion angustifolium (L.) Scop	3	2	4	1
Chenopodium album L.	-	<1	-	-
Crepis tectorum L.	-	-	1	-
Dactylis glomerata L.	-	-	<1	1
Deschampsia cespitosa (L.) Beauv.	-	-	<1	<1
Empetrum hermaphroditum (Lange) Hagerup	<1	<1	<1	<1
Equisetum arvense L.	3	2	1	<1
Festuca ovina L.	10	16	41	20
Hieracium umbellatum L.	1	3	<1	2
Solidago virgaurea L.	<1	<1	<1	<1
Tripleurospermum perforatum (Merat.) M.Lainz	<1	<1	-	-
Vaccinium uliginosum L.	-	<1	-	-
Vaccinium vitis-idaea L.	<1	<1	<1	<1
Mosses				
Ceratodon purpureus (Hedw.) Brid	<1	<1	<1	5
Polytrichum juniperinum Hedw.	<1	<1	1	<1
Polytrichum piliferum Hedw.	<1	<1	1	1
Moss protonema	-	21	32	30
Total projective cover	**30**	**48**	**75**	**70**
Number of herb species	**18**	**18**	**17**	**17**
Number of moss species	**3**	**3**	**3**	**3**

Table 12. Development characterization of herbaceous cover in the optimization experiment (projective cover by years, %).

The changes in ground vegetation cover provoked changes in substratum. Substratum surface was identified for a loose layer of dead plant remnants (litter). On the fourth-fifth restoration year it became underlain by a weakly-compact soddy layer up to 3(5) cm thick. Slow dead plant material decomposition in the North causes slow organic carbon accumulation in substratum (Table 13, Fig. 3). This fact was proven by other scientists (Abakumov, 2008). There is an existed positive trend in content of biogenic elements (Figs. 4,5,6) related to the already started organic matter biological cycle.

Plot, №	Sampling depth, cm	pH_{water}	C, %	$N_{hydr.}$	P_2O_5	K_2O	Ca^{2+}	Mg^{2+}
				Mg/100g a.d.s.			mM/100g a.d.s.	
initial substratum								
control	0-10	5.7	0.1	0.2	6.6	2.1	0.6	0
	20-30	5.7	0.1	0.4	8.1	2.2	0.5	0.1
trial	0-10	6.5	0.2	0.3	5.7	2.9	0.9	0.0
on the second year after planting								
control	0-5	5.8	0.2	0.3	6.3	3.8	1.1	0.5
	5-10	5.7	0.2	0.3	6.5	3.2	0.9	0.3
trial	0-5	6.1	0.3	1.3	5.6	3.0	1.1	0.1
	5-10	6.1	0.2	0.3	3.4	1.8	1.0	0.4
on the fourth year after planting								
control	0-5	5.8	0.1	0.4	10.5	4.7	0.6	0.1
	5-15	5.9	0.1	0.3	9.5	2.7	0.8	0.1
trial	0-3	6.1	0.3	1.0	16.2	7.4	0.8	0.1
	3-15	6.0	0.1	0.2	5.7	3.9	1.3	0.2
	15-30	6.4	0.1	0.5	5.6	2.5	1.3	0.4
on the fifth year after planting								
control	0-5	5.3	0.1	0.4	7.2	5.1	0.4	0.3
	5-10	5.2	0.1	0.3	7.9	5.4	0.3	0.2
	10-20	5.2	0.2	0.4	6.9	5.5	0.3	0.2
	20-30	5.2	0.1	0.3	7.0	5.3	0.3	0.2
trial	0-5	5.1	0.2	1.5	11.1	13.4	0.7	0.2
	5-10	5.5	0.1	0.7	5.0	4.7	0.9	0.3
	10-15	6.0	0.1	1.0	6.5	4.1	0.9	0.3
	15-30	6.0	0.1	0.7	5.0	3.4	0.7	0.3
on the sixth year after planting								
control	0-5	5.2	0.1	0.4	7.2	2.2	0.6	0.2
	5-10	5.2	0.1	0.2	7.9	2.9	0.8	0.2
	10-20	5.2	0.1	0.2	8.8	3.2	0.8	0.3
	20-30	5.3	0.1	0.3	9	2.9	0.8	0.4
trial	Ад 0-2	5.2	0.2	2.2	32.1	10.2	0.6	0.3
	АдА₁ 2-5	5.4	0.1	2.5	13.8	12.3	0.4	0.1
	A// 5-10	5.4	0.2	0.7	9.5	10.1	0.6	0.2
	AB 10-20	5.6	0.2	1.9	12.1		0.8	0.2
	B 20-30	6.1	0.1	0.3	9.1	2.9	1.3	0.3

Table 13. Changes in substratum agrochemical parameters in the optimization experiment.

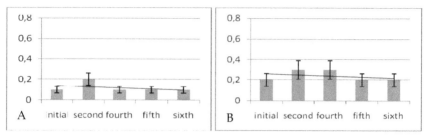

Fig. 3. Organic carbon content dynamics (%) by years in upper substrata layer in the background (A) and experimental (B) plots.

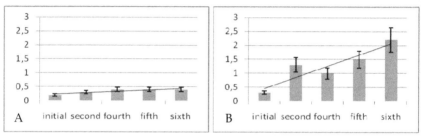

Fig. 4. Hydrolizable nitrogen content dynamics (mg/100 g a.d.s.) in upper substrata layer in the background (A) and experimental (B) plots.

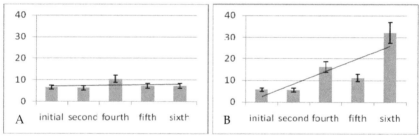

Fig. 5. Phosphorus oxide content dynamics (mg/100 g a.d.s.) in upper substrata layer in the background (A) and experimental (B) plots.

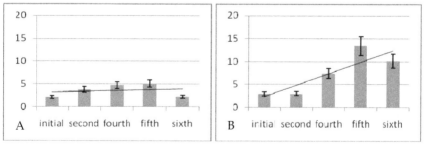

Fig. 6. Potassium oxide content dynamics (mg/100 g a.d.s.) in upper substrata layer in the background (A) and experimental (B) plots.

By the obtained experiment results, usage of intensive agromethods combined with high-quality planting material, big-size *Pinus sylvestris* plantings with ground clot, ensures the high survival rate and active growth of the planted conifers together with herbaceous layer formation. This is important because it causes simultaneous transformation of technogenic substratum and formation of soil as a forest ecosystem component. Soil development is behind plant community development; soil formation in morphological and chemical senses can be only after accumulation of some "critical" plant mass and its transformation products (humus) in substratum.

The popular opinion about the necessity of herb cover destruction while planting of trees in order to improve their growth (competition for nutrients) appeared to be questionable, especially on usage of big-size plantings.

This experiment has shown possibility of accelerated formation of forest ecosystem already on the first ("intensive") restoration stage. Further observations will allow for more recommendations on the optimized experiment.

6. Conclusion

The usage of two-stage "nature restoration" approach ensures active self-restoration of forest ecosystem in far-north taiga (on forest distribution border). It is vital to mention that "nature restoration" conception and its application in practice widen the traditional sense of the term "recultivation" not only by geographical point but also by understanding the functional interdependence between ecosystem components linked together by organic matter biological cycle.

Taking into consideration the serious ecological situation not only in the North, very significant is to revise our opinion on recultivation. Common sense of recultivation is returning lands into repeated agricultural usage. But nowadays there is the need to revise this term. In this view, "nature restoration" conception has a deep sense with its system approach aiming at accelerating restoration of nature ecosystems on disturbed areas exerting important biosphere functions. The system of "nature restoration" can be widely used, including tropical forests, however with some corrections in respect to particular climatic conditions. It is important to mention that oil-polluted lands' restoration at "intensive" stage requires usage of special purifying preparations followed by agrothechnical methods. In view of progressive development of economics, intensive "nature exploitation" should be accompanied by full-scale accelerated (managed) restoration of zonal ecosystems on disturbed areas, proportional to disturbance extent. Imbalance in "human-nature" system produces ecological critical situations (Ecological principles…, 2010).

In relation to the above-said, in newly-published work of K.S. Losev (2010) it says that only some part of natural ecosystems on Earth can be replaced by artificial ecosystems (agrarian or technogenic) without hampering the biological regulation mechanism, responsible for biosphere balance. He calls territories under such artificial ecosystems as ecological (economic) biosphere parts. The importance of natural ecosystems as a biosphere stability factor is now underestimated. The aftereffects cause a row of nature medium changes visualized in climate change, progressive environmental pollution, soil poorness, poor human's health, etc. There is a need of changing the understanding and treat the nature with more responsibility for conservation the environment and wild world in its initial diversity. One simple rule should be followed – whatever we took from nature (disturbed) we are to recover by means of additional work and financial expenses. This is the closest link between ecology and economics.

This work was financially supported by RAS Presidium Program № 23 "Biological diversity" within the topic "Biodiversity formation rules of plant communities in restoring and transforming ecosystems in different types of technogenic objects in the North-East of European Russia", № 09-П-4-1028.

7. References

[1] Abakumov E.V. Accumulation and transformation of organic matter on different-aged waste piles of sand pit // Soil Science J., 2008. №8. p.955-963

[2] Agrochemical methods of soil study. M.: USSR Science Academy Publishing House, 1960. p.556

[3] Archegova I.B. Nature restoration efficient system – the base of promising nature management in the Far North, Syktyvkar. 1998. 12 p. (Scientific reports / Komi SC UrD RAS; Iss. 412).

[4] Ecological principles of nature management and nature restoration in the North. Co-authorship. Editor-in-chief I.B. Archegova. Syktyvkar. 2009. p.176

[5] Field geobotany. M.-L.: *Nauka*, 1964. p.532

[6] Guidance on humus content and composition estimation in (mineral and peat) soils. Leningrad, 1975. p.106

[7] Larin V.B., Pautov Yu.A., Pruchkin V.D. Stable development strategy of forest region in the North // Finno-Ugric World: Nature Health Status and Regional Strategy of Environment Protection: Proc. Int. Conf. Syktyvkar, 2000. p.93-98

[8] Losev K.S. Myths and mistakes in ecology. M.: *Nauchnyi mir*, 2010. p.224

[9] Losev K.S., Mnatsakanyan R.A., Dronin N.M. Consumption of renewed resources: ecological and social-economic consequences (global and regional aspects). M.: GEOS, 2005, p.158

[10] Ogievskiy V.V., Khirov A.A. Observation and investigation of forest cultures. M.: *Lesnaya promyshlennost'*, 1964. p.48

[11] Parfenyuk V.I. The method of oil-polluted soil recultivation. Patent №20009626 (patented in the State Register of Inventions February 19, 1991).

[12] Podzolic soils of the central and eastern parts of the European USSR. – L., 1981. p. – 200

[13] Ponomareva V.V. Forest as elluvial-resistant vegetation type // Botanical Journal, Band 55, № 11. 1970. p.1585-1595.

[14] Ponomareva V.V., Plotnikova T.A. Humus and soil formation (study methods and results). L.: *Nauka*. 1980. p.222

[15] Scientific-applied reference book on the USSR climate. Series 3, Perennial data. P. 1-6. Iss. 1. Archangelskaya and Vologodskaya regions, Komi ASSR. L.: *Hidrometeoizdat*, 1989, book 1. p.483

[16] Semenkova I.G., Sokolova E.C. Forest phytopathology. M.: *Ecology*, 1992. p.352

[17] Shennikov A.P. Introduction to geobotany. – L.: Leningrad Uni. Publishing House, p.447

[18] Theory and practice of soil chemical analysis / Ed. by Vorobyeva L.A. M.: *GEOS*, 2006. p.400

[19] Yudin Yu.P. Geobotanical zoning // Productive forces of the Komi ASSR, M.-L. 1954, Band. 3, P. 1. p.323-359

9

Moving from Ecological Conservation to Restoration: An Example from Central Taiwan, Asia

Yueh-Hsin Lo[1], Yi-Ching Lin[2], Juan A. Blanco[3],
Chih-Wei Yu[4] and Biing T. Guan[1,*]
[1]National Taiwan University,
[2]Tunghai University,
[3]University or British Columbia
[4]New Taipei City Government
[1,2,4]Taiwan
[3]Canada

1. Introduction

The concept of "natural conservation" has been evolving since the beginning of the first efforts to preserve the natural landscape. The creation of the first national parks in the 19th century was originated by the belief that landscapes of exceptional beauty should be preserved from human influence and maintained in their current state for the enjoyment of future generations (Runte, 1997). During the first years of the establishment of national parks around the world, defining these "exceptional landscapes" was usually based on the static beauty of the area: majestic mountains, glaciers, old forests, gorges, canyons, waterfalls, etc. The protection in these areas was basically achieved through the prevention of creating human structures in the sites, reducing and controlling human activity and, in practice, maintaining the areas as they look at the time when they were declared as protected. Therefore, this protection was not based on ecological considerations, but on a human-centered view of natural sites.

The type of protected ecosystems varied widely among regions, depending on the history of human impact on them. For example, national parks in North America were created to protect largely untouched, almost pristine landscapes practically unaffected by the low populations of native peoples previous to European contact (Runte, 1997). Similarly, in South America and Africa, large natural areas could still be found during the 19th and 20th centuries were the human impact was thought to be minimal. However, in Europe or Asia, were the history of urban development can be traced back for millennia and the density of population is also higher, it was more difficult to found those untouched areas. As a consequence, national parks were created to protect landscapes of indisputable beauty but usually with a noticeable human influence on them. Ecosystems at this time of early

---

*Corresponding Author

conservation efforts were seen as static entities, that should not be altered or they would lose their integrity. However, with the arrival of modern ecology, this static view was gradually substituted by the classic view of gradual linear change along a continuum, arriving to a single climax state (Clements, 1916; Odum, 1969; Pickett and McDonnell, 1989). This climax state was usually identified with the state that nature reaches in absence of human influence, and therefore the efforts were oriented into maintaining it. As a consequence of this human-centered vision to define the areas to protect, regions such as oceans, deserts, swamps, shrublands and other similar ecosystems were usually considered "badlands" and unworthy of legal protection. On the other hand, very few protected areas have been established in productive landscapes with clear economical potential for agriculture or forest management (Scott et al., 2001). Even today, these regions are still underrepresented in the protected areas around the globe (Noss et al., 1995). Since the Rio Earth Summit in 1992, the global network of protected areas has continued to grow steadily, increasing yearly by an average 2.5% in total area and 1.4% in numbers of sites, and by 2006 covering more than 24 million km^2 in about 133,000 designated sites (Butchard et al., 2010). Protected areas overall remain a core element of biodiversity conservation (Andam et al., 2008; Gaston et al., 2008).

During the 20th century urban development was extended to all the regions of the world. With a booming human population and the intensification of economic development (first in Europe and North America and lately in the rest of the world) practically all the ecosystems in the world were impacted in one way or another. Therefore, it was just a matter of time that some of the iconic wildlife species of the world started to suffer from fast reductions in their populations, or even facing extinction. The danger of losing species such as whales, lions, tigers, elephants, panda bears, gorillas, brown bears, buffalos, sequoias, etc. was and still remains very real (Laliberte & Ripple, 2004; Sanderson et al., 2008). This danger was highlighted by scientists and environmental managers around the world, and the society responded with the creation of environmentalist groups, whose social pressure helped to create lists of endangered animal and plant species needing specific actions for conservation. This was the base to develop programs and activities focused on the protection of individual high-profile species. Many of these campaigns were supported by the public due to the easy sympathy or spiritual connection with some of these majestic species, and as a consequence, natural conservation was seen by the main public as "avoiding things getting worse". Some of these activities have achieved important successes, such as the halt in commercial hunting of whales (Stevick et al., 2003), the breeding programs of panda bears (Peng et al., 2001) or the increase in numbers of American buffalos (Waldman, 2001). However, in other cases the protection of the target species was not enough to prevent its decline or extinction (e.g. the Yangtze River dolphin, Turvey et al., 2007; or the Pyrenean wild goat, Folch et al., 2009), or just the species were not interesting enough for the public opinion and therefore not the main focus of protection efforts, such in the case of "ugly" species as it is amphibians, reptiles, insects, cacti, etc.

The improvement of this species-oriented conservationism from the first days of creation of protected areas is that it recognizes individual species as worth of the preservation, even if they are not in "beautiful landscapes" with some sort of legal protection. Therefore, it moves one step from the human-centered conservation of some specific favourite areas to protect species and control the factors that affect their populations. However, the main drawback of this type of ecological conservation is that it is targeted to one species, not to the ecosystem that supports that species. This species-oriented conservation followed the theory that if the causes of non-natural mortality are controlled (i.e. hunting, harvesting, poaching, poisoning, clear-cutting, etc.), and the availability of resources increased, the target species could survive or even increase its population. Therefore, actions such as banning hunting, stopping illegal logging, controlling access to the areas were the species is distributed can be part of this strategy (Folch et al., 2009). In addition, zoos, herbariums, arboretums and other centers where collections of plant and animals are kept under controlled conditions are an important part of this strategy, as they provide research insights in the biology of the species and they can specially increase the population sizes of plants and animal species (Bagarinao, 1998).

However, actions that could be beneficial for the target species are not necessary relevant for other species in the ecosystem, and they could be ineffective if the ecosystem is too altered to keep the target species, even after removing the human factors directly affecting it. Ultimately, any plant, animal or microorganism species will survive in a given ecosystem as long as the right conditions exist to support the niche that the species inhabits. Conservation paradigms, practices, and policies have shifted over time recognizing this need to preserve the ecosystem and not just the target species (Adams, 2004). As a consequence, a more holistic approach to conservation has emerged since the last quarter of the 20th century. Within this approach, the actions in the conservation effort will be directed to keep the integrity (bio-physical diversity) and the functionality of the ecosystem. This new approach is in the origin of the last trend in conservation: ecological restoration.

Ecological restoration involves assisting the recovery of an ecosystem that has been degraded, damaged, or destroyed, typically as a result of human activities (Sala et al., 2000). Ecological restoration is based on the new view of ecosystems as biological communities established on a geophysical substrate that can develop into alternative stable states rather than into a single climax state (Lewontin 1969). As a consequence, the idea of the balance of nature has been replaced with the flux of nature (Wu & Loucks, 1995; Pickett & Ostfield, 1995; Wallengton et al., 2005), and ecosystems are thought to be mostly in non-equilibrium. Their dynamics are not only complex but also dependent on the spatial context and the history of natural disturbance and human influence (Hobbs & Cramer, 2008). The main implication of this conceptual model is that ecosystems that have been altered by human activity may not revert back to its original state if left alone. On the contrary, these altered ecosystems could just reach a different stable state defined by the actions of human management on them (i.e. soil alteration and erosion, invasive species, lost of native species, changes in hydrological regime, etc.). Examples of such alternative states are grasslands or forest dominated by invasive species or shrublands that substitute forests. The goal of ecological restoration is therefore the reestablishment of the characteristics of an ecosystem, such as biodiversity and ecological function that were prevalent before degradation (Jordan et al., 1987), and that will not be reached (or if so, in very long time scales) by the ecosystems if left alone.

Ecological restoration is different from the earlier protection of specific areas because "restoration" means human intervention to bring the ecosystem back to a state different from the one in which it is currently. Therefore, it is not a passive conservation effort in which humans are just consider outsiders that should not be "in the way" of Nature. Quite differently, ecological restoration needs direct human actions (i.e., modification of the physic-chemical environment, introduction of lost species, removal of invasive species, plantation of trees and plants, etc.). In addition, ecological restoration is different from conservation of emblematic species in that restoration targets the whole ecosystem, assuming that if the correct ecological conditions are maintained, the emblematic species (and their companion species) will be preserved in the restored area.

Ecological restoration has developed quickly since the first meetings of the Society of Ecological Restoration International in the early 1990s (Greipsson, 2011). Ecological restoration, both within and outside protected areas, is being increasingly applied worldwide (Clewell & Aronson, 2007; Nelleman & Corcoran, 2010), and it is increasingly found as part of natural resource management plans. Actions such as targeted habitat management, removal of invasive species, captive breeding, seed production and species reintroduction have yielded notable successes: among many examples, at least 16 bird species extinctions have been prevented by such means between 1994 and 2004 (Butchart et al., 2006).

Large-scale ecosystem restoration is needed to arrest and reverse the degradation of landscapes around the world (Manning et al., 2006). However, restoration efforts to date have been criticized for being ad hoc, site and situation specific (Hobbs & Norton, 1996), or focusing on small, protected nature reserves (Naveh, 1994; Soulé & Terborgh 1999). Hence, the effectiveness of restoration actions in increasing the provision of both biodiversity and ecosystem services has not been evaluated systematically. A meta-analysis of 89 restoration assessments in a wide range of ecosystem types across the globe indicated that ecological restoration increased the provision of biodiversity by 44% and ecosystem services by 25%, but values of both remained lower in restored versus intact reference ecosystems (Rey-Benayas et al., 2009). Although small-scale restoration projects can be valuable, there is an urgent need to greatly expand the scale of ecosystem restoration for both conservation and production (Naveh, 1994; Hobbs & Norton, 1996).

From these results, it is clear that detailed field research is needed to guide restoration efforts. Field experiments can be a useful guide as to which restoration practices are the most useful for achieving this goal (Kimmins et al., 2010). In this chapter we describe the field research done to guide the restoration of native conifer forests in central Taiwan. Our research has as main objective the identification of the best conditions for seedling establishment of two tree native species: the evergreen Fabaceae *Lithocarpus castanopsisifolius* and *Lithocarpus kwakamii* (stone oak) in a former plantation of Japanese cedar (*Cryptomeria japonica*), an species alien to Taiwan. In this study, we seek to evaluate whether the combination of selective cutting, direct seeding, and understory vegetation control can be a cost-effective method to gradually restore plantation forests to native forests. We also seek to identify the potential barriers that hinder seed and seedling survival.

2. An example from central Taiwan

2.1 Historical background

Taiwan covers an area of 36,000 km^2 and is located at the fringe of the Asian continental shelf at the western rim of the Pacific Basin, and separated from the main continent by a strait of 130 km in its narrowest point. The island has a very complex terrain, with about two thirds of Taiwan's land area at slopes over 10% and almost half of the island with slopes over 40%, and with Jade Mountain (the highest peak), reaching 3952 m a.s.l. (Hsu & Agaramoorthy, 1999). As a consequence of its insularity, closeness to the continent and wide gradient of altitudes, Taiwan harbours over 4,000 vascular plants in six different forest types (Boufford et al., 1996). Wildlife resources are also abundant with 61 species of mammals, 400 species of birds, 92 species of reptiles, 30 species of amphibians, 140 species of fresh-water fish, and the estimated 50,000 species of insects including 400 species of butterflies (Hsu & Agoramoorthy, 1997). To protect this rich biodiversity, the first national park was created in 1984 in Kenting (south Taiwan). Till date, 6 national parks, 18 nature reserves and 24 nature protected areas have been designated to ensure protection for wildlife and their habitats. The protected area covers 12.2% of the total land area of Taiwan (Hsu & Agoramoorthy, 1999).

The history of forest conservation and restoration in Taiwan is closely linked to the economic development of the island. Timber harvesting peaked during the Japanese colonial period and immediately following World War II. Large areas of valuable timber, primarily cypress, spruce, and camphor, were cut and shipped primarily to Japan. Economic pressures led to an aggressive management, with plantations of native species and timber harvesting program through the 1950s to the 1970s, with an average of 1,552,600 m^3 harvested from 1965 to 1975, corresponding to about 18,000 ha cut annually (Lu et al., 2001). These levels of harvesting brought petitions from citizens and environmental protection groups urging forest protection. This intensive level of exploitation was essentially halted with the national forestry management policy of 1976 (Wang, 1997). Since then, the emphasis of forest management in Taiwan has shifted almost entirely from timber production to forest protection. After 1977, timber was harvested mainly from forest plantations with an annual cut of about 100,000 m^3 and by 1990, 99% of Taiwan's timber supply was imported (Wang, 1997; Lu et al., 2001). Currently, national forest lands are managed almost exclusively for the purposes of streamflow regulation, erosion control, and conservation of biological diversity. Under this new approach, the harvesting-reforestation approach is no longer viable and alternatives need to be devised.

The interest on conservation is not limited to natural forests, but it is also extending into plantation forests, especially in the marginal plantations created during the 50s and 60s, at the peak of exploitative management in the island. To restore and promote biodiversity, the current management directives mandate the restoration of plantations no longer serving for timber production back to native forests, in a gradual manner. One example is the important number of existing Japanese cedar (*Cryptomeria japonica* D. Don) plantations that were established in sites now considered as unsuitable for harvesting, mainly due to soil and slope protection concerns. This species was introduced from Japan with the start of the Japanese colonial rule at the end of the 19th century. It has become the most widely planted tree species in Taiwan, covering about 1.1% (41,132 ha) of the island's total land area (Taiwan Forestry Bureau, 1995). However, due to increasing production costs and declining timber prices, most of Taiwan's Japanese cedar plantations either are approaching or have passed the prescribed rotation age.

Knowledge on how to use current forestry practices to accelerate and support the conversion from plantations into native forests is needed to design successful restoration plans in these plantations. Among other concerns, it is necessary to understand the best ways of promoting native trees establishment. Seed and seedling survival are limited by multiple biotic and abiotic factors (Beckage et al., 2000; Fenner & Thompson, 2005), making these stages the bottleneck of ecological restoration (Fenner & Thompson, 2005; Leck et al., 2008). Drought, herbivory, and light are the three most important causes for seedling mortality (Leck et al., 2008)

To improve seed establishment rates, seedling planting and direct seeding are two common tools used in forest restoration. The former has the advantage of high success rate, but it is also more expensive than the later (Bullard et al., 1992). Direct seeding has the advantage in term of cost, but it usually has low success rates. Thus, the creation of an environment that enhances the survival of tree seeds and seedlings is a key element of a successful gradual forest restoration strategy. Selective cutting and thinning are common forestry practices that can also be used for restoration. During these procedures, only a portion of trees is removed, and the overall stand abiotic environment is not greatly altered. Therefore, partial removal of trees can enhance local light availability and create physical environments similar to natural gaps that are essential for seedling survival (Augspurger, 1984; Brokaw & Busing, 2000; Masaki et al., 2007). In addition, the presence of understory vegetation may reduce seedling survival by reducing light availability or increasing competition between seedlings and understory vegetation (Leck et al., 2008). On the other hand, understory vegetation may reduce seedling predation by providing protection (Smit et al., 2006). Therefore, understory vegetation control may cast both positive and negative effects on the survival of seeds and seedlings (Beckage et al., 2000; Fenner & Thompson, 2005; Leck et al., 2008).

2.2 Material and methods

2.2.1 Experimental site

This study was carried out in a 10-ha Japanese cedar plantation in the Heshe District of the National Taiwan University Experimental Forest, central Taiwan (120° 52' E, 23° 37' N, 1442-1602 m a.s.l.; Fig. 1). Mean annual temperature of the study site is 19.8°C, with a mean annual rainfall of 1500 mm (NTUF, 2011). Originally an evergreen broad-leaf forest dominated by *Fagaceae* and *Lauraceae* species, the site was clear-cut in 1958 and planted with

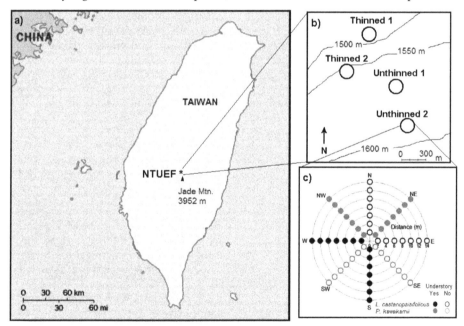

Fig. 1. Location of the experimental site, National Taiwan University Experimental Forest – NTUEF (a), spatial arrangements of the treatment combinations (b), and seed transects (c).

Chinese fir (*Cunninghamia lanceolata*). Due to typhoon damages in 1969, the stand was re-planted with Japanese cedar in 1971. In 2005, the plantation was selected as a demonstration site to study how to gradually restore Japanese cedar forests back to native vegetation communities. Within 500-m from the edges of the plantation, remnants of the original vegetation can still be found. We regard those edge areas as the reference for the restoration project and to set the initial restoration goal: the successful establishment of late succession components of the reference stands in the plantation.

2.2.2 Materials and experimental design

An initial inventory found that while the saplings of late succession *Lauraceae* species (mainly dispersed by birds) were relatively abundant, only a few saplings of *Fagaceae* species were present. Thus, we focused only on the reintroduction of the main *Fagaceae* species. *Lithocarpus castanopsisifolius* (Hayata) Hayata and *Pasania kawakamii* (Hayata) Schottky were selected as the target species for reintroduction as they are late succession species and relatively abundant in the surrounding areas from were collected seeds. All fresh seeds used in this study were collected between October and November 2008 from the reference stands in the edges of the Japanese cedar plantation, and they were stored at 4 °C until they were used in January 2009.

As a part of the experiment, 20% of the standing volume was harvested to create gaps of different sizes. We established four plots, two thinned and two unthinned, within the plantation (Fig. 1b). The canopy openness of the two thinned plots was 27% and 29%, whereas canopy openness of the two unthinned plots was 13% and 11%. For each plot, a 10-meter transect was set at each of the 8 cardinal and inter-cardinal directions (Fig. 1c). The seeds of *L. castanopsisifolius* were placed along the cardinal direction transects, whereas the seeds of *P. kawakamii* were placed along the remaining four transects (Fig. 1c). For each group of transects in each plot (cardinal or inter-cardinal), we randomly selected two transects from where we removed the ground vegetation in a strip of 1-m wide along the entire transect (devegetated transects), whereas the ground vegetation of other two transects was left untouched (vegetated transects, Fig. 1c). Thus, the entire experiment consisted of 4 treatments for each species.

In January 2009 we placed 30 fresh seeds every 2.5 m along each transect, starting and ending at the 2.5-m and 10-m marks, respectively, for a total of 120 seeds per transect and 1920 seeds per species for the entire experiment. After placing the seeds, the number of seeds still present was counted every day during the first 35 days. After that, we went back on day 140 as the final checking time. At day 140, almost all the seeds were removed or consumed, therefore the experiment ended at that time.

Eight infrared automatic cameras, one for each species-treatment combination, were also set up to capture how the seeds were removed or consumed and by which animal species under different treatment conditions.

2.2.3 Data analysis

Cox regressions were used to analyze the survival and seedling establishment, with the hazard defined as the instantaneous mortality risk of a seed (Cox, 1972). The thinning and understory vegetation removal treatments were used as the explanatory variables. Species were analyzed separately. We used R to conduct all statistical analyses, with survival analysis using the R package Survival (R Core Team, 2010).

2.3 Results

2.3.1 Field observations

Seeds were first removed from transects with no ground vegetation cover (Fig. 2). Twenty days since the beginning of the observation, 82% and 48% of *L. castanopsisifolius* seeds were missing in the unthinned and thinned plots, respectively. These results were similar for the *P. kwakamii* seeds, with 95% and 48% seeds disappearing in the unthinned and thinned plots, respectively.

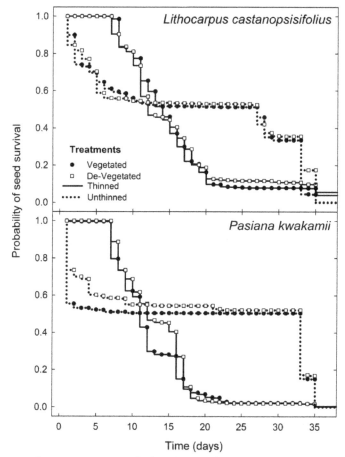

Fig. 2. Seed removal rates over a period of 61 days in different thinning and understory vegetation treatments for *Lithocarpus castanopsisifolius* (upper panel) and *Pasania kawakamii* (lower panel). Solid lines indicate thinned treatment, and dotted lines represent unthinned treatment. Solid black dots indicate ground vegetation intact, whereas opened squares represent without ground vegetation.

Seeds of *P. kwakamii* were removed more slowly from the devegetated plots for both thinned and unthinned treatments, but this effect disappeared after 23 days of exposure. This pattern was not clearly identified for *L. castanopsisifolius* seeds, but after 15 days there seemed to be a tendency for a slightly higher probability survival in devegetated plots for both canopy types. Ninety-four percent of *L. castanopsisifolious* seeds had disappeared from the vegetated plots after 35 days, versus 91% from devegetated plots. However, the treatment producing the biggest differences after 35 days for *P. kwakamii* was thinning, with 98% of seeds disappearing from thinned, a much lower probability of survival than in close canopy plots (84% seeds disappeared after 35 days).

From the images captured by the automatic cameras, we identified two mammal species as the acorn consumers/removers during the observation period. These two species were red-bellied squirrel (*Callosciurus erythraeus*) and Owslon's long-nosed tree squirrel (*Dremomys pernyi owstoni*). These two species can be seen as potential dispersers of large seeds in the late succession period.

2.3.2 Seed removal

The results from Cox regressions indicated that, for both species, removing part of the canopy significantly influenced seed removal (mortality risk) in the study site (Table 1, Fig.2). Results indicated that, for *P. kwakamii*, seeds in the thinned treatments suffered the highest removal risk (Table 1). Similar results were found for *L. castanopsisifolius* (Table 1). In addition to the main effects, the interaction between canopy type and understory vegetation cover was non-significant (Table 1).

Treatment	df	Hazard ratio[1]	Z[4]	P
Lithocarpus castanopsisifolius				
Thinning[2]	1	0.70	-5.11	<0.001
Understory removal[3]	1	1.05	0.73	0.466
Thinning × Vegetation	1	1.18	1.69	0.091
Pasania kwakamii				
Thinning	1	0.65	-6.18	<0.001
Understory removal	1	1.04	0.66	0.509
Thinning × understory removal	1	1.09	0.93	0.351

Table 1. Effects of thinning and ground vegetation treatments on the seed hazard (instantaneous mortality risk) based on Cox regressions. Notes: 1) Hazard ratio is defined as the ratio of mortality risk between two factor levels. If hazard ratio = 1, it indicates equal mortality risk; 2) Risk ratio of thinned relative to unthinned treatment; 3) Risk ratio of devegetated relative to vegetation treatment; 4) Cox regression coefficient.

2.3.3 Seedling establishment

Probabilities of successful seedling establishment at day 140 differed among species and treatments. For *P. kwakamii*, seedlings only successfully established in the unthinned plots (Fig. 3 left panel). For *L. castanopsisifolious*, there was a small seedling establishment probability in thinned plots, but the success rate was much higher for unthinned plots. (Fig. 3 right panel). No significant difference was detected between vegetated and devegetated plots.

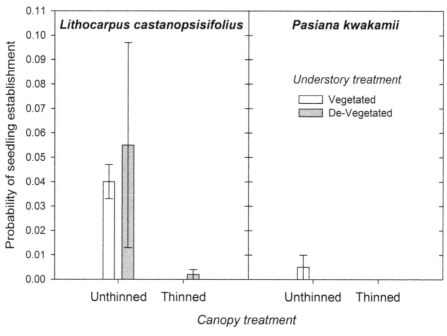

Fig. 3. Probabilities of successful seedling establishment after 140 days of seed planting for *Lithocarpus castanopsisifolius* (left panel) and *Pasania kawakamii* (right panel). Error bars represent the mean ± standard error.

2.4 Discussion

Differential seed survival and seedling establishment among species, canopy and ground vegetation conditions were observed. The thinning treatment reduced seedling establishment in both species and significantly reduced seedling survival in both species. In addition, seedling establishment was not significantly affected by removing ground vegetation.

Despite that our results are species-specific, they suggest that keeping canopy vegetation intact may be more effective than removing understory for the two species studied. Survival rates for the unthinned plots were consistently higher than for the thinned plots in either vegetated or devegetated after two weeks of seeding (Figure 2). In contrast, differences in seed survival between the two vegetation types did not show consistent patterns and the difference was rather small. Such results suggest that removing the understory layer may not be as effective as keeping the canopy cover, although understory removal showed a tendency to increase seed survival in *L. castanopsisifolius*, although statistically not significant.

The observed differences in seed survival may arise from changes in foraging behaviours of seed predators as a result of the forest management treatments. The understory vegetation was rather dense in the study site, especially in the thinned plots. The understory grew rapidly after selective thinning was conducted in 2005. The improved light conditions due to thinning were believed to facilitate the rapid growth, which was mostly composed by broadleaf species, such as *Schefflera actophylla,* and *Machilus thunbergii,* which are shade-tolerant but can take advantage of increased light levels (C.-W. Yu, personal observation). Therefore, thinning could reduce seed survival by favouring understory growth and then providing more protection for seed predators.

In addition, the openness of the canopy layer can also influence the behaviour of seed predators (Boman and Casper, 1995; Schnurr et al., 2004), as well as the effects of understory vegetation (Chambers and MacMahon, 1994; Fenner and Thompson, 2005). The effect of both canopy structure or understory vegetation density on seed survival, however, is somewhat context-dependent (den Ouden, 2004; Hulme & Kollmann, 2004; Tamura & Katsuki, 2004). Seed predation may increase or decrease among different microhabitats depending on the composition of seed predators (Schupp et al., 2002; den Ouden, 2004; Hulme & Kollmann, 2004; Tamura & Katsuki, 2004). For instance, seed predation by squirrels is reduced in canopy gaps, while predation rates by field mice, *Peromyscus mexicanus* are higher in canopy gaps (Tamura & Katsuki, 2004). This may be the case in our study, which showed increases in seed predation in conditions with higher canopy openness after thinning. To try to understand the relationship between seed predators and seedling establishment, the movements of mammals were monitored in a nearby plantation.

Mammalian activities were decreased immediately after thinning treatments (Lin & Bridgman, 2010). In addition, the photos taken from our automatic cameras in our plots also suggested that major seed predators in the study sites were granivorous animals, including red-bellied Squirrel and Owslon's long-nosed tree squirrel. These animals may also become seed dispersers via scatter-hoarding behaviour. Therefore, they can be reducing the seed survival but increasing seed dispersal rates at the same time. Scatter-hoarding behaviour has been widely observed in squirrels and jays in various forest ecosystems, where instead of consuming the seeds, they moved seeds away and cached the seeds (Vander Wall, 1994; Forget et al., 2004; Zhang et al., 2004). The cached seeds may germinate at later dates. Some of the animals observed by the automatic camera may also function as seedling predators. For example, many rodent species, such as red-bellied squirrels or the Formosan field mouse have been observed to eat seedlings (Young, personal observation). Seed tracking techniques are required to study the details of scatter-hoarding behaviours (Forget et al., 2004), and they could be a future research line at these sites.

In addition of these two species, we are carrying similar experiments for other tree species at the same plots, and our preliminary results indicate that for other species the effects of thinning and understory removal can be the opposite (data not shown). Our experiments are the first ones of these kind in Taiwan, and the species-specific responses of seedling establishment to management activities provide a clear indication of the need to shift from a static conservation in which the "no human action" approach is favoured into a more active restoration strategy. If no action were taken, it can be expected that *L. castanopsisifolious* will successfully regenerate, and in a lesser way *P. kwakamii.* Under this scenario, the future tree composition of these stands could be a multi-story canopy in which the Japanese cedar (an alien species) dominates the canopy, with lower layers of *L. castanopsisifolius* and other similar native species, but in which *P. kwakamii* will remain mostly suppressed. As a consequence, the new state of this forest under a conservation-only strategy would not be either the Japanese cedar plantation or the original mixed forest, but a hybrid of both. Only after a major natural disturbance (i.e. a typhoon or a stand-replacing wildfire) this hybrid stand could be transformed into the original mixed forests. Therefore, if this "new ecological state" is to be avoided and the recovery towards the original mixed forest is needed, a program of seeding combined with additional research to improve the regeneration of *P. kwakamii* should be implemented in these stands to ensure that *P. kwakamii* finds favourable conditions to regenerate. However, these activites should be localized and not general through the stands to avoid the inhibitory effect that close canopy or exposed forest floor could have for the regeneration of other important native species.

3. Conclusions

The native forests within the elevational range in this experimental area were dominated by *Fagaceae* and *Lauraceae* species. These families were, however, rare in the seed rain of the planted forest after thinning (Sun, 2010). Such results suggested a high degree of recruitment limitation in the plantation. Therefore, to facilitate the transition from planted forest to native forest, it is essential to develop management strategies to overcome recruitment limitation of native species in the plantation forest. Our study indicated that for the two species studied, keeping the canopy cover could be an effective management tool to overcome recruitment limitation, suggesting an easy and inexpensive mean for forest restoration. With extensive plantation forests in Taiwan, the management practices could be widely applied to facilitate the regeneration of native species. The next step is to apply such treatments to a broader area to assess the operational costs of such management techniques. The efficiency of these management techniques, however, seems to be species-specific. Other research has shown opposite effects of thinning and vegetation removal in other species at the same sites (data not shown). Therefore, we warn the readers from assuming that the results presented here could be applied to other forests types or regions. The interaction between light availability, soil moisture and species-specific factors for trees (seed size, seed dispersal) and seed predators (foraging behaviour, seed preference) can make the same management have very different results in restoring different sites.

All things considered, our research shows how the shift from passive ecological conservation towards a more active ecological restoration can be successful if enough ecological information on ecosystem structure and function is available. Also, if no active restoration is implemented, the ecological barriers for seedling establishment could prevent these ecosystems from recovery for a long time, generating a new "hybrid" state with elements from both the human-altered (plantation) and original (mixed forest) ecosystems. We suggest that other similar programs monitoring seed survival and seedling establishment should be enacted in other forest regions around the world, especially in tropical forest where little is known about many of the native tree species.

4. Acknowledgements

The authors thank Dr. Po-Jen Jiang for his assistance in setting up automatic cameras. We also thank National Taiwan University Experimental Forest for their help during the field work period. This study was supported by a grant to Dr. Biing T. Guan and Dr. Yiching Lin by the Taiwan National Science Council (NSC 97-2313-B-002-041-MY3).

5. References

Adams, W.M. (2004). *Against Extinction: The Story of Conservation*, Earthscan, ISBN 9781844070565, London, UK.

Andam, K.S.; Ferraro, P.J.; Pfaff, A.; Sanchez-Azofeifa, G.A.; Robalino, J.A. (2008). Measuring the effectiveness of protected area networks in reducing deforestation. *Proceedings of the National Academy of Science USA*, Vol. 105, pp. 16089-16094, ISSN-0027-8424.

Augspurger, C.K. (1984). Light requirements of neotropical tree seedlings: A comparative study of growth and survival. *Journal of Ecology*, Vol. 72, pp. 777-796, ISSN: 1365-2745.

Baskin, C.C.; Baskin, J.M. (1998). *Seeds: ecology, biogeography, and evolution of dormancy and germination*. Academic Press, ISBN 978-0120802609, San Diego, USA.

Bazzaz, F.A., (1996). *Plants in Changing Environments: Linking Physiological, Population, and Community Ecology.* Cambridge University Press, ISBN 978-0521398435, Cambridge, UK.

Beckage, B.; Clark, J.S. (2005). Does predation contribute to tree diversity? *Oecologia*, Vol. 143, pp. 458-469. ISSN 0029-8549.

Beckage, B.; Clark, J.S.; Clinton, B.D.; Haines, B.L. (2000). A long-term study of tree seedling recruitment in southern Appalachian forests: the effects of canopy gaps and shrub understories. *Canadian Journal of Forest Research*, Vol. 30, pp. 1617-1631, ISSN 1208-6037.

Boman, J.S.; Casper, B.B. (1995). Differential Postdispersal Seed Predation in Disturbed and Intact Temperate Forest. *American Midland Naturalist*, Vol. 134, pp. 107-116, ISSN 0003-0031.

Boufford, D.E.; Hsieh, C.F.; Huang, T.C.; Ohasi, H.; Yang, Y.P.; Lu, S.Y. (eds.) (1996). *Flora of Taiwan*, 2nd Ed. Editorial Committee of the Flora of Taiwan, ISBN 957-02-7534-0, Taipei, Taiwan, ROC.

Brockerhoff, E.G.; Jactel, H.; Parrotta, J.A.; Quine, C.P.; Sayer, J. (2008). Plantation forests and biodiversity: oxymoron or opportunity? *Biodiversity and Conservation*, Vol. 17, pp. 925-951. ISSN 0960- 3115.

Brokaw, N.; Busing, R.T. (2000). Niche versus chance and tree diversity in forest gaps. *Trends in Ecology and Evolution, Vol.* 15, pp. 183-188. ISSN 0169-5347.

Butchart, S.H.M.; et al. (2010) Global biodiversity: indicators of recent declines. *Science*, Vol. 328, pp. 1164-1168, ISSN 0036-8075.

Butchart, S. H. M.; Stattersfield, A. J.; Collar, N. J. (2006). How many bird extinctions have we prevented? *Oryx*, Vol 40, pp. 266-278, ISSN: 0030-6053.

Caccia, F.D.; Ballare, C.L. (1998). Effects of tree cover, understory vegetation, and litter on regeneration of Douglas-fir (*Pseudotsuga menziesii*) in southwestern Argentina. *Canadian Journal of Forest Research*, Vol. 28, 683-692, ISSN 1208-6037.

Chambers, J.C.; MacMahon, J.A. (1994). A day in the life of a seed: movements and fates of seeds and their implications for natural and managed systems. *Annual Review of Ecology and Systematics*, Vol. 25, pp. 263-292, ISSN 0066-4162.

Chazdon, R.L. (2008). Beyond deforestation: restoring forests and ecosystem services on degraded lands. *Science*, Vol. 320, pp. 1458-1460, ISSN 0036-8075.

Clements, F.E. (1916). *Plant succession: an analysis of the development of vegetation*. Carnegie Institute of Washington, Publication No. 242. Washington DC, USA.

Clewell, A.F.; Aronson, J. (2007). *Ecological Restoration: Principles, Values, and Structure of an Emerging Profession*. Island Press, ISBN 978-1597261692. Washington DC, USA.

Connell, J.H. (1971). On the role of natural enemies in preventing competitive exclusion in some marine animals and in rain forest trees. In: *Dynamics of populations*, den Boer, P.J.; Gradwell, G.R. (eds.), pp. 298-310, Centre for Agricultural Publishing and Documentation, Wageningen, The Netherlands.

Cox, D.R. (1972). Regression models and life tables. *Journal of the Royal Statistical Society. Series B (Methodological)*, Vol. 34, pp. 187–220, ISSN: 0035-9246.

den Ouden, J. (2004). Jays, Mice and Oaks: Predation and Dispersal of *Quercus robur* and *C. petraea* in North-Western Europe In: *Seed Fate: Predation, Dispersal and Seedling Establishment*, Forget, P.M.; Lambert, J.E.; Hulme, P.E.; Vander Wall, S.B. (Eds.), CABI International, Wallingford, UK, ISBN 978-0851998060.

Fenner, M.; Thompson, K. (2005). *The ecology of seeds*. Cambridge University Press, Cambridge, UK. ISBN 978-0521653688.

Forget, P.M., Lambert, J.E., Hulme, P.E., Vander Wall, S.B. (Eds.), 2004. *Seed Fate: Predation, Dispersal and Seedling Establishment*. CABI International, Wallingford, UK, ISBN: 978- 0851998060.

Gaston, K.J.; Jackson, S.E.; Cantu-Salazar, L.; Cruz-Pinon, G. (2008). The Ecological Performance of Protected Areas. *Annual Review of Ecology, Evolution and Systematics*, Vol. 39, pp. 93-113, ISSN 1545-2069.

Gove, A.D.; Majer, J.D.; Rico-Gray, V. (2005). Methods for conservation outside of formal reserve systems: The case of ants in the seasonally dry tropics of Veracruz, Mexico. *Biological Conservation*, Vol. 126, pp. 328-338, ISSN 0006-3207.

Hobbs, R.J.; Norton, D.A. (1996). Towards a conceptual framework for restoration ecology. *Restoration Ecology*, Vol. 4, pp. 93–110, ISSN 1061-2971.

Hulme, P.E.; Kollmann, J. (2004). Seed Predator Guilds, Spatial Variation in Post-Dispersal Seed Predation and Potential Effects on Plant Demography - a Temperate Perspective. In: *Seed Fate: Predation, Dispersal and Seedling Establishment*. Forget, P.M.; Lambert, J.E.; Hulme, P.E.; Vander Wall, S.B. (Eds.), CABI International, Wallingford, UK, ISBN: 978- 0851998060.

Hsu, M.J.; Agoramoorthy, G. (1997). Wildlife conservation in Taiwan. *Conservation Biology*, Vol. 11, pp. 834-836, ISSN 0888-8892.

Hsu, M.J.; Agoramoorthy, G. (1999). Conserving the biodiversity of Kenting National Park, Taiwan: Present status and future challenges. In: *Proceedings of the Symposium on Biodiversity – 1999*. Lin, Y.S. (ed.) pp. 62-72, Council of Agriculture, Executive Yuan, Taipei, Taiwan.

Janzen, D.H. (1970). Herbivores and the number of tree species in tropical forests. *American Naturalist*, Vol. 104, pp. 501-528, ISSN 0003-0147.

Janzen, D.H. (1971). Seed predation by animals. *Annual Review of Ecology and Systematycs*, Vol. 2, pp. 465-493, ISSN 0066-4162.

Jordan, W.R.; Gilpin, M.E.; Aber, J. (Eds.) (1987) *Restoration Ecology: A Synthetic Approach to Ecological Research*. Cambridge University Press, ISBN 978-0521337281, Cambridge, UK.

Kimmins, J.P.; Blanco, J.A.; Seely, B.; Welham, C. & Scoullar, K. (2010). Forecasting Forest Futures: A Hybrid Modelling Approach to the Assessment of Sustainability of

Forest Ecosystems and their Values. Earthscan Ltd., ISBN 978-1-84407-922-3, London, UK.

Laliberte, A.S.; Ripple, W.J. (2004). Range contractions of North American carnivores and ungulates. *BioScience*, Vol. 54, pp. 123–138, ISSN 0006-3568.

Leck, M.A.; Parker, V.T.; Simpson, R.L. (Eds.) (2008). *Seedling Ecology and Evolution*, Cambridge University Press Cambridge, UK, ISBN: 978-0521873055.

Lewontin, R.C. (1969). Meaning of stability. *Brookhaven Symposia in Biology*, Vol. 22, pp. 13-24, ISSN 0068-2799.

Lin, L.-K.; Bridgman, C.L. (2010). Effect of differential thinning on habitat selection of birds and mammals in a *Cryptomeria japonica* plantations In: *Symposium of thinning effects on biodiversity and ecosystem functions*. Taiwan Forestry Research Institute, Taichung, Taiwan (R.O.C.).

Lu, S.-Y.; Cheng, J.D.; Brooks, K.N. (2001). Managing forests for watershed protection in Taiwan. *Forest Ecology and Management*, Vol. 143, pp. 77-85, ISSN 0378-1127.

Masaki, T.; Osumi, K.; Takahashi, K.; Hoshizaki, K.; Matsune, K.; Suzuki, W. (2007). Effects of microenvironmental heterogeneity on the seed-to-seedling process and tree coexistence in a riparian forest. *Ecological Research*, Vol. 22, pp. 724-734, ISSN 0013-9351.

National Taiwan University Experimental Forest (NTUEF) (2011). *The Experimental Forest College of Bioresources and Agriculture* (Weather Data. Retrieved on 2011/4/8 from http://www.exfo.ntu.edu.tw/cht/05teaching/

Naveh, Z. (1994). From biodiversity to ecodiversity: a landscape-ecology approach to conservation and restoration. *Restoration Ecology*, Vol. 2, pp. 180–189, ISSN 1061-2971.

Nellemann, C.; Corcoran, E. (Eds.) (2010). *Dead Planet, Living Planet — Biodiversity and Ecosystem Restoration for Sustainable Development: A Rapid Response Assessment*. United Nations Environment Programme, ISBN 978-82-7701-083-0, Nairobi, Kenya.

Noss, R.F.; LaRoe, E.T.; Scott. J.M. (1995). *Endangered ecosystems of the United States: a preliminary assessment of loss and degradation*. Biological Report 28, U.S. Department of the Interior, National Biological Service, Washington, D.C., USA.

Odum E.P. 1969. The strategy of ecosystem development. *Science*, Vol. 164, pp. 262-270, ISSN 0036-8075.

Pickett, S.T.A.; McDonnell, M.J.M. (1989). Changing perspectives in community dynamics – atheory of successional forces. *Trends in ecology and evolution*, Vol. 4, pp. 241-245, ISSN 0169-5347.

Pickett, S.T.A.; Ostfield, R.S. (1995). The shifting paradigm in ecology. In: *A New Century for Natural Resources Management*, R.L. Knight, S.F. Bates (Eds.), 261–278. Island Press, ISBN: 978-1559632621, Washington DC, USA.

R Development Core Team, 2010. R: A language and environment for statistical computing. R Foundation for Statistical Computing, Vienna, Austria.

Runte, A. (1997). *National Parks: the American experience* 3rd Ed. University of Nebraska Press, ISBN: 978-0803289635, Lincoln, USA.

Sala O. E. et al. (2000). Global biodiversity scenarios for the year 2100. *Science*, Vol. 287, pp. 1770-1774, ISSN 0036-8075.

Schnurr, J.L.; Canham, C.D.; Ostfeld, R.S.; Inouye, R.S. (2004). Neighborhood analyses of small-mammal dynamics: Impacts on seed predation and seedling establishment. *Ecology*, Vol. 85, pp. 741-755, ISSN 0012-9658.

Schupp, E.W.; Milleron, T.; Russo, S.E. (2002). Dissemination limitation and the origin and maintenance of species-rich tropical forests. In: *Seed Dispersal and Frugivory: Ecology, Evolution and Conservation,* Levey, pp. 19-34. D.J.; Silva, W.R.; Galetti, M. (Eds.). CABI International, Wallingford, UK, ISBN: 978- 0851998060.

Scott J.M.; Davis, F.W.; McGhie, R.G.; Wright, R.G.; Groves, C.; Estes, J. (2001). Nature reserves: do they capture the full range of America's biological diversity? *Ecological applications,* Vol. 11, pp. 999-1007, ISSN 1051-0761.

Smit, C.; Gusberti, M.; Müller-Schärer, H. (2006). Safe for saplings; safe for seeds? *Forest Ecology and Management,* Vol. 237, pp. 471-477, ISSN: 0378-1127.

Soulé, M.E.; Terborgh, J. (1999). Conserving nature at regional and continental scales—a scientific program for North America. *BioScience,* Vol. 49, pp. 809–817, ISSN 0006-3568.

Sun, I.-F. (2010). Effects of thinning on forest regeneration and restoration In: *Symposium of thinning effects on biodiversity and ecosystem functions.* Taiwan Forestry Research Institute, Taichung, Taiwan (R.O.C.).

Tamura, N.; Katsuki, T. (2004). Walnut Seed Dispersal: Mixed Effects of Tree Squirrels and Field Mice with Different Hoarding Ability. In: *Seed Fate: Predation, Dispersal and Seedling Establishment,* Forget, P.M., Lambert, J.E., Hulme, P.E., Vander Wall, S.B. (Eds.), CABI International, Wallingford, UK, ISBN: 978- 0851998060.

Vander Wall, S.B. (1994). Seed fate pathways of antelope bitterbrush: Dispersal by seed-caching yellow pine chipmunks. *Ecology,* Vol. 75, pp. 1911-1926.

Waldman, N. 2001. *They came from the Bronx: how the buffalo were saved from extinction.* Boyds Mill Press, ISBN 978-1563978913. Honesdale, USA.

Wallington, T.J.; Hobbs, R.J.; Moore, S.A. (2005). Implications of current ecological thinking for biodiversity conservation: a review of the salient issues. *Ecology and Society,* Vol. 10, article 15, ISSN 1708-3087.

Wang, D.H. (1997). *Perspectives on forestry resources managemen. Part III. Republic of China, Vol* 2. Asian Productivite Organization, Tokyo, Japan, pp. 153-165.

Wu, J.; Loucks, O.L. (1995). From balance of nature to heirarchical patch dynamics: a paradigm shift in ecology. *The Quarterly Review of Biology,* Vol. 70, pp. 439–466, ISSN 0033-5770.

Zahawi, R.A.; Augspurger, C.K. (2006). Tropical forest restoration: Tree islands as recruitment foci in degraded lands of Honduras. *Ecological Applications,* Vol. 16, pp. 464-478, ISSN 1051-0761.

Zahawi, R.A.; Holl, K.D. (2009). Comparing the Performance of tree stakes and seedlings to restore abandoned tropical pastures. *Restoration Ecology,* Vol. 17, pp. 854-864, ISSN 1061-2971.

Zhang, Z.-B.; Xiao, Z.-S.; Li, H.-J. (2004). Impact of Small Rodents on Tree Seeds in Temperate and Subtropical Forests, China. In: *Seed Fate: Predation, Dispersal and Seedling Establishment,* Forget, P.M., Lambert, J.E., Hulme, P.E., Vander Wall, S.B. (Eds.) CABI International, Wallingford, UK, ISBN: 978- 0851998060.

10

Interactions of Forest Road, Forest Harvesting and Forest Ecosystems

Murat Demir

Istanbul University, Faculty of Forestry,
Department of Forest Construction and Transportation
Turkey

1. Introduction

Forestry developed with the goal, ultimately, of maximizing the long-term economic return from the forest, a goal that has remained virtually unchanged to the present day, despite the growth in understanding of ecosystem function (Farrell et al. 2000). Forest ecosystems supply a wide range of commodities sought by an expanding human population, including structural materials, fuels, and medicines, along with a wide range of critical ecosystem services including nutrient cycling, climate regulation, maintaining water balances and carbon sequestration (Klenner, 2009). The concept of sustainable forest management, which may be defined as the use and regulation of forests and forest areas, at local, national and global levels, in such manner and to such extent as to protect their biological diversification, their productivity, rejuvenation capacity and survival energy as well as their potential to fulfill their ecological, economic and social functions, both at present and in the future, while not causing any harm to other ecosystems, is well recognized by all countries in the world (Demir, 2007).

The use of the forests and other elements in the landscape is driven by our human needs in local, regional and global perspectives. The way we use the forest as a natural resource is determined by a number of factors. These factors, which in their character are social, economic, biological and ecological, can be seen as forces and constraints (Anderssona et al. 2000). Forests, which are renewable natural assets, are formed by gathering of a large number of living and non-living creatures. However, this formation is not a random mass, but whole, a system. When making use of the forest ecosystem for various purposes, care must be taken not to spoil the forest structure. As it is the case in every engineering activity, in carrying out the road planning and construction works, the requirements regarding compatibility with nature and safety and economy must be met. The compatibility with nature, that is, the requirement that the road to be constructed as a result of the works carried out should have the characteristics enabling it to perform its expected functions is thus recognized to be of primary concern. To meet this requirement, first the purpose of construction of planned facility must be precisely defined. Meeting the second requirement regarding safety involves the construction of planned facilities according to relevant standards within the prescribed period to enable them to serve in line with the contemplated purpose (Hasdemir and Demir, 2005).

Developing and maintaining the economic activity that is vital for the quality of modern life would be difficult without roads. Roads are critical component of civilization. Roads provide access for people to study, enjoy, or contemplate natural ecosystems. In fact, the development of human civilization has benefited from transportation systems that evolved from root trails to complex highway systems (Crisholm, 1990; Grübler, 1994). Building and maintaining roads have become controversial, however, because of public concerns about their short and long term effects on the environment and the value that society now places on road less wilderness (Cole and Landres, 1996). Oppositions to road building and pressure to decommission roads in rural landscapes will continue to increase as road less areas decrease in relation to roaded ones. Decisions about road alignment, building, maintenance, or decommissioning are complex because of the many trade off involved (Lugo and Gucinski, 2000).

Traditionally, the planning of rural road network is based on economic and social considerations. In the last years, traffic volumes showed a considerable growth, despite an extension of the road networks. Meanwhile, some harmful effects of these networks and their traffic flows appeared. Traffic unsafety, emissions and noise affect local people, flora and fauna (Jaarsma, 1994; Jaarsma and van Langevelde, 1996; Jaarsma, 1997). Evaluating the ecological effects of roads requires rigorous analysis and an understanding of the ecology of roads, that is, the interplay between all of the living components, the function of roads, and the environmental factors that regulate processes along the road corridor (Forman et.al. 1997).

2. Forest road ecosystems

Forest road can be defined as ecosystems because they occupy ecological space (Hall et.al.1992), have structure, support a specialized biota, exchange matter and energy with other ecosystems, and experience temporal change. Forest road ecosystem are built and maintained by people (Haber, 1990). Forest road ecosystem includes both the paved and unpaved rights of way and adjacent structure, including other infrastructure, ditches, drainage features, and other components that provide the means for vegetation to establish and provide habitat for associated plants and animals (Fig.1). Forest roads are crucial for effective forest management, regardless of its main objectives. Forest maintenance, wood harvesting, game control, recreational activities - all require the accessibility provided by a suitable road network. Forest roads, in former times planned and constructed for the needs of wood harvesting and transport, are the key factor for recreational access to and activities in forest environments. Leisure activities in urban forests include hiking, biking, horse-riding, jogging and inline-skating (Janowsky and Becker, 2003). The opening of forests to exploitation is usually realized by means of well-planned forest road networks. The parameters and location requirements of forest road networks vary depending on variations in landscape conditions and according to the technology used and administrative activities. These requirements and planning approaches may be related to economic, ecological and management characteristics (Potocnik, 1996). The road network is a form of land use, which planning strongly depends on for other land uses. These decide the desirable density of the network (mesh size) and the capacity of the road links (pavement width). Simultaneously, all human land uses are strongly dependent on this network. Economic developments, and efficient use of land resources and, as asocial aim, accessibility of rural areas, need a well-developed road network. Most regions in industrialized countries have, from a quantitative point of view, a sufficient rural road network (road density, mesh size) (Jaarsma, 1997). The model in Fig.1 highlights the six-way flow of materials, energy, and organisms along the road corridor; vegetation zone; the interaction with the human economy and human activity; external forces that converge on the road corridor. The structure and functioning of a road varies according to its design, use, type of surface, and location (Lugo and Gucinski, 2000). Forest roads are also corridors that can connect contrasting ecosystem types. Since forest roads provide a fairly homogeneous condition through the length of the corridor, they provide opportunity for organisms and materials to move along the corridor, thus increasing the connectivity among those ecosystems that interface with the forest road (Lugo and Gucinski, 2000; Merriam, 1984).

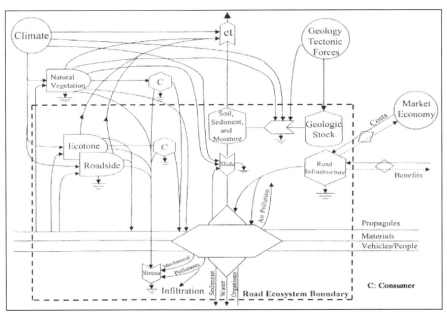

Fig. 1. Model of road ecosystem (Lugo and Gucinski, 2000).

3. Forest road impacts on forest ecosystems

Forest roads create many collateral problems adversely affecting the conservation of ecosystems and landscape integrity (Smith and Wass, 1980; Thompson, 1991). Like all ecosystems, roads are constantly changing as do the relations between the road ecosystem and adjacent ecosystems. The major phases of road development are building, operating, maintaining, and abandonment. Forest road building is often the most environmental traumatic to adjacent ecosystems because earth movement and other activities can disturb whole watersheds. Changes -mechanical, geochemical, hydrologic, biotic, and so on- to the immediate land area and any adjacent upstream and downstream ecosystem affected by building activities can be predicted. During this phase, the road is primarily a disturbance and agent of change (Lugo and Gucinski, 2000).

3.1 Wildlife

Forest roads segments can be part of forest road networks criss-crossing the landscape. A forest road network system has environmental effects and ecosystem properties that appear to transcend those of its individual segments. For example, some wildlife species, such as bear, wolf, or mountain lion, respond more to forest road density than to individual road segments (Forman et.al.1997). Similarly, forest road networks are more relevant to issues of forest fragmentation or to hydrological effects than are isolated forest road segments (Jones, 1998). Roads create barriers and additional that in turn causes fragmentation of the landscape and its populations (Jaarsma and Willems, 2002).

3.2 Vegetation

Maintaining forest roads, particularly if improperly done, act as periodic disturbances to both the road biota and landscape as a whole. Maintenance activities can approximate building activities in the amount and extent of disturbance, and they can prolong environmental effects to adjacent ecosystems. Not maintaining forest roads, however, can hinder the primary function of the road and also significantly affect the environment. For example, poorly maintained drainage systems in wet montane roads can induce mass-wasting events large enough to destroy the road and affect adjacent forest and aquatic systems. Such events sometimes exceed those observed during forest road building (Larsen and Parks, 1997).

Forest road use itself affects the landscape, for example through spills of toxic substances, pollution, dust, or effects on plants and animals by the presence of people. Forest roads as part of long-range transportation networks are likely to introduce alien species. The type and intensity of use are associated with particular environmental effects. For example, logging truck traffic is known to facilitate the transport of fungal root diseases and heavy vehicular traffic increases the risk of dispersing roadside weeds and different types and intensities of pollution (air, soil, or water) or chemical spills (Lugo and Gucinski, 2000). Furthermore, a dense forest road network for example, has a more likely effect on fragmentation than a low density network (Forman et.al.1997; Forman and Hersperger, 1996). High density road networks are more likely to affect hydrological parameters than low density ones. However, forest road density is less important to fragmentation of forest where topography dominates the structure and size of vegetation stands (Miller et al. 1996).

Road abandonment allows successional processes to recapture the road corridor. The speed and direction of succession after a road is abandoned depends on the type of road, landscape, and environment. Some road segments may be overgrown with vegetation quickly, but the pavement can arrest succession in others. Rehabilitation techniques are usually needed to accelerate succession to reach management goals after abandonment (Luce, 1997). With time, the road ecosystem ages and matures. As it does, and regardless of disturbances, segments of the road can adjust to conditions, blend with the landscape, and reach a new ecological and hydrological state (Olander et al.1998).

Finally, like other ecosystems, roads produce long term legacies on the landscape (Hutchinson, 1973). The environmental gradients believed to be most important in describing the ecological space in which roads function as ecosystem are shown in Fig 2.

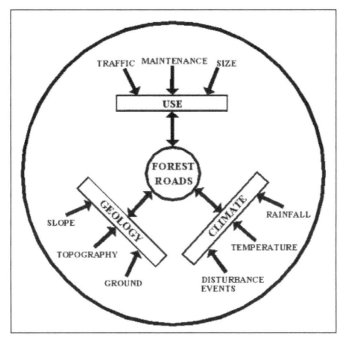

Fig. 2. Parameters of forest road ecosystems (Demir, 2007).

Forest road effects on surrounding environments and road function as ecosystem are mainly influenced by climate, geologic conditions, and uses or functions of the road. Climatic conditions are mainly the precipitation and temperature regime, and the frequency and intensity of climatic disturbance events. Geologic side is the type of substrate such as volcanic, limestone, or alluvial and the topography (Fig.3).

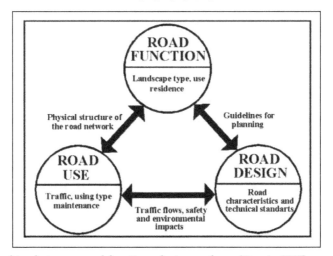

Fig. 3. Relationships between road function, design and use (Demir, 2007).

3.3 Soil

Forest roads produce the highest production of sediment yield to streams from forest lands (Binkley and Brown, 1993; McClelland at al. 1999; Reid and Dune 1984). Road construction removes the forest vegetation, disturbs forest floor, and damages soil structure, which dramatically increases the sediment yield (Grace, 2002; Megahan 1974). Sediment delivered to streams from road sections leads to number of dramatic effects on water quality (Megahan 1974). On unsealed roads, road surface erosion is generally the dominant source of sediment (Ramos Scharrón and MacDonald, 2007). Road surface and ditch areas still continue to deliver sediment to the streams as long as the road is used. The production of sediment produced from road surface highly depends on traffic density, road surface type, road dimensions, and road gradient. The ditches receive the sediment yield from the cut-slope areas, depending on road section length, ground slope, and vegetation and rock cover density. In slope roads with ditch keep the runoff water away from the fill-slope which may cause much smaller sediment yield than road surface and cut-slope areas (Akay et al. 2008). Sediment can be eroded from all road features. The factors affecting surface erosion from roads include rainfall intensity and duration, snowfall, the characteristics of surface materials, the hydraulic characteristics of the road surface, road slope, traffic, construction and maintenance, and the contributing road area (MacDonald and Coe, 2008). Previous studies indicated that sediment production rates from unpaved road surfaces were several orders of magnitude higher than undisturbed hill slopes (MacDonald et al. 1997; MacDonald 2001). Changes in vegetation cover also might have played a role in declining sediment production (Ramos Scharrón, 2010). Ditches are potentially important sediment sources particularly when erosion is caused by scour from road runoff. Ditches may deepen or widen, or be filled with deposited sediment during rainfall events (Croke et al. 2006; Lane and Sheridan 2002). Several studies found the sediment production between 0.01 and 105 kg m^{-2} yr^{-1} depending on the several observation time, different cut slopes characteristics, cover density and parent material (Table 1).

Sediment production from unpaved roads was significantly related to total rainfall, road segment slope and graded or no graded (Ramos Scharrón and MacDonald 2006). Ramos Scharrón and MacDonald (2006) obtained a different result measured sediment production rates for graded roads ranged from 5.7 to 580 Mg ha^{-1} yr^{-1} for roads with slopes of 2% and 21%, respectively. The sediment production rate for ungraded roads was about 40% lower than for comparable graded roads. Abandoned road segments had a mean erosion rate of 12 Mg ha^{-1} yr^{-1}. Reid and Dunne (1984) report that a heavily used road contributes 130 times as much sediment as an abandoned road and a paved road, along which cut slopes and ditches are the only sources of sediment, yields less than 1% as much sediment as a heavily used road with a gravel surface.

Kartaloğlu (2011) has been determined that through sediment traps established on unpaved (UPFR) and paved forest road (PFR) ditch and in an undisturbed (UA) area at Belgrad forest-Istanbul, Turkey between date of November 2009 to October 2010 (12 months) (Fig.4 and 5). In this research reported that significantly differences were found on sediment productions among experiment sites (UPFR, PFR and UA). Annual sediment production was 0.654 t ha^{-1} yr^{-1} on UPFR, 0.334 t ha^{-1} yr^{-1} on PFR and 0.056 t ha^{-1} yr^{-1} on UA. Total sediment production of UPFR was 1.96 times higher than to PFR and 11.68 times higher than UA. Kartaloğlu (2011) stated that monthly sediment production on UPFR significantly higher than the PFR an UA for each month and it clearly shows that stabilizing cover on forest road led to less sediment production and more soil protection. Significantly differences on monthly sediment production were determined among experiment sites in all observation period (12 months).

Location	Cutslope description	Reported sediment production rate	Normalized sediment production (kg m^{-2} yr^{-1})	Reference
Georgia, USA	Unvegetated	102–230 Mg ha^{-1} yr^{-1}	5,1-11	Diseker and Richardson (1962)
Oregon, USA	6–7 yr old cutslopes	153 Mg ha^{-1} yr^{-1}	15	Wilson (1963)
	New cutslopes	370 Mg ha^{-1} yr^{-1}	37	Wilson (1963)
Oregon, USA	5 yr old cutslopes	0.5 cm yr^{-1}	7,5	Dyrness (1970 and 1975)
	1 yr old cutslopes	0.7 cm yr^{-1}	10	Dyrness (1970 and 1975)
Idaho, USA	45 yr old cutslopes, soil	0.01 m^3 m^{-2} yr^{-1}	15	Megahan (1980)
	45 yr old cutslopes, granite	0.011 m^3 m^{-2} yr^{-1}	16	Megahan (1980)
Washington, USA	55-70 degrees	16.5 mm yr^{-1}	25	Reid (1981)
Papua New Guinea	NA	70 mm yr^{-1}	105	Blong and Humphyreys (1982)
Idaho, USA	NA	11 mm yr^{-1}	16	Megahan et al. (1983)
New South Wales, Australia	NA	2.4–3.9 mm yr^{-1}	3,6-5,8	Riley (1988)
New Zeland	Unvegetated, granite	NA	5,2-15	Fahey and Coker (1989 ve 1992) Smith and Fenton (1993)
Idaho, USA	Cover density 0.1–89%, 55–104% gradient	0.1–248 Mg ha^{-1} yr^{-1}	0,01-25	Megahan et al. (2001)
St. John, USA	Unvegetated, 2–5 m high	NA	2-17	Ramos-Scharrón and MacDonald (2007)

Table 1. Estimated cutslope contribution to sediment yields at the road segment (Ramos-Scharrón ve MacDonald, 2007).

Fig. 4. Sediment traps on forest road and collected sediments on sediment traps at Belgrad Forest-Istanbul, Turkey (Kartaloğlu, 2011).

Fig. 5. Collected sediments on sediment traps at Belgrad Forest-Istanbul, Turkey (Kartaloğlu, 2011).

4. Forest harvesting impacts on forest ecosystems

4.1 Vegetation

Production work being carried out in the forest have many negative impact on the forest ecosystem is well known. Skidding or yarding on terrain requires the construction of relatively dense network of forest roads including skid roads, haul roads and landings (Demir et al. 2007a; Demir et al. 2008; Ketcheson et al., 1999; Makineci et al. 2007a; Makineci et al. 2007b; Swift, 1988). It has also been determined that the forest harvesting, timber production and timber skidding negatively affect the amount and variety of forest floor and herbaceous understory as well as youth development and living conditions of the soil organisms (Arocena, 2000; Bengtsson et al., 1998; Gilliam, 2002; Godefroid and Koedam, 2004; Johnston and Johnston, 2004; Marshall, 2000; Messina et al., 1997; Wang, 1997; Williamson and Neilsen, 2003).

Several studies have documented varying degrees of reduced tree growth on non-rehabilitated skid roads, ranging from 15% to 59% averaged over the trail, when compared to trees grown on undisturbed soil in the same cut block (Smith and Wass, 1980; Thompson, 1991). The wide variation in previous findings may be due to species and site-specific responses to soil disturbance, variations in the severity of disturbance, or other growth limiting factors that may magnify or alleviate the impacts of soil disturbance (Lewis, 1991; Dykstra and Curran, 2000). Sat Gungor et al. (2008) stated that ground based skidding destroyed the soil and ecosystem and the timber skidding limits recovery and growth of plant cover on skid roads. However, some herbaceous plant species show healthy habitat, and they can revegetate and survive after the extreme degradation in study area. Earlier results clearly show that skidding has particularly negative impacts on herbaceous cover, forest floor and soil. These effects of skidding on skid road have been demonstrated to have detrimental impacts on native flora and herbaceous plant establishing and maintaining were limited. Similarly, Mariani et al. (2006) reported that organic layer removal reduced abundance of herbs and shrubs.

Soil compaction can also severely reduce plant growth by restricting root growth may be due to oxygen stress and lower the percentage of water and air space in the soil (Berzegar et al. 2006). Also, Kozlowski (1999) mentioned a reduced total photosynthesis when soils become increasingly compacted, as a result of smaller leaf areas.

Yilmaz et al. (2010) reported that significant differences on widths of annual rings, dbh growth and increment values are found same tree species when in an undisturbed area and adjacent to a road under the same microclimate and site conditions, supporting the notion that long term timber skidding reduces the annual ring width, dbh growth and increment of nearby trees. The trees growing there commonly display better growth and diameter increment than the trees growing on the skid road. Tree growth and increment on the undisturbed area was found to be about 60% greater than the skid road.

4.2 Soil

Logging operations can cause significant and wide spread soil disturbance, including removal, mixing and compaction of the various soil layers (Demir et al. 2007b ; Demir et al. 2010; ; Makineci et al. 2007c; Makineci et al. 2007d; Makineci et al. 2008) . Ground based skidding, timber harvesting and logging operations in forest ecosystems cause the reduction and redistribution of organic matter, changes in plant cover, organic layer and soil properties, and modification of microclimate (Buckley et al., 2003). Timber harvesting can adversely affect both soil physical properties and soil nutrient levels. Logging can cause diminished growth of subsequent tree rotations, significant increase in runoff and sediment loads (Laffan et al., 2001). Erosion of organic and nutrient rich surface soil and compaction decrease forest productivity (Pritchett and Fisher, 1987) and the transport of sediment to streams and subsequent sedimentation lead to loss of stream habitat and altered stream hydrology. The soil micro flora and fauna complement each other in the commination of litter, mineralization of essential plant nutrients and conservation of these nutrients within the soil system. Harvesting directly affects these processes through the reduction and redistribution of organic matter, compaction, changes in plant cover, and modification of microclimate (Marshall, 2000).

The extent of severe disturbance from ground based timber harvesting systems varies due to slope and terrain, timber harvesting machines, methods of designating skid roads and harvesting season. Forest harvesting and ground based skidding may result in soil compaction and other soil structural changes, influencing soil water retention, and reducing soil aeration, drainage and root penetration (Froehlich et al., 1986). Soil damage on forest roads, skid roads and landings includes the removal of the organic layer and topsoil, soil compaction and erosion of the exposed soil. The soil damage affects hill slope infiltration and surface and subsurface flows (Binkley, 1986).

5. Conclusion

Sustainable forest management requires an adequate understanding not only of the forest ecosystem but also of the interaction between different disciplines. The planning process and finding the appropriate balance of different interests for the use of our forest and landscape resources will be vital to the achievement of the goal of sustainable forest management. The main purpose in the planning of forest roads is that, when then unfavorable effects of planned and constructed forest roads on the forest ecosystem are compared with the benefit to be derived from the roads constructed as a result of planning within the concept of sustainable forest management, such benefit must be within the acceptable limits. In this context, it has become evident that the density and road space criteria presently employed to provide each piece of a forest area with a systematic forest road network in order to enable it to fulfill its planned functions shall not be equally applicable to every area. In summary, it is recommended that the purposes of management of forests should be put forth in detail, and the road density and road space values to enable the realization of these purposes should be determined separately (Demir, 2007).

Roads affect both the biotic and the abiotic components of landscapes by changing the dynamics of populations of plants and animals, altering flows of materials in the landscape, introducing exotic elements, and changing levels of available resources, such as water, light and nutrients (Coffin, 2007).

In the recent years, an ever-increasing trend has been observed in public consciousness regarding environment. This has led to the creation of a medium of constant controversy between the foresters and environmentalists. The main issue of controversy is centered on the argument that the construction of forest roads destroys the natural environment to a great extent, causes soil erosion, completely destroys the habitat and impairs the integrity of landscape. As stated also in the declaration issued by UNCED (United Nations Conference on Environment and Development), the utilization of nature's renewable resources is a key component of development based on environment. It is, however, an essential requirement that access to relevant areas be provided for utilization of resources concerned. Therefore the construction of forest roads can in no case be abandoned. It follows, however, that relevant forestry organizations are obliged to figure out new ways that could be approved by the public and would cause no harm to the environment (Heinimann, 1996).

Coffin (2007) stated that roads have many direct ecological effects on adjacent aquatic and terrestrial systems, as network structures, they also have far reaching, cumulative effects on landscapes (Riitters and Wickham, 2003). Some major effects to landscapes that directly relate to roads include the loss of habitat through the transformation of existing land covers to roads and road-induced land use and land cover change (Angelsen and Kaimowitz, 1999) and reduced habitat quality by fragmentation and the loss of connectivity (Theobald et al., 1997; Carr et al., 2002). Together they point to the larger issue of the synergistic effects of roads and road networks on ecosystems at broader scales (Forman et al., 2003).

In tropical forested areas, econometric models of land use and land cover change have revealed important relationships between biophysical and economic variables relative to roads. In rural areas, particularly in developing countries, the presence of roads has been most strongly correlated with processes of land cover change by facilitating deforestation (Chomitz and Gray, 1996; Angelsen and Kaimowitz, 1999; Lambin et al., 2001; Mertens and Lambin, 1997). There are a variety of different effects of roads in boreal, temperate, Mediterranean, tropical and sub-tropical or alpine forest ecosystems.

In this study, tried to give interactions of forest roads, forest harvesting and forest ecosystems. Furthermore, the study results were tried to give related to the subject in the world and Turkey. Despite the negative impacts of forest roads, forest harvesting and wood transports the realization of sustainable forestry on these structures, and studies are needed. Forest roads construction, forest harvesting and wood transports eliminating the negative effects to the realization of appropriate technique.

6. References

Akay, A.E.; Erdas, O.; Reis, M. & Yuksel, A. (2008). Estimating sediment yield from a forest road network by using a sediment prediction model and GIS techniques. *Building and Environment*, Vol.43, pp. 687-695, ISSN 0360-1323

Anderssona, F.O.; Fegerb, K.H.; Hüttlc, R.F.; Krauchid, N.; Mattssone, L.; Sallnasf, O. & Sjöbergg, K. (2000). Forest ecosystem research - priorities for Europe. *Forest Ecology and Management*, Vol.132, pp. 111-119, ISSN 0378-1127

Angelsen, A. & Kaimowitz, D. (1999). Rethinking the causes of deforestation: lessons from economic models. *The World Bank Research Observer*, Vol. 14, pp. 73–98.

Arocena, J.M. (2000). Cations in solution from forest soils subjected to forest floor removal and compaction treatments. *Forest Ecology and Management*, Vol.133, pp. 71-80, ISSN 0378-1127

Barzegar, A.R.; Nadian, H.; Heidari, F.; Herbert, S.J. & Hashemi, A.M. (2006). Interaction of soil compaction, phosphorus and zinc on clover growth and accumulation of phosphorus. *Soil and Tillage Research*, Vol.87, pp. 155–62, ISSN 0167-1987

Bengtsson, J.; Lundkvist, H.; Saetre, P.; Sohlenius B. & Solbreck, B. (1998). Effects of organic matter removal on the soil food web: Forestry practices meet ecological theory. *Applied Soil Ecology*, Vol.9, pp. 137-143, ISSN 0929-1393

Binkley, D., 1986. *Forest Nutrition Management*, Wiley, ISBN 978-0-471-81883-0, New York.

Binkley, D. & Brown, T.C. (1993). Forest practices as nonpoint sources of pollution in North America. *Water Research Bulletin* Vol.29, pp. 729-740, ISSN 0043-1370

Blong, R.J. & Humphreys, G.S. (1982). Erosion of road batters in Chim Shale, Papua New Guinea. *Civil Engineering Transactions, Institution Engineers Australia*, Vol.CE24 (1), pp. 62–68, ISSN 0020-3297

Buckley,D.S.; Crow, T.R.; Nauertz E.A. & Schulz, K.E. (2003). Influence of skid trails and haul roads on understory plant richness and composition in managed forest landscapes in upper michigan, USA. *Forest Ecology and Management*, Vol.175, pp. 509-520, ISSN 0378-1127

Carr, L.W.; Fahrig, L.; Pope, S.E. (2002). Impacts of landscape transformation by roads. In: Gutzwiller, K.J. (Ed.), Applying Landscape Ecology in Biological Conservation. Springer-Verlag, New York, pp. 225–243.

Chomitz, K.M., Gray, D.A., 1996. Roads, land use and deforestation: a spatial model applied to belize. *The World Bank Economic Review*, Vol. 10, pp. 487–512.

Cole, D.N. & Landres, P.B. (1996). Threats to wilderness ecosystems: impacts and research needs. *Ecological Applications*, Vol.6, No.1, pp. 168-184, ISSN 1051-0761

Coffin, A.W. (2007). From roadkill to road ecology: A review of the ecological effects of roads. *Journal of Transport Geography*, Vol. 15, pp. 396-406, ISSN 0966-6923

Crisholm, M. (1990). The increasing separation of production and consumption. In: *The earth as transformed by human action*, B. L. Turner II, William C. Clark, Robert W. Kates, John F. Richards, Jessica T. Mathews, William B. Meyer, (Ed.), pp. 87-101, Cambridge University Press, ISBN 978-0-521-36357-0, Cambridge

Croke, J.; Mockler, S.; Hairsine, P. & Fogarty, P. (2006). Relative contributions of runoff and sediment from sources within a road prism and implications for total sediment delivery. *Earth Surface Processes and Landforms*, Vol.31, No.4, pp. 457–468, ISSN 1096-9837

Demir M. & Hasdemir, M., (2005). Functional planning criterion of forest road network systems according to recent forestry development and suggestion in Turkey. *American Journal of Environmental Science*, Vol. 1, No.1, pp. 22-28, ISSN 1553-345X

Demir, M. (2007). Impacts, management and functional planning criterion of forest road network system in Turkey. *Transportation Research Part A: Policy and Practice*, Vol.41, No.1, pp. 56-68, ISSN 0965-8564

Demir, M., Makineci E. & Yilmaz, E. (2007a): Harvesting impacts on herbaceous understory, forest floor and top soil properties on skid road in a beech (*Fagus orientalis* Lipsky.) stand. *Journal of Environmental Biology*, Vol. 28, pp. 427-432, ISSN 0254-8704

Demir, M., Makineci E. & Yilmaz, E. (2007b). Investigation of timber harvesting impacts on herbaceous cover, forest floor and surface soil properties on skid road in an oak (*Quercus petrea* L.) stand. *Building and Environment*, Vol. 42, pp. 1194-1199, ISSN 0360-1323

Demir, M., Makineci, E.; Şat Güngör, B. (2008). Plant Species Recovery on a Compacted Skid Road. *Sensors*, Vol. 8, No.5, pp. 3123-3133, ISSN 1424-8220

Demir, M.; Makineci, E.; Comez, A. & Yilmaz, E., (2010). Impacts of Repeated Timber Skidding on the Chemical Properties of Topsoil, Herbaceous Cover and Forest Floor in an Eastern Beech (*Fagus orientalis* Lipsky) Stand. *Journal of Environmental Biology*, Vol. 31, No. 4, pp. 477-482, ISSN 0254-8704

Diseker, E.G. & Richardson, E.C., (1962). Erosion rates and control methods on highway cuts. *Transactions of the American Society Agricultural Engineers*, Vol. 5, pp. 153–155, ISSN 0001-2351

Dykstra, P.R. & Curran, M.P. (2000). Tree growth on rehabilitated skid roads in south British Columbia. *Forest Ecology and Management*, Vol. 133, pp. 145–156, ISSN 0378-1127

Dyrness, C.T. (1970). *Stabilization of newly constructed road back-slopes by mulch and grass-legume treatments*. USDA Forest Service Research Note PNW-123, Corvallis, OR.

Dyrness, C.T. (1975). Grass-legume mixtures for erosion control along forest roads in western Oregon. *Journal of Soil and Water Conservation*, Vol. 30, pp. 169–173, ISSN 0022-4561

Fahey, B.D. & Coker, R.J. (1989). Forest road erosion in the granite terrain of southwest Nelson, New Zealand. *Journal of Hydrology (N.Z.)*, Vol.28, No.2, pp. 123–141, ISSN 0022-1694

Fahey, B.D. & Coker, R.J., 1992. Sediment production from forest roads in Queen Charlotte Forest and potential impact on marine water quality, Marlborough Sounds. *New Zealand Journal of Marine and Freshwater Research*, Vol. 26, pp. 187–195, ISSN 0028-8330

Farrell, E.D.; Führer, E.; Ryan, D.; Andersson, F.; Hüttl, R. & Piussi, P., (2000). European forest ecosystems: building the future on the legacy of the past. *Forest Ecology and Management* Vol. 132, pp. 5-20, ISSN 0378-1127

Forman, R.T.T. & Hersperger, A.M., (1996). *Road ecology and road density in different landscapes, with international planning and mitigation solutions*. Florida Department of Transportation, pp.1-22, Tallahassee.

Forman, R.T.T.; Friedman, D.S.; Fitzhenry, D.; Martin, J.D.; Chen, A.S. & Alexander, L.E. (1997). *Ecological effects of roads: Toward three summary indices and an overview for North America, habitat fragmentation and infrastructure*. Ministry of Transport, Public Works and Water Management, Delf, pp. 40-54.

Forman, R.T.T.; Sperling, D.; Bissonette, J.A.; Clevenger, A.P.; Cutshall, C.D.; Dale, V.H.; Fahrig, L.; France, R.; Goldman, C.R.; Heanue, K.; Jones, J.A.; Swanson, F.J.; Turrentine, T.; Winter, T.C. (2003). Road Ecology: Science and Solutions. Island Press, Washington.

Froehlich, H.A.; Miles D.W.R. & Robbins, R.W. (1986). Soil bulk density recovery on compacted skid trails in central Idaho. *Soil Science Society of America Journals*, Vol.49, pp. 1015-1017, ISSN 0361-5995

Gilliam, F.S. (2002). Effects of harvesting on herbaceous layer diversity of a central appalachian hardwood forest in west Virginia, USA. *Forest Ecology and Management* Vol. 155, pp. 33-43, ISSN 0378-1127

Godefroid, S. & Koedam, N., (2004). The impact of forest paths upon adjacent vegetation: Effects of the paths surfacing material on the species composition and soil compaction. *Biological Conservation*, Vol. 119, pp. 405-419, ISSN 0006-3207

Grace, J.M. (2002). Control of sediment export from the forest road prism. *Transactions of the American Society Agricultural Engineers*, Vol. 45, No. 4, pp. 1-6, ISSN 0001-2351

Grübler, A., (1994). *Technology. Changes in land use and land cover: A global perspective.* Cambridge University Press, with Clark University, ISBN 978-0521470858, Cambridge

Haber, W., (1990). Using landscape ecology in planning and management. In: *Changing Landscapes: An Ecological Perspective.* Zonneveld, I. S., Forman, R.T.T., (Ed.), pp.217-232, Springer, ISBN 3-540-97102-5, Newyork

Hall, C.A.S.; Stanford, J.A. & Hauer, R., (1992). The distribution and abundance of organisms as consequence of energy balances along multiple environmental gradients. *Oikos,* Vol. 65, pp. 377-390, ISSN 0030-1299

Heinimann, H.R., (1996). Openning-up planning taking into account environmental and social integrety. *The seminar on environmentally sound forest roads and wood transport proceedings,* Sinaia, Romania, June 1996

Hutchinson, G.E. (1973). Eutrophication. *American Scientist,* Vol. 61, pp. 269-279, ISSN 0003-0996

Jaarsma, F.C. (1994). Rural low-traffic roads (LTRs): The challenge for improvement of traffic safety for all road users. *27ᵗʰ ISATA, dedicated Conference on Road and Vehicle Safety,* Aachen, Germany,

Jaarsma, F.C. & van Langevelde, F. (1996). The motor vehicle and the environment: Balancing between accessibility and habitat fragmentation. *29ᵗʰ ISATA, dedicated Conference on the Motor Vehicle and the Environment,* Florence, Italy

Jaarsma, F.C., (1997). Approaches for the planning of rural road network according to sustainable land use planning. *Landscape and Urban Planning,* Vol.39, pp. 47-54, ISSN 0169-2046

Jaarsma, F.C. & Willems, G.P.A. (2002). Reducing habitat fragmentation by minor rural roads through traffic calming. *Landscape and Urban Planning,* Vol. 58, pp. 125-135, ISSN 0169-2046

Janowsky, D. & Becker, G. (2003). Characteristics and needs of different user groups in the urban forest of Stuttgart. *Journal of Nature Conservation,* Vol. 11, pp. 251-259, ISSN 1617-1381

Johnston, F.M. & Johnston, S.W. (2004). Impacts of road disturbance on soil properties and exotic plant occurrence in subalpine areas of Australian Alps. *Arctic Antarctic and Alpine Research,* Vol. 36, No.2, 201-207, 1523-0430

Jones, J.A., (1998). Roads and their major ecological effects. Effects of forest roads on water and sediment routing. *Ecological Society of America Annual Meeting Abstracts,* Vol. 29, pp.

Kartaloğlu, M. (2011). Temporal variation of sedimentation on paved and unpaved forest roads. Master of Science Thesis (Advisory of Dr.Murat Demir and Dr.Ender Makineci), Istanbul University, Science Institute, Istanbul.

Ketcheson, G.L., Megahan W.F. & King J.G., (1999). R1-R4 and Boised sediment prediction model tests using forest roads in granitics. *J. Am. Wat. Resour. Ass.,* 35(1), 83-98.

Klenner, W., Arsenault, A., Brockerhoff, E.G. & Vyse, A. (2009). Biodiversity in forest ecosystems and landscapes: A conference to discuss future directions in biodiversity management for sustainable forestry. *Forest Ecology and Management* Vol. 258S, pp. S1–S4, ISSN 0378-1127

Kozlowski, T.T. (1999). Soil compaction and growth of woody plants. *Scandinavian Journal of Forest Research,* Vol. 14, pp. 596–619, ISSN 0282-7581

Laffan, M., Jordan, G. & Duhig, N. (2001). Impacts on soils from cable logging steep slopes in Northeastern Tasmania, Australia. *Forest Ecology and Management,* Vol. 144, pp. 91-99, ISSN 0378-1127

Lambin, E.F.; Turner, B.L.; Geist, H.J.; Agbola, S.B.; Angelsen, A.; Bruce, J.W.; Coomes, O.T.; Dirzo, R.; Fischer, G.; Folke, C. (2001). The causes of land-use and land-cover change: moving beyond the myths. *Global Environmental Change,* Vol. 11, pp. 261–269.

Lane, P.N.J. & Sheridan G.J. (2002). Impact of an unsealed forest road stream crossing: water quality and sediment sources. *Hydrological Processes,* Vol. 16, pp. 2599–2612, ISSN 1099-1085

Larsen, M.C. & Parks, J.E. (1997). How wide is a road? The association of roads and mass-wasting in a forested montane environment. *Earth Surface Processes and Landforms,* Vol. 22, pp. 835-848, ISSN 1096-9837

Lewis, T. (1991). Developing *Timber Harvesting Prescriptions to Minimize Site Degradation.* B.C. Ministry for Land Management, Report No. 62, Vancouver.

Luce, C.H. (1997). Effectiveness of road ripping in restoring infiltration capacity of roads. *Restoration Ecology,* Vol. 5, pp. 265-270, ISSN 1061-2971

Lugo, A.E. & Gucinski, H. (2000). Function, effects, and management of forest roads. *Forest Ecology and Management,* Vol. 133, pp. 249-262, ISSN 0378-1127

McClelland, D.E.; Foltz, R.B.; Falter, C.M.; Wilson, W.D.; Cundy, T. & Schuster, R.L. (1999). Relative effects on a low-volume road system of landslides resulting from episodic storms in Northern Idaho. In: *Proceedings of the seventh international conference on low-volume roads,* vol. 2 (1999), pp.235-243, Washington, DC

MacDonald, L.H.; Anderson, D.M. & Dietrich, W.E. (1997). Paradise threatened: Land use and erosion on St John, US Virgin Islands. *Environmental Management,* Vol. 21, pp. 851–863, ISSN 0364-152X

MacDonald, L.H.; Sampson, R.W. & Anderson, D.M., (2001). Runoff and road erosion at the plot and road segment scales, St John, US Virgin Islands. *Earth Surface Processes and Landforms,* Vol. 26, pp. 251-272, ISSN 1096-9837

MacDonald, L.H. & Coe, D.B.R. (2008). Road sediment production and delivery: processes and management. In: *Proceedings of the First World Landslide Forum, International Programme on Landslides and International Strategy for Disaster Reduction.* United Nations University, pp. 381–384, Tokyo.

Makineci, E.; M. Demir & Yilmaz, E. (2007a). Long term harvesting effects on skid road in a fir (*Abies bornmulleriana* Mattf.) plantation forest. *Building and Environment,* Vol. 42, pp. 1538-1543, ISSN 0360-1323

Makineci, E.; Demir, M.; Comez A. & Yilmaz, E. (2007b). Effects of timber skidding on chemical characteristics of herbaceous cover, forest floor and topsoil on skidroad in an oak (*Quercus petrea* L.) forest. *Journal of Terramechanics,* Vol. 44, No.6, pp. 423-428, ISSN 0022-4898

Makineci, E.; Demir, M.; Comez, A. & Yilmaz, E. (2007c). Chemical characteristics of the surface soil, herbaceous cover and organic layer of a compacted skid road in a fir

(*Abies bornmulleriana* Mattf.) forest. *Transportation Research Part D: Transport and Environment*, Vol. 12, No. 7, pp. 453-459, ISSN 1361-9209

Makineci, E.; Demir, M. & Yılmaz, E. (2007). Ecological effects of timber harvesting and skidding works on forest ecosystem (Belgrad Forest case study). *International Symposium on Bottlenecks, Solutions, and Priorities in the Context of Functions of Forest Resources, Proceedings of Poster Presentation*, ISBN 978-975-9060-45-9, Page: 131-142, October 17-19th, 2007, Istanbul University, Faculty of Forestry, Istanbul, Turkey.

Makineci, E.; Şat Güngör, B. & Demir, M. (2008). Survived herbaceous plant species on compacted skid road in a fir (*Abies bornmulleriana* Mattf.) forest - A note. *Transportation Research Part D: Transport and Environment*. Vol. 13, No. 3, pp. 187-192, ISSN 1361-9209

Mariani, L.; Chang, S.X. & Kabzems, R. (2006). Effects of tree harvesting, forest floor removal, and compaction on soil microbial biomass, microbial respiration, and N availability in boreal aspen forest in British Colombia. *Soil Biology and Biochemistry*, Vol. 38, pp. 1734–44, ISSN 0038-0717

Marshall, V.G. (2000). Impacts of forest harvesting on biological processes in northern forest soils. *Forest Ecology and Management* Vol. 133, pp. 43-60, ISSN 0378-1127

Megahan, W.F. (1974). *Erosion over time: a model*. US Department of Agriculture Forest Service, Intermountain Res Stn, Ogden, Utah Res Paper INT-156, 14 p.

Megahan, W.F. (1980). Erosion from roadcuts in granitic slopes of the Idaho batholith. *Proceedings Cordilleran section of the Geological Society of America, 76th Annual Meeting*, Oregon State University, Corvallis, OR, p. 120.

Megahan, W.F.; Seyedbagheri, K.A. & Dodson, P.C. (1983). Long term erosion on granitic roadcuts based on exposed tree roots. *Earth Surface Processes and Landforms*, Vol. 8, pp. 19–28, ISSN 1096-9837

Megahan, W.F.; Wilson, M. & Monsen, S.B. (2001). Sediment production from granitic cutslopes on forest roads in Idaho, USA. *Earth Surface Processes and Landforms*, Vol. 26, pp. 153–163, ISSN 1096-9837

Merriam, G. (1984). Connectivity: A fundamental ecological characteristic of landscape pattern. *Proceedings of the First International Seminar on Methodology in Landscape Ecological Research and Planning*, Roskilde University Center, pp. 5-15, Denmark

Mertens, B. & Lambin, E.F. (1997). Spatial modelling of deforestation in southern Cameroon: spatial disaggregation of diverse deforestation processes. *Applied Geography*, Vol. 17, pp. 143–162.

Messina, M.G.; Schoenholtz, S.H.; Lowe, M.W.; Wang, Z.; Gunter D.K. & Londo, A.J. (1997). Initial responses of woody vegetation, water quality and soils to harvesting intensity in a texas bottomland hardwood ecosystem. *Forest Ecology and Management*, Vol. 90, pp. 201-215, ISSN 0378-1127

Miller, J.R.; Joyce, L.A.; Knight, R.L. & King, R.M. (1996). Forest roads and landscape structure in the southern Rocky Mountains, *Landscape Ecology*, Vol. 11, pp. 115-127, ISSN 0921-2973

Olander, L.P.; Scatena, F.N. & Silver, W.L. (1998). Impacts of disturbance initiated by road construction in a subtropical cloud forest in the Luquillo Experimental

Forest, Puerto Rico. *Forest Ecology and Management,* Vol. 109, pp. 33-49, ISSN 0378-1127

Potocnik, I. (1996). The multiple use of forest roads their classification. *The Seminar on Environmentally Sound Forest Roads and Wood Transport,* Sinaia, Romania.

Pritchett, W.L. & Fisher, R.F. (1987). *Properties and management of forest soils.* Wiley, ISBN 978-0471895725, New York.

Ramos Scharrón, C.E. & MacDonald, L.H. (2007). Measurement and prediction of natural and antropogenic sediment sources, St.John, US Virgin Islands. *Catena,* Vol. 71, pp. 250-266, ISSN 0341-8162

Ramos Scharrón, C.E. (2010). Sediment production from unpaved roads in a sub-tropical dry setting - Southwestern Puerto Rico. *Catena,* Vol. 82, pp. 146-158, ISSN 0341-8162

Reid, L.M. (1981). *Sediment production from gravel-surfaced roads, Clearwater Basin, Washington.* University of Washington Fisheries Research Institute, Publ. FRI-UW-8108, Seattle, WA,

Reid, L.M. & Dunne, T. (1984). Sediment production from forest road surfaces. *Water Resources Research,* Vol. 20, pp. 1753-1761, ISSN 0043-1397

Riley, S.J. (1988). Soil loss from road batters in the Karuah State Forest, eastern Australia. *Soil Technology,* Vol. 1, pp. 313–332, ISSN 0933-3630

Riitters, K.H. & Wickham, J.D. (2003). How far to the nearest road? *Frontiers in Ecology and the Environment,* Vol.1, pp. 125–129, ISSN 1540-9295

Sat Gungor, B.; Makineci, E. & Demir, M., 2008. Revegetated herbaceous plant species on a compacted skid road. *Journal of Terramechanics,* Vol. 45, No. 1-2, pp. 45-49, ISSN 0022-4898

Smith, R.B. & Wass, E.F. (1980). *Tree growth on skidroads on steep slopes logged after wildfires in Central and Southeastern British Columbia.* Canadian Forest Service, Pacific Forestry Centre Information Report. BC-R-6, Vancouver.

Smith, M. & Fenton, T. (1993). *Sediment yields from logging tracks in Kaingaroa Forest.* New Zealand Logging Industry Research Organisation Report, Vol. 18, No. 18, 10 pp.

Swift Jr., L.W. (1988). Forest access roads: Design, maintenance and soil loss. In: *Forest Hydrology and Ecology at Coweeta,* Eds: W.T. Swank and D.A. Crosley, pp.325-338, Springer-Verlag, New York

Theobald, D.M., Miller, J.R., Hobbs, N.T., 1997. Estimating the cumulative effects of development on wildlife habitat. *Landscape and Urban Planning,* Vol. 39, pp. 25–36, ISSN 0169-2046

Thompson, S.R. (1991). Growth of juvenile lodgepole pine on skid roads in Southeastern B.C. Synopsis of results. Prepared for Crestbrook Forest Industries Ltd.

Wang, L. (1997). Assessment of animal skidding and ground machine skidding under mountain conditions. *Journal of Forest Engineering,* Vol. 8, No. 2, pp. 57-64, ISSN 1913-2220

Williamson, J.R. & Neilsen, W.A. (2003). The effect of soil compaction, profile disturbance and fertilizer application on the growth of eucalypt seedlings in two glasshouse studies. *Soil Tillage and Research,* Vol. 71, pp. 95-107, ISSN 0167-1987

Wilson, R.L. (1963). *Source of erosion on newly constructed logging roads in the H.J. Andrews experimental forest.* Bureau of Land Management, University of Washington, Seattle, WA.

Yilmaz, E.; Makineci, E. & Demir, M. (2010). Skid road effects on annual ring widths and diameter increment of fir (*Abies bornmulleriana* Mattf.) trees. *Transportation Research Part D: Transport and Environment,* Vol. 15, No. 6, pp. 350-355, ISSN 1361-9209

11

Close to Nature Management in High-Mountain Forests of Norway Spruce Vegetation Zone in Slovakia

Martin Moravčík[1], Zuzana Sarvašová[1],
Ján Merganič[2,3] and Miroslav Kovalčík[1]
[1]*National Forest Centre – Forest Research Institute Zvolen*
[2]*Czech University of Life Sciences in Prague, Faculty of Forestry,*
Wildlife and Wood Sciences, Department of Forest Management, Praha
[3]*Forest Research, Inventory and Monitoring (FORIM), Železná Breznica*
[1,3]*Slovak Republic*
[2]*Czech Republic*

1. Introduction

The Slovak Republic is one of the most forested countries in Europe. Forest covers about 20,000 km^2 (41%) of the total area of the country, a substantial part of which is occupied by the mountains of the Carpathian Arch (highest peak: Gerlachovsky Peak, 2655 m). Forests in Slovakia have commercial functions as well as functions of benefit to the public, for example: timber production, water management, soil erosion control, avalanche control, nature conservation, tourism, and aesthetic value. Many rivers that are important for neighbouring countries spring from the Slovak mountains; Slovakia is therefore sometimes called the roof of Central Europe.

The Slovak forests are classified according to dominating tree species into eight vegetation zones: 1. Oak (located on altitudes below approximately 300 m), 2. Beech–oak (about 200-500 m), 3. Oak–beech (300-700 m), 4. Beech (400-800 m), 5. Fir–beech (500-1000 m), 6. Spruce–beech–fir (900-1300 m), 7. Spruce (1250-1550 m) and 8. Mountain pine (over 1,500 m). One of the most important forest ecosystems in terms of benefit to the public is Norway spruce forests, located at an altitude of 1250-1550 m in the so-called Norway spruce vegetation zone (SVZ).

Mission of these forests is fulfilment of their important protective functions and specific social needs which govern also the way of their management. A substantial part of them are situated in protected territories pursuant to the Act on nature and landscape conservation in national parks or protected landscape areas. Thus forests of the SVZ fulfill in addition to ecological functions (water management, soil protective, avalanche control functions) also significant social functions, especially nature protective function and recreational function. By Greguš (1989) the mentioned forests with prevailing ecological and social functions fulfill in the best way all required functions in such a state, which corresponds to the state of stands not affected by human activity. The aim of management should be regeneration and preserving sustained forest existence with functionally effective stand structure of natural or primeval character with preserved self-regulating capabilities and good health condition. In this way the management and care about such forests is very effective as the treatments of manager are not required or they are minimal.

On the other side forest stands with substantially changed tree species, age and spatial structure, usually artificially established or disintegrating and disintegrated stands without natural regeneration cannot be retained for self-regulating processes without intervention. Required functionality can be secured in an acceptable time horizon only by means of reconstruction management measures with resultant creation of desired differentiated structure. In the forests with partially changed structure (semi-natural and natural) according to criteria of Zlatník (1976) only necessary correction measures should be carried out to direct their development toward target state.

According to Korpeľ (1990) we have only little experience with regulating the structure of forest stands with prevailing ecological and environmental functions, so there prevail considerable caution to complete passivity in this approach. Only few authors have been dealing with forest management of protection and special purpose forests in Slovakia. Because of such reasons, there is an urgent need for gathering of objective knowledge about the functionally desirable condition of these forests, patterns of their existence, development, risk factors that damage its stability and functionality as well as deepening the connections between forest management planning and work on patterns of dynamics of natural forests and nature friendly silviculture. This need led to the development of proposal presented in this paper focused on objectified practice of framework and detailed forest management planning depending on the degree of conservation of natural structure (naturalnee class) in forests with prevailing ecological or social functions.

The knowledge about the naturalness class of forest ecosystems is therefore of great importance. Its objective assessment is essential in the decision-making process dealing with forest utilisation and subsequent forest management. According to Hoerr (1993) and Schmidt (1997), naturalness is the most significant and widely applied criterion for the evaluation of nature conservation, and serves as a key tool in analyses and as a support in planning nature conservation measures. Unfortunately, the assessment of the forest naturalness class lacks the application of the complex objective procedures and methods not only in Slovakia, but also in other countries. This situation results from the facts that research has not provided the practice with any suitable methodological mechanisms that would enable its scientifically based and statistically provable determination. The same fact has been reported by Bartha and others (2006) who mentioned that in the last decades, a number of authors developed procedures for the assessment of forest naturalness. However, in all these schemes subjective elements have been included. The assessed values of the indicators depend partially on the expert judgement and partially on their estimation. In addition, the experts make decisions, which attributes are to be assessed and what their weight is. The classification of forest naturalness proposed by Zlatník (1976) for Slovakia is also primarily based on subjective expert evaluation of the extent of human influence on forests (Table 2).

In Slovakia, several authors dealt with the evaluation of forest naturalness using typological surveys (Šmídt 2002; Glončák 2007; Viewegh and Hokr 2003; Bublinec and Pichler 2001; Polák and Saxa 2005). These works are characterised by insufficiently complex evaluation of forest naturalness, since the authors primarily assess the suitability of tree species composition. For example, Glončák (2007) identified areas which require active management of forest ecosystems in protected areas by comparing real tree species composition with model using GIS tools. The disadvantage of this method is a high level of subjectivity needed for the development of the model of natural tree species composition. On the other hand, precise distribution of the values of naturalness of tree species composition in GIS environment is a practical advantage of this method. The proposal of the network Natura 2000 in Slovakia was based on the assessment of qualitative attributes of forest ecosystems using numerical quantifiers (Šmelko ex Polák and Saxa 2005; Šmelko and Fabrika 2007). However, this system assessed also features which were not directly connected to forest naturalness (e.g. forest health status, adverse external influences), and when evaluating the majority of attributes, artificial securing of forest status needed from the point of nature conservation was accepted. Hence, this system was more likely aimed at the assessment of nature conservation values than at naturalness of ecosystems.

Naturalness is also a pan-European indicator of sustainable forest management (SFM) belonging to the set of criteria and indicators for sustainable forest management (No. 4.3) proposed within the framework of the Ministerial Conference on the Protection of Forests in Europe (MCPFE (Ministerial Conference on the Protection of Forests in Europe) 2002). In this context, forests are divided into forests undisturbed by man, which encompass forests with least human interventions; modified natural forests, seminatural forests and plantations (productive and protective), which cover man-made (artificial) forests. According to the Global Forest Resources Assessment 2010 (FAO 2007), forests are distinguished into primary forests defined as naturally regenerated forests of native tree species with no clearly visible indications of human activities and with not significantly disturbed ecological processes; other naturally regenerated forests which are also regenerated naturally but the indications of human activities are clearly visible; and planted forests, where the trees established through planting or seeding prevail.

The degree of forest naturalness is assessed through various indicators, mainly: nativeness of species and genotypes, differentiation of stand structure (e.g. diameter frequency distribution, vertical and age structure, occurrence of deadwood, natural regeneration of forests and coverage of ground vegetation), as well as the existence and extent of human influence in particular forest ecosystems (e.g. occurrence of timber felling and forest re-establishment and the applied methods, soil scarification, existence of forest roads, recreational activities, grazing, forest damage). (e.g. McComb and Lindenmayer 1999; Müller-Starck 1996; Peterken 1996; Scherzinger 1996; Frank 2000). Some European countries assess forest naturalness at a sample plot level within the framework of their national forest inventories. However, such an assessment provides summary information on individual forest naturalness classes only at national or regional levels.

Since the assessment of forest naturalness is very demanding from the points of methodology, applied techniques and funding, its realisation is reasonable if this indicator is an essential element in a specific decisionmaking process. In forestry, forest naturalness is of the greatest significance in the decision-makings that deal with the designation of forests as protected areas, and as a tool for determination the need and the urgency of management (cultivation, tending) in such a way, which will secure the protection of biological diversity, ecological stability and other natural values in forest with prevailing ecological and social functions – protective and protected forests. For these purposes, it is required to perform detailed surveys of forest naturalness.

In contrast to the above-mentioned methods, our proposal (presented in this chapter) is based on more precise data gathering methods, it deals with exclusive relationship with forest naturalness, and allows to account for the specifications of particular biotopes. And above all, it presents the proposal of mathematical and statistical assessment, formulation and presentation of results. In this point the significance of this work has a great international value. It can be used as a basis for efficient application of differentiated methods of utilisation and subsequent forest management. It fits also for application of Assessment Guidelines for Protected and Protective Forests and Other Wooded Land in Europe (MCPFE 2003) which can be regarded as one tool for differentiated management of protected forests. In Guidelines, three classes of forests, in which biodiversity is the main management objective, were defined. Class 1.1 comprises the forests where no active direct human interventions can take place. In class 1.2, only minimum human interventions are permitted. Class 1.3 comprises the forests designated for biodiversity conservation through active management.

2. Characteristics and main problems of forests in the Norway spruce vegetation zone

In the SVZ, total annual precipitation ranges between 1000 and 1300 mm, mean annual temperature ranges between 2°C and 4°C, and the vegetation period lasts 70–100 days. The SVZ forests cover about 40,000 ha or 2% of the total forest area and are located in the central and northern parts of the country, some of them in national parks.

The original SVZ forests were made up mostly of sparse stands or groups of trees with Norway spruce as a dominant species. Some forests also have European larch, European beech, mountain ash, and individual stands of dense mountain pine. Silver fir, cembra pine, and sycamore maple can also be found. The most frequent forest type groups (original species composition before human influence) are *Sorbeto-Piceetum* (mountain ash-spruce) and *Lariceto-Piceetum* (Larch-spruce).

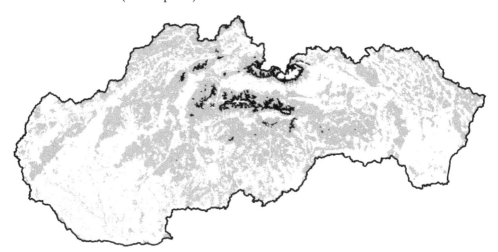

Similar forest types occurred in other European mountain ranges (e.g. Alpine and Carpathian regions in Romania, Ukraine, Poland, Austria, Germany, Switzerland, France and Czech Republic.

Fig. 1. Distribution of spruce vegetation zone over the area of Slovakia.

The age, diameter, and height structure of the forests in SVZ should be highly (horizontally and vertically) diversified to ensure the fulfillment of important ecological and social functions. Their static stability and the continuous influence of forest stand structure on forest functions are significant. Some authors (Korpeľ 1978, Turok 1990, 1991) stated that in spite of the existence of the trees that live to a greater age, the upper age of the mountain Norway spruce primeval forests is approximately 250 years. Almost identical forest types are spread over the whole Alpine and Carpathian region, less frequently they can be found also in other European mountain ranges (Palearctic habitat 42,21: Alpine and Carpathian subalpine spruce forests).

In spite of significant ecological and social functions of these forest ecosystems, their actual condition is not favourable. Moravčík and others (2005) presented the following reasons of the current, not always favourable state of the forests:

- Natural conditions: the SVZ forests are situated at an elevation from approximately 1250 up to almost 1600 m above sea level on long and steep slopes; growing in shallow, skeletal, drying-out (mainly due to the climate change), and nutrient-poor soils; on the sites with high potential and real soil erosion; on remote and technologically inaccessible locations.

- Climatic conditions: extreme temperature, moisture, and wind conditions; frequent intensive precipitation with occurrence of storm rainfalls, which is in the last years intensified by the climate change; and short vegetation season (70 to 100 days).

- Another negative ecological factor significantly influencing the health of high-mountain forests is unfavourable climatic conditions (lack or unsuitable distribution of annual precipitation, temperature extremes, etc). Formerly, the Slovak high-mountain forests were considered to have sufficient precipitation and favourable soil moisture. However, recent studies showed a dramatic change in the water regime in mountain forest soils, especially in sparse spruce stands. Soil acidification and lack of soil moisture are considered the most negative factors — worsening, or on some sites even disabling, natural regeneration of high-mountain forests.

- Monitoring of forest health in certain areas within the SVZ showed that about 90% of SVZ forest can be considered to be affected by air pollution. A rise in ozone concentration with altitude has been proven within the SVZ. Furthermore, a significant decrease of soil pH values was recorded (0.5–1.0 unit since the 1960s). Although sulphur and nitrogen emissions were considerably reduced during the past two or three decades, these substances are still accumulated in the soils.

- Age, vertical and horizontal stand structure of a large proportion of these forests is altered and little/unsufficiently differentiated. This state is the result of the strong colonisation pressure and clear-cutting management in the past.

- A significant proportion of these stands are over-mature (average age 105 years), disintegrating or disintegrated; average stocking 0.63 is significantly lower than the target stocking (0.7).

- Influence of injurious agents. The forests of SVZ are exposed to a complex of negativelly influencing factors, particularly at the upper limit of their occurence. This refers to the influence of air pollution in conjunction with natural factors (insects, fungi, wildlife), with climatic effects (windthrows and snow breakage), and with the impact of tropospheric ozone.

- These forests have been seriously damaged by storms. Trees damaged by wind or physiologically weakened by climatic extremes create suitable conditions for bark beetle outbreaks. Whereas in the past such outbreaks occurred only up to 1000 m, presently this limit is at 1300 m and in certain areas even at the timberline. All these factors cause weakening or even collapse of forest ecosystems. Forest stands become sparse and fragmented. This phenomenon is most evident on mountain ridges at 1300–1600 m.

- Long-term tendency of leaving forests in SVZ without any treatments (since the first half of the last century), which was caused by the fact that their management was unprofitable if assessed solely from the point of management costs and returns obtained from selling the wood. Considering lower wood quality and long extraction distances, both tending and regeneration measures are loss-making.

- Lack of objective knowledge about the functionally desired state of forests in SVZ and about the regulation or management of the structure of the stands with prevailing ecological and social functions. This has resulted in considerable cautiousness or even in passivity in their management (Korpeľ 1989).

- The attitude of state administration of nature conservation and some organisations of nature conservationists, who support a so-called passive conservation, i.e. against any treatments of these forests regardless their altered origin and their actual state.

3. Objectives, materials and methods

The purpose of this scientific paper is to improve the practices of forest management and tending of forests with particularly ecological and social functions on the example of SVZ. It contains development of objectified processes of forest tending according to their naturalness. The primary instruments of systematic forest tending in Slovakia are models of forest management (Fig. 2) that determines for specific natural conditions and stand conditions: management targets, basic management decisions and forest management principles. Our primary objective was to develop differentiated models of forest management for conditions in SVZ forests with added differentiating measure – naturalness class, which has not been applied systematically in the models so far.

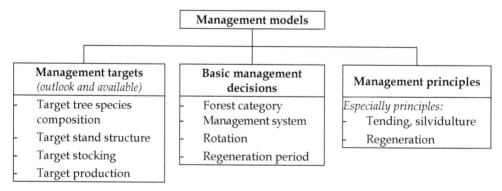

Fig. 2. Framework classification and content of management models.

For mentioned research we selected the SVZ ecosystem because of these reasons: 1) This forest community is very valuable but also vulnerable with significant ecological and social functions including nature-conservation functions. 2) Current condition of these forests (see previous section) is characterized by inappropriate structure of forest stands on a considerable part of them, exposure to harmful agents and adverse changes of the environment, which cause the urgent need to implement the measures to restore and improve their functional efficiency. 3) Because extensive national and international scientific activities have been carried out in the forests of SVZ (e.g. 4-year research project at the Zvolen Forest Research Institute dealing with methods for high-mountain forest management based on principles of sustainable development; S4C Initiative, Mountain Research Initiative, International Scientific Committee on Research in the Alps ISCAR).

Greguš (1989) formulated the general principles of forest management as follows: The mission of forest management should be: (1) achieving the maximum benefits from forest, (2) their permanent provision, (3) minimum risk, and (4) efficiency of providing these benefits. Once formalized these principles to the conditions of SVZ forests, in terms of the forest tending it is essential to achieve:

• The maximum observation of ecological and environmental functions, in particular through the functionally efficient forest stand structure, equivalent or approaching the state of natural anthropicly unaffected stands.
• Ensuring the permanent existence of the forest with good health condition and with corresponding forest stand structure according to which stands are capable of self-regulatory processes.
• Minimal risk of benefits in the given natural conditions through healthy, statically and ecologically stable forests with high-differentiated structure.
• Effectiveness of provision of benefits, through conservation and cultivation of natural forests to primeval forests with preserving their self-regulating features, which will reduce intervention of the forest manager to a minimum.

From the above-mentioned principles, the passive approach for SVZ forests with changed forest stand structure is unacceptable, respectively retaining them without the intervention (the self-development), because there is a risk of their subsequent destruction in large areas. The time horizon of the natural return into the stage of climax forest with the desired functionality through the stages of pioneer and intermediate forests is as a result of the passive approach unacceptable.

Proposing the methodology of solving the problem area was based on the following principles:

• Due to variety of natural conditions and the SVZ stand conditions, forest management models should be differentiated according to:
    - The naturalness class of forest stands (primeval forests, natural and semi - natural forests, man-made forests) and development stages of nature friendly forests.
    - Specific natural conditions characterized by present groups of forest site types (GFT) *Sorbeto-Piceetum* (SP) *Lariceto Piceetum* higher degree (LP hd) *Cembreto-Piceetum* (CP) *Acereto-Piceetum* higher degree (AcP hd) *Fageto Piceetum* higher degree (FP hd) and *Pineto-Laricetum* higher degree (PiL hd).

- Height zones of SVZ (lower zone – to an altitude of about 1400 m and high zone – from about 1400 m up to the upper limit of tree vegetation).
- With regard to the mission of SVZ forests, the basic objectives of management are the following: the target tree species composition, target stand structure and the target stocking.
- Available management targets should be proposed for man-made forests in which the target status can not be achieved during one rotation period.
- Identify the need and urgency of the implementation of forestry measures in forests with different naturalness.

To achieve planned objectives the following procedure was chosen:

- To obtain and evaluate own empirical material from the permanent research plots (PRP) established, so that:
  - each of the PRP represents a particular naturalness class,
  - PRP represents significant typological units: groups of forest site types and the both vertical zones (hight zones) (upper and lower) in SVZ,
  - they were etsablished in the forest areas with a significant presence of SVZ.
- To find a detailed information on natural conditions on PRP and on forest stand condition of SVZ forests using indicators appropriate for expression of condition of structurally differentiated forests – according to their naturalness.
- Deduce the average values of the indicators on forest condition in the PRP classified by the aggregated naturalness classes and development stages so that they could be used retroactively in forest management practice to identify the given forest types and to precise and add the existing management models.
- To use and to process data available from literature and documents of Forestry Information Centre.

We used data of detailed analysis of SVZ forest relized on the basis of data from 122 PRP and published knowledge from authors dealing with the issue of mountain forests as a background material to provide the above-mentiond activities. Empirical material was collected in PRPs by preferential and non-random sampling. The PRPs were established as circle plots of a size of 100-1,000 m² in order to meet the prerequisite that a minimum of 25 trees occur within each plot. The PRPs were localized using the global positioning system (GPS). The methodological intention was to establish PRPs in such a manner that detailed information about the natural and stand conditions (inclusive of forest naturalness) of forests in SVZ could be obtained. In the process of the methodology preparation, indicators suitable for the description of the state of structurally differentiated forests that were assumed to be related to the forest naturalness class were identified and proposed.

To find out natural conditions we monitored on PRP the status of these indicators: exposure, slope, altitude, relief, geological parent rock, thickness and form of humus layer, surface skeleton, forest type, soil type and the soil was also sampled. To characterise the state of the forest stocking and canopy were monitored, basic mensurational parameters were taken and development stage and naturalness class were determined on each PRP. All trees were localised as regards position and visualised by means of *Stand Visualisation System* (SVS), version 3.36. Then damage to trees, loss of assimilatory organs and social status were determined, crown length was measured and a necessary number of bores for age analyses was taken. Assimilatory organs were sampled for laboratory analyses. Ground vegetation was assessed as well as conditions for natural regeneration of Norway spruce and existing natural regeneration. The database system „*Mountainous Forests*" was constructed in MS Access 2000 for the processing and assessment of empirical material. An overview of the classification of 122 PRP according to natural and stand conditions under which they were established (forest eco-region, group of forest site types, naturalness classes and elevation) is given in Table 1.

Aggregated naturalness classes, n / %							
Primeval forest			*Natural forest*			*Man-made forest*	
17/13.9			94/77.1			11/9.0	
Of it the stage of			Of it the stage of			Of it the phase of	
Growth	Optimum	Decline	Growth	Optimum	Decline	Tending	Regeneration
2	9	6	32	36	26	2	9
Forest eco-region, n / %							
Veľká Fatra		Poľana		Nízke Tatry		Vysoké Tatry	
7/5.7		12/9.8		85/69.7		18/14.8	
Group of forest site types, n / %							
SP, LP hd		AcP hd		FP hd		CP	
84/68.9		22/18.0		9/7.4		7/5.7	
Elevation (meters above sea level), n / %							
Up to 1,350	1,351–1,400		1,401–1,450	1,451–1,500		1,501–1,550	1,551 and above
14/11.5	21/17.2		29/23.8	32/26.2		19/15.6	7/5.7

SP – *Sorbeto-Piceetum*, LP hd – *Lariceto-Piceetum* higher degree, AcP hd – *Acereto-Piceetum* higher degree, FP hd – *Fageto*
Piceetum higher degree, CP – Cembreto-Piceetum.

Table 1. Data structure with regard to natural and stand conditions.

The classification of forest into forest naturalness classes in each PRP in the field was based on the categorisation of Zlatník (1976) (Table 2). The assessed forest naturalness classes resulted from the detailed, though subjective evaluation of the forest status. Naturalness was assessed as a rate of human influence on a forest on the base of visual features that indicate human interventions (inclusive of forest management), which affect tree species, spatial and age structure (Fleischer 1999) of forests in SVZ. Each PRP was assigned one of forest naturalness class from the scale A to G (Zlatník 1976).

Forest naturalness classes (NC) by Zlatník (1976) were further aggregated into three degrees: Primeval forest, Natural forest, Man-made forest (Moravčík and others 2003; Moravčík and others 2005; Moravčík 2007a, b) prior to data processing. This was done due to insufficient number of plots in the degrees of the finer scale from A to G, and also from the reason of the need their practical application. The aggregated degrees of naturalness were complemented by the classification according to basic development stages defined by Korpeľ (1989) (Table 3).

NC	Name	Signs of anthropic effect; signs of stand structure
A	Primeval forest	without any effect of human activity
B	Natural forest	appearance of primeval forest without obvious signs of anthropic activity, possible selective felling in the past, natural forests affected by natural disasters left to natural development are included as well
C	Semi-natural forest	natural tree species composition, altered spatial structure due to extensive human activity
D	Predominantly natural forest	natural signs predominate over anthropic signs
E	Slightly altered forest	forest with natural as well as anthropic signs, the latter ones prevail
F	Markedly altered forest	forests only with anthropic signs but of natural appearance
G	Completely altered forest	forest stand only with anthropic signs of its origin or formation

Table 2. Criteria for the classification of stands by the naturalness classes.

1 – primeval forests (A)	2 – natural and semi-natural forests (B, C)	3 – man-made forests (D-G)
11 – in the stage of growth	21 – in the stage of growth	34 – tending phase
12 – in the stage of optimum	22 – in the stage of optimum	35 – regeneration phase
13 – in the stage of decline	23 – in the stage of decline	–

Table 3. Overview of aggregated naturalness classes and their classification by development stages.

These development stages of naturalness classes 1 and 2 in SVZ can be characterized as follows:

- Stage of growth – this stage is characterized by the largest diameter, height and area (vertical and horizontal) differentiation of stands. Canopy is graded to vertical, with a significant participation of trees in the middle and lower layer. There is characterized high vitality of trees and slight tree mortality in upper layer. Smaller gaps as results of tree falling from the previous cycle or accidental death of a strong tree from a new cycle are rapidly canopying.

- Stage of optimum – due to a longer life than height growth, the forest adjusts in height despite the large all-age. The maximum growstock is reached. Characteristic is: small number of trees per area unit and loss of foliation. Construction of stand is graded in height

- Stage of decline – overaged trees in good health condition begin to die in numbers at the end of stage of optimum and the forest is getting to the stage of decline. Growing stock rapidly decreases due to mortality of numerous large trees and is distributed very irregularly. Squads and groups of trees from the old generation are altered by gaps or incoming forest regeneration. Individuals of natural regeneration from the end stage of optimum merge into a continuous regeneration. Usually, there is regeneration of climax (target) tree species, only after fast (calamity) damage also the regeneration of preparatory tree species.

Characteristics of the development stages of man-made forests in SVZ:

- Forests in a period of forest tending - Horizontal involved, even-aged mostly spruce stands in the growth phase of cultures, providing cultures, youg wood, pole young forest and pole mature forest that require forest tending interventions.

- Forests in the regeneration period – even-aged spruce stands in various stages of thinning, or even locally disrupted, in different ages and unstocked areas that require regeneration.

Considering the structure and the type of data stored in the database system "Mountainous forests", a number of indicators that were assumed to be related to a degree of forest naturalness were proposed. In total, 25 different indicators of naturalness of forest ecosystems in SVZ were quantified, while tree species diversity was represented with 10 indicators, and structural diversity with 15 indicators (Table 4a, 4b). Tree species diversity was quantified with five indices of species richness, two indices of species heterogeneity, and three indices of species evenness. The indices of species heterogeneity were calculated from the proportion of basal area of particular tree species from the total basal area in a sample plot. The indicators of structural diversity reflect the diversity of structural elements of a forest ecosystem in horizontal and vertical directions. From 15 proposed structural indicators, two characterise vertical diversity (number of tree layers determined on the base of the sociological position of trees, and "Arten Profil" (species profile) index (Pretzsch 1996), while horizontal diversity is quantified by an aggregation index (Clark and Evans 1954). The remaining structural indicators are relatively simple and easy to be quantified, and are also related to static stability, stand density, and site quality. The average ratio of crown length to tree height, and the average ratio of tree height to tree diameter were calculated from the trees ranked in 1st to 3rd sociological layers. The indicators describing the coverage of herbs, grasses, mosses and lichens, shrubs and subshrubs; the coverage of phases describing the conditions for natural regeneration (juvenile, optimal, senile); the coverage of natural regeneration were visually estimated in the field and are given in relative values (%) (Moravčík and others 2005).

Structural diversity			
Indicator	**Formula**	**Units**	**Reference**
Number of tree layers (Z)	$Z = j$	DIM	
Arten profil index (A)	$A = -\sum_{i=1}^{S}\sum_{j=1}^{Z} p_{ij} \cdot \ln p_{ij}$	DIM	Pretzsch 1996
Aggregation index (R)	$R = \dfrac{\dfrac{1}{M} \cdot \sum_{i=1}^{M} r_i}{0.5 \cdot \sqrt{\dfrac{M}{A}}}$	DIM	Clark and Evans 1954
Coefficient of variation of tree diameter ($CV\_D1.3$)	$CV\_D1.3 = \dfrac{\bar{d}}{SD_d}$	%	Šmelko 2000
Coefficient of variation of height (CV_H)	$CV\_H = \dfrac{\bar{h}}{SD_h}$	%	Šmelko 2000
Average ratio of crown length to tree height (AM_K)	$AM\_K = \dfrac{\sum_{i=1}^{M}\frac{cl_i}{h_i}}{M}$	%	Šmelko 2000
Average height / diameter (h/d) ratio (AM_HDR) (Slenderness quotient)	$AM\_HDR = \dfrac{\sum_{i=1}^{M}\frac{h_i}{d_i}}{M}$	DIM	Šmelko 2000
Coverage of grasses (PK_T)	$PK\_T = p_i$	%	
Coverage of herbs (PK_B)	$PK\_B = p_i$	%	
Coverage of mosses and lichens (PK_M)	$PK\_M = p_i$	%	
Coverage of shrubs and subshrubs (PK_K)	$PK\_K = p_i$	%	
Coverage of juvenile regeneration stage (PK_JS)	$PK\_JS = p_i$	%	
Coverage of optimum regeneration stage (PK_OS)	$PK\_OS = p_i$	%	
Coverage of senile regeneration stage (PK_SS)	$PK\_SS = p_i$	%	
Coverage of natural regeneration (PK_NR)	$PK\_NR = p_i$	%	
Deadwood volume (MOD)	$MOD = \dfrac{\sum_{i=1}^{m} v_i}{A\,/\,10000}$	m³/ha	

Table 4a. Calculated indicators of structural diversity of forest ecosystems.

Tree species diversity				
Category	Indicator	Formula	Units	Reference
Species richness	Index N0 – living trees	$N0 = S$	DIM	Hill 1973
	Index N0 – mosses and lichens	$N0 = S$	DIM	Hill 1973
	Index N0 – shrubs and subshrubs	$N0 = S$	DIM	Hill 1973
	Index R1	$R1 = (S\text{-}1)/\ln(M)$	DIM	Margalef 1958
	Index R2	$R2 = S/\sqrt{M}$	DIM	Menhinick 1964
Species heterogeneity	Index λ	$\lambda = 1 - \sum_{i=1}^{S} p_i^2$	DIM	Simpson 1949
	Index H´	$H` = -\sum_{i=1}^{S} p_i \cdot \ln(p_i)$	DIM	Shannon and Weaver 1949
Species evenness	Index E1	$E1 = H`/\ln(S)$	DIM	Pielou 1975, 1977
	Index E3	$E3 = (e^{H`}\text{-}1)/(S\text{-}1)$	DIM	Heip 1974
	Index E5	$E5 = ((1/\lambda)\text{-}1)/(e^{H`}\text{-}1)$	DIM	Hill 1973

Legende for Tables 4a and 4b:
S – number of species; M – number of individuals, number of living trees in a sample plot; m – number of deadwood individuals (stumps, lying deadwoood); pi – probability, proportion of ith species or category in a sample plot; pij– proportion of trees of ith tree species in jth stand layer; Z – number of layers – stories of the stand; ri – distance between ith tree and its closest neighbour (m); A – area of a sample plot (m2); d – tree diameter; SDd – standard deviation of tree diameters in a sample plot; cl – crown length; h – tree height; v – volume.

Table 4b. Calculated indicators of tree species diversity of forest ecosystems.

4. Results and discussion

4.1 Management targets

4.1.1 Target stand structure

To derive the target structure we used data from literature and the values of selected indicators of spatial structure obtained from the assessment of empirical material, namely from PRP classified into the highest naturalness class (primeval forests) for the derivation of outlook target structure and into the 2nd naturalness class (natural and semi-natural forests) for achievable target structure. The results of testing statistical significance of the differences in diameter variability, height variability, slenderness quotient, crown length (in %) between individual naturalness classes showed to be statistically significant (*) up to highly significant (**). But mostly no statistical significant differences were confirmed between individual altitudinal zones (lower zone – *lz* and upper zone – *uz)* in the same naturalness classes.

The objective is to achieve and keep the structure of forest stands with markedly differentiated age, diameter and height (horizontal and vertical), which ensures the fulfillment of their significant protective (ecological and social) functions. Static stability of these forest stands is of primary importance. Their target structure is not connected with a single moment of the forest stand life. A permanent effect of target structure mainly on soil protective function (soil erosion control, avalanche control) and water management function is desirable.

The threshold values of selected indicators for target stand structure were derived from data collected on the PRP classified in the 1st naturalness class (primeval forest). They characterize the most original SVZ forest stands and were therefore considered as a benchmark for the desired outlook stand structure. Primeval forests have 3 developmental stages – growth, optimum, and disintegration – characterized by adjusted average values of the following indicators: degree of diameter dispersion (to assess tree diameter variability); share of canopy level (to assess tree height variability); ratio between crown length and tree height, and tree height and tree diameter; and mosaic of stand clusters.

Indicator		1st natural– ness class	Development stage (adjusted average values)		
			Growth	Optimum	Decline
Tree diameter variability (Sx%)		50 ± 15	60	45	50
Degree of diameter dispersion		3	3	2–3	3
Tree height variability (Sx%)		40 ± 20	50	30	40
Share of canopy level (%)	Upper	55 ± 15	45	65	60
	Middle	25 ± 15	30	25	20
	Downer	20 ± 15	25	15	20
Crown length / tree height (%)		75 ± 10	80	75	75
Tree height / tree diameter (slenderness quotient)		0,6 ± 0,1	0,55	0,60	0,55
Texture: mosaic of stand clusters and groups of the area 0.5 hectare max.					

Table 5. Model of outlook target structure derived from the values of indicators of forest state on permanent research plots classified into the 1st naturalness class.

However, it will not be possible to reach the desired stand structure even in the next generation because of large areas of artificially formed stands where management has been neglected. The characteristics of a realistic (achievable) target stand structure were therefore derived from the data representing the 2nd degree of naturalness (natural forest).

Indicator		2nd natural-ness class	Development stage (adjusted average values)		
			Growth	Optimum	Decline
Tree diameter variability (Sx%)		35 ± 15	45	30	35
Degree of diameter dispersion		2	2-3	2	2
Tree height variability (Sx%)		30 ± 15	40	20	30
Share of canopy level (%)	Upper	65 ± 20	50	75	65
	Middle	20 ± 15	30	15	20
	Downer	15 ± 15	20	10	15
Crown length / tree height (%)		70 ± 10	75	67,5	72,5
Tree height / tree diameter (slenderness quotient)		0,6 ± 0,1	0,65	0,6	0,55
Texture:	area form of structural types mixture (above 0,5 ha)				

Table 6. Model of achievable target structure derived from the values of indicators of forest state on permanent research plots classified into the 2nd naturalness class.

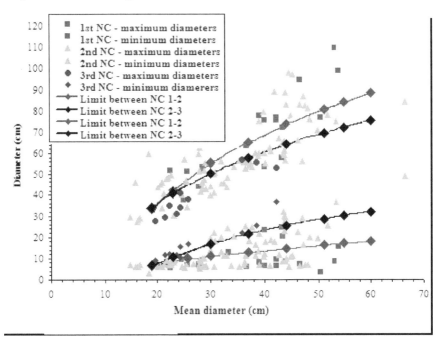

Fig. 3. Graph of diameter variance for Norway spruce in the SVZ in dependence on the naturalness classes.

To simplify the evaluation of diameter variability it is possible to use the degree of diameter variance as a practically usable indicator. In dependence on the mean diameter given on x axis the values of minimal and maximal diameter (y axis) of each PRP were illustrated. Minimal and maximal values of diameters were separately equalled graphically for all three naturalness classes. Equalling was made by means of logarithm curves. As the obtained curves represented only average values for the naturalness classes, the curves of limit values between the 1st and 2nd naturalness class, and the 2nd and the 3rd naturalness class were put between them. In this way 4 limit curves were constructed (Fig. 3) determining the variances of diameters for all three naturalness classes. The highest degree of variance 3 corresponds to the 1st class of naturalness, the 2nd degree of variance corresponds to the 2nd class of naturalness and the lowest degree of variance 1 corresponds to the 3rd class of naturalness.

4.1.2 Target stocking

As a rule stocking is defined as an indicator of the growth space utilization by a forest stand. Traditional way of its determination is the share of considered trees and the sum of considered trees and missing trees to the full stocking. According to Greguš (1976) target stocking is the stocking when the stand fulfils the determined functions in the best way. In commercial forests it is mainly production of wood and simultaneously fulfillment of other functions; in protective forests mainly fulfillment of publicly beneficial (ecological and social) functions (Midriak 1994). Greguš (1989) considered target stocking as an important component of management objectives especially because it informs us, though indirectly, but clearly about the fulfillment of desired functions and about the phase of regeneration. Especially by a change in stocking the manager can influence the development in forests. Derivation of target stocking is therefore a significant prerequisite to ensure professional care of forests, including those in the SVZ with the objective of achievement of their maximum functional utility. Assmann (1961) defined these concepts: optimum stocking with optimal stand basal area in which the forest stand produces maximum volume increment; maximum stocking with maximum stand basal area formed by living trees; critical stocking with critical stand basal area in which the forest stand still produces 95% of its maximum increment. In Slovakia mainly these authors dealt with issues related to target stocking: Halaj (1973, 1985), Faith and Grék (1975, 1979), Korpeľ (1978, 1979,1980), Šmelko et al. (1992), Korpeľ and Saniga (1993), Kamenský et al. (2002), Fleischer (1999), Moravčík et al. (2002).

Target stocking in the forests of the SVZ was derived on the basis of an original procedure as optimum stocking with harmonization of the requirements for the fulfillment of ecological functions, securing static stability and the existence of adequate conditions for formation and development of natural regeneration. To achieve this objective our own empirical material (122 PRP) was analyzed. Research was aimed at the investigation of relations between stocking and indicators (Table 4a) of static stability (slenderness coefficient and ratio of crown length to tree height), conditions for the formation and development of natural regeneration, coverage of natural regeneration and coverage of ground and non-wood vegetation in natural and semi-natural stands of the SVZ.

- *Ground vegetation* was found out as the percent of coverage of non-wood and shrubby vegetation on PRP; percent of coverage was determined in the groups: grasses, herbs, mosses and lichens, shrubs and semi-shrubs and total coverage.
- *Young regeneration and thicket* on PRP were found out as the percent of coverage by tree species in respective development stages; current year seedlings, natural seeding being high 50 cm, advance growth being high 1 m and thicket within diameter $d1.3 < 6$ cm were distinguished.

- Conditions for natural regeneration of the Norway spruce were evaluated according to Korpeľ (1990), Vacek et al. (2003) in three phases (juvenile, optimal and senile).
  - *Juvenile (early/premature) phase* – it is characterized by the almost closed canopy of stand with a marked microclimate buffering climatic extremes and by low coverage of ground vegetation. In the forests of the SVZ the soil is usually covered by a layer of forest floor, and low herbs and mosses with total coverage 30–40% prevail in the ground vegetation. The parent stand is capable to ensure natural seeding of the plot being regenerated by a sufficient amount of seeds that can germinate but the conditions of the stand environment are not suitable for the growth of natural seeding and formation of advance growth.
  - *Optimal phase* – it is characterized by the relatively open canopy, and thus by an increased access lof light, warmth and moisture to the soil surface. Climatic extremes are alleviated by the stand. Thin ground vegetation with prevalence of herbs over grasses occurs on the whole plot. In the forests of the SVZ this phase is frequently characterized also by the whole-area occurrence of mosses (more than 20%). Conditions of the stand environment enable the stages of germination, natural seeding, as well as advance growth on the same plot.
  - *Senile (late) phase* – it has the markedly open canopy of parent stand that enables almost a full access of light, warmth and moisture to the soil surface. In the dense ground vegetation grasses and high herbs prevail markedly. Ferns can be dominant in the stands of the SVZ at northern exposures as well. Conditions for the stages of seedling germination and their growth are not favourable any more. Providing there are natural seedlings or advance growth in the stand they can develop successfully.

Actual stocking on PRP was analyzed in the forests of the SVZ in relation to the degrees of naturalness classes, development stages, altitude and groups of forest site types. Average stocking on PRP established in primeval forests reached the value 0.61, in natural and semi-natural forests 0.62 and in artificial man-made forests 0.76. The lowest values of stocking were found in the decline stage (0.52 in NC 1 and 0.45 in NC 2). In the growth stage these values are 0.55 in NC 1 and 0.65 in NC 2. In the stage of optimum the values 0.69 and 0.72 were found. In average data on stocking there were not any statistically significant differences between stocking in the upper and lower altitudinal zone. Forests of the SVZ are permanently naturally open and thin by their appearance, towards the timberline the stands are thinner. Along the timberline they have a character of thin park forests.

In extreme site conditions the density of stands is lower. Trees in extreme conditions need a relatively greater growth area. Using the traditional way of stocking determination we estimate its value to be lower than 1.0 though it is frequently only the result of natural growth processes not influenced by man or injurious agents and its higher value under the given conditions (with regular spacing of trees) is not possible. In this case reduced clearing is unproductive clearing. Its reforestation is impossible. It is a part of the natural growth process and natural stocking of stands below the timberline also according to Assmann (1961).

Optimal stocking in the forests of SVZ was derived so as it would correspond in the best possible way to requirements for the fulfillment of ecological functions (soil protection, hydrological function), securing static stability and the existence of conditions for the formation and development of natural regeneration. It follows from the analysis of the relation between the ratio of crown length to tree height and stocking that with lower stocking the ratio is increasing, up to stocking about 0.7. Further drop of stocking is not reflected significantly in the increase in the ratio (Fig. 4).

It follows from the analysis of the relation between slenderness coefficient and stocking that with lower stocking the value of slenderness coefficient is lower as well. It drops to the value about 0.7. Further drop of stocking is not reflected significantly in the drop of the slenderness coefficient (Fig. 5).

It follows from the analysis of the relation between the conditions for natural regeneration and stocking that the most suitable combination of all three phases of preconditions for natural regeneration (juvenile, optimal, senile) is with stocking 0.7 (Fig. 6). At this value there are the most suitable conditions for the formation and development (advance) of natural regeneration as well as adequate coverage of ground and non-wood vegetation (Fig. 7).

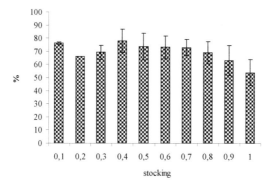

Fig. 4. A relation between the ratio of crown length to tree height (%) and stocking.

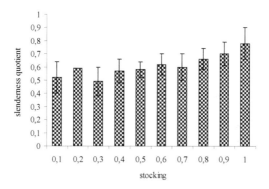

Fig. 5. A relation between slenderness coefficient and stocking.

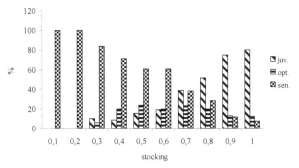

Fig. 6. A relation between natural regeneration phases (%) and stocking.

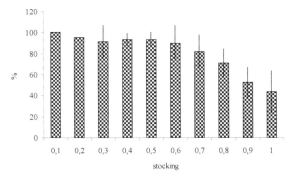

Fig. 7. A relation between ground and non-wood vegetation coverage (%) and stocking.

The optimum values of stocking with regard to the state of evaluated indicators are for stocking 0.7 or 0.7+. It follows from this finding that on average target stocking is about 0.7 for the forests of SVZ. It can differ slightly in dependence on the altitudinal zone or group of forest site zones. More significant differentiation can occur in dependence on the development stage but the objective of the care of forests of SVZ is to prevent the occurrence of the development stage "decline" on large areas. It is a desirable permanent (continuous) effect of this indicator of stand structure on forest functions. We can consider the given stocking rounded to 0.7 as Assman's natural stocking of the stands of SVZ below the timberline being evaluated by a practical manager with a traditional attitude. The values of stocking lower than 0.7 but within 0.7 determine the area share to complement or regenerate the stand.

4.1.3 Target tree species composition

Norway spruce has an absolute dominante in autochtonous stands of SVZ. It is the only tree species, which bears the harsh and extreme existential conditions of this vegetation zone. Therefore, spruce is the doninant tree species in current stands. Current tree species composition in groups of forest site types does not differ from the desired target tree species composition (such as in the sixth vegetation zone, or in other); on the contrary, it is very similar in the main tree species.

In neighbour lower sixth vegetation zone in the original stands of individual forest site type groups tree species as spruce, fir and beech (possibly along with other coniferous and broad-leaved tree species such as pine, larch, sycamore maple, ash, elm) formed various mixtures. Currently, tree species composition is altered in a large part of these forests. It consists of spruce monocultures and often is different compared to desired target diverse tree species composition. This has led to introduction of the so-called achievable tree species composition into forest management planning (achievable target), which reflects the current, altered tree species composition and it disintegrates (into the phases) the complex process of a desired target status into shorter periods, respectively to the possible achievable change in the current rotation period. This process is not necessary in SVZ with respect to closely existing and target tree species composition.

Table 7 shows the target tree species composition for groups of forest site types of SVZ. Selection and shares in forest site type groups is based on the original tree species composition as introduced by Zlatník (1956, 1957, 1959). There is an expected arborescent growth of spruce, in LP also of larch and swiss pine. Other tree species may have a height of a tree or shrub, according to altitude and locality conditions of a specific site. Desirable are all original, also rarely growing shrub tree species. Representation of mixed tree species can be provided by permanent, constant cluster and individual natural regeneration, or filling, in the emerging stand spaces.

Forest site type groups	Target tree species composition, %	Detailed specification considering the height degree
SP	Spruce 70 – 90, rowan, Silesian willow, betula carpatica, sallow, dwarf pine 10 – 30	In the higher degree towards the upper forest limit (which is mostly irregular and continuous), dwarf pine should be more applied in tree species composition for a continuous transition to 8. vegetation zone.
LP hd	Spruce 70 – 90, larch, swiss pine, rowan, rowan, Silesian willow, betula carpatica, sallow, dwarf pine 10 – 30	In the higher degree towards the upper forest limit (this has a continuous transition), individually and cluster mixed swiss pine, larch and dwarf pine should be more applied in tree species composition.
AcP hd	Spruce 70 – 90, sycamore maple, rowan, beech, Silesian willow, dwarf pine 10 – 30	Beech can be applied in a lower degree (referring to 6. vs), there is a lack in higher degree, to apply more dwarf pine towards the upper forest limit.
FP hd	Spruce 70 – 90, beech 0 – 10, white beam, rowan, larch, Silesian willow, dwarf pine 10 – 30	Beech mainly in a lower degree, there is a lack in higher, to apply more dwarf pine towards the upper forest limit.
PiL hd	Larch, dwarf pine, spruce, fir, pine, white beam, Silesian willow (individual tree species on rocks and cliffs)	Small areas of extreme localities of different tree species composition.

SP – *Sorbeto-Piceetum*, LP hd – *Lariceto-Piceetum* higher degree, AcP hd – *Acereto-Piceetum* higher degree, FP hd – *Fageto Piceetum* higher degree, PiL hd - *Pineto-Laricetum* higher degree.

Table 7. Target tree species composition by groups of forest types and height degree.

4.2 Basic management decisions

Basic management decisions in the Slovak forest management practice concern forest category (forests: commercial, protective, special purposes), silvicultural system (clear-cutting, shelter-wood, selective cutting), rotation, and regeneration period.

4.2.1 Functional focus

In terms of significance of SVZ forests when carrying out the environmental and social functions, their current categorization as protective forests is correct and there is no need to change it. Highly functional potential in providing natural and protective functions is in many cases seen in their declaration as a special purpose forests.

4.2.2 Rotation period (non-production rotation maturity)

The derivation of rotation period in SVZ forests was based on rotation maturity of the main tree species – spruce, life of forest stands and functions that are provided by these forests. Rotation period in protective forests results from their ability to carry out the required functions and it should not be higher than life of forest stands. Life of forest stand is deemed as the maximum age to which the stand retains in the given tree species composition and structure the character of canopy forest, not weedy by foreign species of forest phytocoenoses, preventing the forest regeneration.

These principles were the basis for evaluating the backround material with aim to derivate the rotation perion of SVZ forests. If the rotation period does not to exceed the life of forest stand, it must be determined in the age before the onset of stage of decline, i. e. at the end of the stage of optimum. Because the life of man-made stands is shorter due to faster culmination of growth processes compared to stands growing under cover of parent stand (Pliva, 2000), we researched the life of stand with regard to naturalness classes. We selected among the PRP those that are characterized by natural, semi-natural and man-made forests. Forest stand in a stage of optimum were set apart in and natural and semi-natural forests. Despite the fact that in different locality conditions (group of forest site types), altitude and forest area the life of stands may be different, in the framework feature of rotation period are these differences negligible. Therefore, we suggest differentiating the rotation periods only generally by aggregated naturalness classes.

Rotation period should be seen as a benchmark variable set near the maximum life of stands in natural and semi-natural forests. Maximum age of stands in the stage of optimum in these forests is about 180 to 210 years, so we recommend indicating the rotation period at the age of 200 years. Rotation period is only symbolic in the forest stands with a structure corresponding to the state of primeval forests with preserved self-regulatory process and should be close to the maximum physical age of the main tree species – spruce. Based on literature data and own findings, we suggest indicating the age of 250, respectively 300 years in these best-preserved SVZ forests. In man-made forests with the density 0.7 to 1.0 can be found the oldest stands mainly in the age of only 100 to 110 years. There were found no stands of age that exceeded 150 under the given stages of density. Man-made stands in age over 150 years can be usually found only in the advanced stage of decline. Based on these findings, we proposed to indicate the rotation period at maximum of 150 years in man-made forests.

4.2.3 Regeneration period

Stand structure in SVZ forests should be improved, we should prevent its levelling and create conditions for natural regeneration. It is therefore necessary to maintain the current practice of applying a continuous regeneration period in natural and semi-natural forests, when applying the fine methods and regeneration felling of shelter management method, in particular special purpose selection. In man-made forests, which are intended to restructure to more nature friendly forest types, we should bear in mind with respect to the largely neglected forest tending and difficult natural conditions the early onset, low intensity and slow process of regeneration. Therefore, there is also a well-founded application of a continuous regeneration period. The advantage of a continuous regeneration period is that it allows carrying out interventions with aim to improve the functional efficiency of stands and promotion of forest regeneration any time. An excception from these introduced processes are stands with poor health condition and high degree of threat, often thinned and weedy. Felling-regeneration procedures can not be applied in these stands that require a long-term regeneration periods.

4.2.4 Silvicultural methods

The aim of forest management in SVZ forests is to maintain or achieve such condition of stands, when they can carry out the best of the desired ecological and social functions and are able to exist and evolve through their internal self-regulatory abilities, without or with minimal human intervention. In natural and semi-natural forests of SVZ with partly altered structure, this status can be achieved by applying the finest forms and regeneration felling of shelterwood system, mainly special purpose selection. Selection system can be used only in stands with selection structure. The current state of man-made stands does not allow using special purpose selection. Their reconstruction requires using group, respectively marginal shelterwood cutting which will enable to achieve structurally differentiated stands.

Intensified management use of SVZ

In the issue of use of SVZ forests can be noticed an effort to enforce the passive protection by some interested groups of nature protection. However, there are also proponents of more intensive management use, for example Mráz (2001). In assessing the requirements of more intensive management use, we were looking for conjunction between the limit attributes of production capacity of stands and the need to meet the primary protection functions. The suitability of stands for their management use is limited by locality conditions, relief characteristics (slope, sceleton, regularity of micro-relief), transport accessability and an acceptable height stand quality. Locality conditions in SVZ are expressed by five groups of forest site types (SP, LP hd, AcP hd, FP hd, PiL hd). We should exclude from economic use all those forest site types in groups of forest site types FP and PiL becasue of terrain condition resulting from the following conditions:

- Slope should not exceed 40%, in larger slope there is an increased need to realize the soil protection functions, especially erosion and in higher altitudes also avalanche functions.
- Soil and surface sceleton should not exceed 50% for potential management use. SVZ forests are situated on hard sceleton soils. Sceleton directly affects also production capabilities of the site that decrese in greater representation of the sceleton.
- Forests used for the production of wood requires more management measures, so there should be regular slopes with an balanced gradient of slope and micro-relief should not be bouldered or rised.
- With regard to the possibility of use the ecological techniques and technologies, it is essential to provide accessibility to the given localities. This is essential requirement also in terms of preventing the subsequent damage to stands and soil.
- An important indicator of the possibility of management use of stands is the middle height, respectively the absolute height quality. Hančinský (1972) Stands in SVZ groups of forest site types SP, LP a AcP are considered to be possible economically used when they achieve the absolute height quality 24-25 m. Vološčuk (1970) In AcP and its bottom limit in the absolute height quality. According to our results, it could be from the absolute height quality of 24 m. According to this reason, potential economic conditions for management use can be limited by altitude up to about 1400 m, because there is a significant decrease of height quality over this limit and cluster stan structure with a loose canopy.
- With regard to the above conditions and requirements, these forest site types are potential for the management use:
  - 7106 – fertile rowan spruce wood - SP, LP hd
  - 7401 – fertile maple spruce wood – AcP hd
  - 7404 – irrigated maple spruce wood – AcP hd

Based on the analysis of natural, stand and economic conditions of SVZ forests, we concluded that these forests could be increasingly used for timber production after fulfilling the following conditions and criteria:

- They must be situated on forest site types with the potential of economic use (7106, 7401, 7404), indicated by acceptable absolute height quality of 24 meters or more.
- They must be located in the lower zone of SVZ, up to approximately 1400 m.
- They must meet criteria of micro-relief, limited to a maximum allowable slope up to 40% and the proportion of soil and surface skeleton up to 50%.
- They must be accessible by transport and there must be created such conditions for a minimum damage to stands and soil.
- There shall be no deterioration in carrying out the primary ecological and social functions.
- To assign particularly stands with lower naturalness class for intensified economic use (man-made, respectively natural forests).

4.3 Management principles

Until recently, mountain forests were rather domain of natural scientists. With regard to the inefficiency of their management from the momentaly short-term view, they apply a conservative approach. They were retained to self-development regardless of their structure and closely related stability. Reducing of vitality and decline of mountain stands, however, drew the attention of foresters. Extreme climatic and soil conditions along with an unstable structure, which is a natural consequence of the lack of silviculture treatment, create from mountain forests complexes with a low resistance to stress factors and a high probability of catastrophic decline. In cases where this process has already begun, remedial measures are extremely difficult to apply from a technical and economic point of view. Using an appropriate silviculture measures may lead to growing of stands with significantly differentiated structure, which substantially increases their stability. Implementation of silviculture measures under these conditions is very difficult mainly due to their "un-profitability", unaccessability of stands and discrepancies between forestry legislation and environmental legislation.

Aggregated Naturalness Classes / Forest Type Groups	Forest category	Management system	Rotation, year	Regeneration period, year
1 – Primeval forests				
SP, LP, AcP, FP hd, PiL hd	Protective	Retained for self-regulating processes without intervention	Symbolic 250 – 300	Permanent natural regeneration
2 – Natural forests				
SP, LP, AcP, FP hd	Protective	Shelterwood system	200	Continuous
PiL hd		Retained for self-regulating processes without intervention		
3 – Mand-made forests in reconstruction				
SP, LP, AcP, FP hd	Protective	Shelterwood system	150	Continuous
Forests of the SVZ determined for more intensive commercial expolitation				
SP, LP (7106), AcP (7401, 7404)	Protective	Shelterwood system	120 – 130	50 – 60

Table 8. Review of chosen basic management decisions in Norway spruce vegetation zone.

The aim of management in protected forests, including SVZ forests is not the quantity of production as in production forests, but the quality of stands, expressed by target tree species composition, target structure, target stocking and other indicators. It is necessary to focus primarily on use and direction of natural forces towards a low need for additional energy in all phases of management, from establishment through tending and regeneration of stands. Therefore we also propose to differentiate the management principles for the stands with various naturalness classes and development stages. Moreover procedures must be also differentiated with regard to health condition, static stability and the state of natural regeneration of forest stands. From this aspect only preserved forest stands or their parts with parameters corresponding to primeval forests including stand texture, which should be by Plíva (2000) formed of a mosaic of stand clusters, groups and small stands with the area the most 0.5 ha can be retained for self-regulating processes.

Predisposition of mountain forests to forming large-scale horizontal structure Korpeľ (1989, 1995) considers a significant risk factor. Mayer & Ott (1991) state that immediately as there appears a tendency of formation of one-layered stand the spruce stand can be maintained in the state of optimal functional effectiveness only by permanent silvicultural tending. Based on the results of owns research as well as experience from abroad Korpeľ (1990) also notes that high effective differentiated structure cannot be maintained in the altitude below 1400 m for a longer period without intentional silvicultural-logging treatments (with except for extreme soil conditions). He also says there is little experience with regulating desirable structure of stands with prevailing protective (ecological) functions and therefore a very careful almost passive attitude prevails in this field. Korpeľ (1980) evaluated the development and structure of Slovakian natural spruce forests in the SVZ. In the stage of optimum one-layered, height-balanced structure with horizontal canopy is being formed in these forests with long lasting (about 100 years) low resistance potential against wind. Due to fear of weakening the stands are left their natural development frequently, which is ended by calamity.

Due to the mentioned reasons we propose to carry out in the forests with partially altered structure (natural and semi-natural forests) if necessary inevitable correction measures to direct their development towards target state. By KORPEĽ (1980) the most effective and least risky are regulatory treatments through so called purposeful selection felling in advanced phase of growing up or in the initial phase of optimum. Purposeful selection must be aimed at increasing (maintaining) individual stability of trees. Shelterwood regeneration should start in advance on small areas in clusters or groups or in small cleared gaps. The procedure is similar to slow natural disintegration / declina but going on in still resistant stand. In Norway spruce natural forests, where development stages and structurally different parts of stand interchange in a mosaic on plots smaller than 1 ha, regulatory silvicultural-felling treatments are not urgent (especially treatments similar to regeneration felling).

Later when there are still suitable conditions for the germination of seed, survival and growth of spruce seedlings (prior to old-age phase of the conditions of natural regeneration) it is purposeful by Korpeľ (1980) to try to start intentional regeneration. Trees with reduced stability (intermediate with short crown) are removed and the most stable trees as bearers of stand resistance are preserved. By cutting of instable trees concentrated into clusters or groups an irregular regeneration elements rise. It is desirable to use permanently silvicultural and regeneration opportunities for creating strongly differentiated structure of stands and improvement of their static stability, mainly in lower part of the SVZ within the altitude about 1400 m. A great individual stability of trees is conditioned by slow decline of individual trees and thus markedly small-scale regeneration and small area forest texture (Korpeľ 1992). By Míchal (1995) the greater is the area of optimum stage with one-layered stands, little differentiated what concerns height and diameter the faster is their decline and on the greater area. In opposite to that markedly uneven-aged groups decline slowly on a small area.

The difference between actual value of stocking lower than 0.7 and stocking 0.7 determine area proportion to complement or regenerate the stand providing the area has continuous round shape, not very elongated, of minimally 300 m², e.g. 17x18 m, 20x15 m etc., which appears as a marked stand gap after missing trees. Fleischer (1999) states he found only for the plot with area 300 m² more stable progress of natural regeneration. In this sense also Saniga (2000) give the area of 200-300 m² as sufficient also for larch. He states the best conditions for natural regeneration are in the stands with stocking about 0.7 without herbaceous cover (herbs and mosses occur only sporadically). In some places there are small plots with almost 50 about two-year old spruce seedlings per m^2 but the conditions for survival are not suitable and therefore seedlings die (insufficient heat and light).

In structurally altered forests it is impossible to secure in an acceptable time horizon required functionality through retaining the forests for self-regulating processes. Therefore there must be applied reconstruction management measures according to actual state in forest stands with substantially altered age and spatial structure, formed usually as a result of artificial regeneration or in declining and declined stands without natural regeneration. A principal shortcoming is late time of regeneration and state of advanced decline of stands without securing regeneration. In such cases the regeneration can be realized only through artificial or combined regeneration; however only stands with little differentiated age and spatial structure are again created in this manner.

With all this in mind, we propose to plan and carry out any measures in these forests only on the basis of their actual "naturalness" class, which has to be the decisive criterion for determining the urgency of proposed measures. Additional criteria should include an assessment of static stability, natural regeneration, health condition, and stocking, as an indicator of fulfilment of ecological functions (mainly soil and water protection). Basically, it can be stated that the forest stands classified in the 1st naturalness class can be left as is. In such stands, natural regeneration usually fully corresponds to the actual stand structure, and both static stability and health condition are excellent. Forest stands that do not meet these criteria — mostly man-made, even-aged, vertically and horizontally little-differentiated forests, but also natural forests with various development stages whose natural regeneration ability is insufficient — require concrete measures. These measures can be classified according to the degree of urgency, based on the forest's actual status.

Better management of high-mountain forests SVZ will also require building a comprehensive net of forestry roads that are ecologically adapted to the terrain. It will be necessary to adapt all forestry activities in these forests to ecological standards and to introduce the most recent techniques and technologies. Clear-cutting is forbidden in the SVZ and has been fully replaced by shelter-wood and selection (purposefull) systems. On sites with deteriorated soils, recovery measures such as area-wide application of dolomitic limestone by airplane or helicopter, addition of dolomitic limestone and NPK fertilizers in holes, or application of mulching cloths when planting will create suitable growth conditions for subsequent forest stands.

Generally, natural regeneration is preferable. However, on certain sites tree species diversity will be enhanced by planting desired tree species. Mixed stands (especially of Norway spruce, European beech, silver fir, Scots pine, sycamore maple, European larch, and mountain ash) will gradually substitute pure spruce plantations, thus enhancing the ecological stability of the forests (including resistance to ongoing climatic change). The health status of forests and occurrence of harmful agents will continue to be monitored. In the field of forest protection, preventive methods will be given preference over suppressive methods.

4.3.1 Need and urgency of management measures

On the basis of the status of stand structure indicators, the conditions and the state of natural regeneration, static stability, health condition as well as after considering the requirements and conditions being given in basic management decisions and the management targets the manager will decide about the need and urgency of management measures with applying the management principles. The manager will decide whether the stand or its part requires a concrete management measure as well as about the degree of the urgency of management measure based on the fact how the state of stand corresponds to the criteria listed in following table.

Forest stand or its part doesn´t require any measures	*Forest stand or its part requires measures in the 1st degree of urgency (within 3 years)*
1st naturalness class Static stability – excellent Health condition – excellent Natural regeneration – fully corresponding	3rd (2nd) naturalness class Static stability – unsatisfactory Health condition – caduceus or died forest Natural regeneration – slight or minimal
Forest stand or its part requires measures in the 2nd degree of urgency (within 10 years)	*Forest stand or its part requires measures in the 3rd degree of urgency (postponable)*
3rd or 2nd naturalness class Static stability – satisfactory Health condition – mediumly declined Natural regeneration – slight or minimal at the age of forest less than 50 years under rotation	2nd (3rd) naturalness class Static stability – good Health condition – slightly declined Natural regeneration – slight or minimal at the age of forest more than 50 years under rotation

Table 9. Criteria for determination of the need and urgency of the measures.

Stand or its part will be classified into respective naturalness class on the basis of evaluation of diameter and height variability (especially by means of the degree of diameters dispersion and the share of canopy level – Fig. 3 turned into table), crown length and stand texture. Age range may be as an auxiliary indicator. In the following tables (10, 11) are given orientation values of the indicators. They can be used as an aid for assignment of the aggregated naturalness class and development stages of the 2nd NC of respective forest stands or their parts. The values listed in the following tables were derived from empirical material of 122 PRP.

Indicator		Aggregated naturalness classes		
		1	2	3
Degree of diameters dispersion		3	2	1
Share of canopy level; %	1	55 ± 15	65 ± 20	90 ± 10
	2	25 ± 15	20 ± 15	5 ± 5
	3	20 ± 15	15 ± 15	5 ± 5
Crown length; %		75 ± 10	70 ± 10	55 ± 10
Stand texture; ha		> 100	40 – 100	< 40
Age range; years		< 0,2 – 0,5	> 0,5	> 0,5

Table 10. Values of chosen indicators of forest status in naturalness classes 1, 2 and 3.

Indicator		Development stages of naturalness class 2		
		21	22	23
Degree of diameters dispersion		2-3	2	2
Share of canopy level; %	1	50 ± 20	75 ± 15	70 ± 15
	2	30 ± 15	15 ± 10	15 ± 10
	3	20 ± 15	10 ± 5-10	15 ± 10
Crown length; %		75 ± 10	67,5 ± 7,5	72,5 ± 7,5

Table 11. Values of chosen indicators of forest status in development stages of naturalness class 2.

The evaluation of static stability (Konôpka, J., 2002) will be made on the basis of the value of slenderness coefficient, which will be calculated as the proportion of tree height and tree diameter $d_{1,3}$ multiplied by 100. Slenderness coefficient of the stand or its part will be determined as mean value of slenderness coefficients found on respective standpoints. The assessment will be done in four degrees in the dependence on mean diameter and yield class.

Yield class	Degree of static stability	Slenderness coefficients by mean diameter, cm						
		10	15	20	25	30	35	40
≤ 16	1-excellent	0,63	0,60	0,57	0,55	0,52	0,49	0,46
	2-good	0,64-0,73	0,61-0,70	0,58-0,67	0,56-0,64	0,53-0,61	0,50-0,58	0,47-0,55
	3-suitable	0,74-0,82	0,71-0,79	0,68-0,76	0,65-0,73	0,62-0,70	0,59-0,67	0,56-0,64
	4-unsuitable	0,83	0,80	0,77	0,74	0,71	0,68	0,65
18-22	1-excellent	0,67	0,65	0,63	0,60	0,58	0,55	0,53
	2-good	0,68-0,77	0,66-0,74	0,64-0,72	0,61-0,69	0,59-0,67	0,56-0,64	0,54-0,62
	3-suitable	0,78-0,86	0,75-0,83	0,73-0,81	0,70-0,78	0,68-0,76	0,65-0,73	0,63-0,71
	4-unsuitable	0,87	0,84	0,82	0,79	0,77	0,74	0,72
≥ 24	1-excellent	0,68	0,66	0,64	0,62	0,61	0,59	0,57
	2-good	0,69-0,78	0,67-0,76	0,65-0,74	0,63-0,72	0,62-0,70	0,60-0,68	0,58-0,67
	3-suitable	0,79-0,87	0,77-0,85	0,75-0,83	0,73-0,81	0,71-0,79	0,69-0,77	0,68-0,76
	4-unsuitable	0,88	0,86	0,84	0,82	0,80	0,78	0,77

Table 12. Criteria for evaluation of static stability by mean diameter and site classess.

Health condition (Konôpka, J., 2002) is being evaluated according to the damage by identifiable injurious agents (wind, snow, frost, bark beetles, fungal diseases, it means rots, damage by game) and the evaluation of the state of crown, it means the loss of assimilatory organs. Co-dominant and dominant trees are evaluated by means of 5 degree scale – excellent or slightly disturbed, moderately disturbed, heavily disturbed, very heavily disturbed and dying or died stand.

In the assessment of the state of natural regeneration (Jankovič, 2002) actual state of natural regeneration will be estimated and on the basis of actual spatial and age structure of the stand or its parts there will be determined percentage of the area where natural regeneration should occur. It follows from the comparison of actual and required state of natural regeneration that the state being evaluated can be fully corresponding – suitable in 91-100 %, sufficient – 61-90 %, average – 41-60 %, weak – 11-40 % and minimal within 10 %. Only natural regeneration being in accordance with regeneration tree species composition is taken into account.

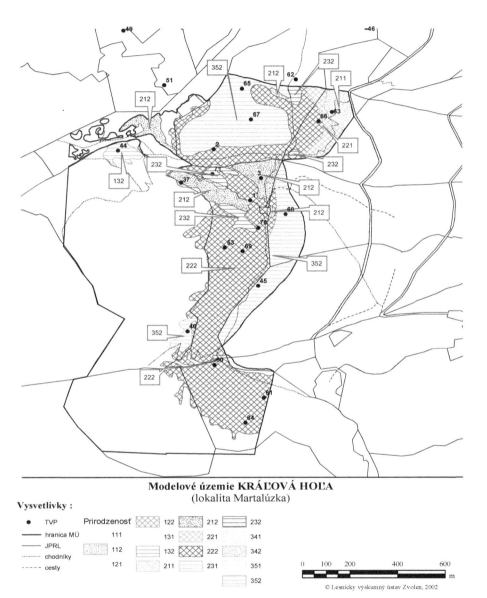

Explanation notes: 1st number – naturalness class, 2nd number – development stage, 3rd number – altitudinal zone.

Fig. 8. Example of forest distribution by naturalness classes and developmet stages in Nature Reserve of Martalúzka.

5. Classification model of a forest naturalness class

Because of the above-mentioned reasons it is required to know the actual forest naturalness class in the forest ecosystems since it can be taken as an objective criterion for decision-making about forest use and consequently about forest management (Hoerr 1993; Schmidt 1997). This is a generally applicable requirement and a need for achieving the optimal and the most effective use of forests. Hence, our goal was to prepare and propose a generally applicable method for the derivation of an integrated indicator and a model of forest naturalness class. Our requirement was to obtain unit values of the indicator and the variability of such a magnitude, that the differences between the individual degrees of forest naturalness would be significant. In order to examine the practical applicability of the proposed method, it was developed for a case of forest ecosystems located in SVZ.

Two variants of the classification model of forest naturalness were proposed, one based on the principles of discriminant analysis, while the second one uses an additive approach to derive the integrated indicator of the forest naturalness class. The discriminant model is derived as an application of multivariate statistical analysis, so-called predictive discriminant analysis (Cooley and Lohnes 1971; Huberty 1994; StatSoft 1996; Merganič and Šmelko 2004). Its role is to classify the sampling unit on the base of several quantitative variables into one of the pre-defined qualitative classes, in our case into one of the three forest naturalness classes. Using the data from the database, three discriminant equations were derived, each for one class of forest naturalness. These discriminant equations serve for the classification of an evaluated forest stand into one of the three forest naturalness classes. Secondly, we proposed an integrated indicator of forest naturalness class. This indicator belongs to complex indicators that combine several diversity components into a single value (Merganič 2008). The indicator is based on an additive approach, while the partial components are given in real measurement units. Mathematical formula of the integrated indicator of the forest naturalness class (IISP) is as follows:

$$IISP = ID_1 + ID_i + \ldots ID_n$$

where ID partial indicator of the forest naturalness class.

5.1 Data adjustment to meet the needs for the derivation of the classification model of the forest naturalness class

The relation between a diversity indicator and an area, for which the indicator was assessed, is known from a number of theoretical and practical studies. Due to the varying area of our sample units, we tested the relationship between the values of the partial indicators of forest naturalness and the area of the sample plot. The analysis revealed that 9 indicators (R1, R2, the average ratio of crown length to tree height, the average ratio of tree height to tree diameter, coverage of herbs and grasses, coverage of juvenile and senile phases and deadwood volume per hectar) had a significant relationship with the plot area ($p<0.05$). This result is logical and is mainly coupled with the effect of the development stages. The significant influence of the development stage on the indicators of forest naturalness was found in 16 out of 25 cases. Since the plots were distributed among the development stages, the varying area of the sample plots should not have a negative influence on subsequent analyses and on the creation of the classification model of the forest naturalness classes. On the contrary, the estimates of the average values and the variation of the indicators derived from tree data (the average ratio of crown length to tree height, aggregation index atc.) are even more representative, since they always represent a similar group of trees (approx. 25 trees).

Numbers of the PRP in individual forest naturalness classes, as well as the numbers of the plots in individual development stages (growth, optimum, decline) within the naturalness classes are imbalanced. Due to this and the above-stated facts, it was required to equalise the number of the sampling units in individual development stages and in individual forest naturalness classes. The missing plots were added by random replication of the existing sample plots using bootstrap technique (Chernick 2008; Yu 2003) until the number of the plots in the most abundant development stages was reached in other stages, too. In this way, the numbers of the plots in less abundant development stages and 1st, 2nd, and 3rd naturalness classes were set to 9, 36, and 9 plots, respectively.

Subsequently two different variants of the integrated complex indicator and the model of the forest naturalness class were proposed, one as a discriminant model, while the other one as an additive model.

Discriminant Model

From a great number of the examined combinations of the indicators (Table 4a, 4b), the best results of the correct classification of the forest naturalness class were obtained using the combination of the following six indicators: the arithmetic mean of the ratio between crown length and tree height (AM_K), the deadwood volume (MOD), the coverage of grasses (PK_T), the coverage of mosses and lichens (PK_M), the aggregation index (R), and the coefficient of variation of tree diameters (CV_D1.3). The general formula of the final discriminant model looks as follows:

$$\text{Discriminant score } j = AM\_K \cdot b_{j1} + MOD \cdot b_{j2} + PK\_T \cdot b_{j3} + PK\_M \cdot b_{j4} + R \cdot b_{j5} + CV\_D1.3 \cdot b_{j6} + b_{j7}$$

where: J = 1st to 3rd forest naturalness class.

The classification of the forest naturalness class is performed in several steps. First, the discriminant score of each naturalness class (1–3) is calculated from the particular discriminant equation using the real values of the partial indicators. An evaluated location, a stand, or in our case a sample plot, is assigned such a forest naturalness class, for which the calculated discriminant score is a maximum.

Forest naturalness class	Correct classification in %	Degree of forest naturalness according to the model			
		1	2	3	Total
		Number of plots			
1	85.2	23*	4	0	27
2	68.5	15	74*	19	108
3	94.4	0	1	17*	18
Total	74.5	38	79	36	153

indicates the cases with correctly classified forest naturalness class.

Table 13. Classification matrix of the discriminant model.

The results of the classification matrix of the parameterisation data set are presented in Table 13. As can be seen in this table, the overall correctness of the classification of the forest naturalness class using the proposed discriminant model is 74.5%. The highest probability of correct classification is in marginal classes (classes 1 and 3), while the lowest probability is in the middle class (class 2, 68.5%).

Following Table 14 presents the statistical characteristics of the model. According to the values of Fischer F and Wilks' Lambda statistics we can, with 99.9% probability, say that the proposed discriminant model is highly significant. The Willks' Lambda can be interpreted in the following manner: if its value is close to 0, the model is appropriate; if, on the other hand, the value approaches 1, the model is not suitable. The partial Lambda values given in the third column of Table 13 provide us with the information about the contribution of each independent variable to the discrimination of the dependent variable. Five out of six selected indicators are significant, which means that their contribution to the discrimination of the forest naturalness class is significant. Although the sixth indicator, the coefficient of variation of tree diameters, was insignificant, its presence in the model improved the classification. The indicators AM_K and MOD have the largest influence on the discrimination of the forest naturalness class.

Discriminant model			
Number of variables: 6		Number of groups: 3	
Wilks' Lambda: 0.43676		$F_{(12,290)}$= 12.401***	
Input variables			
Indicator	Wilks' Lambda	Partial Lambda	$F_{(3,935)}$ **95%, ***99.9%
Arithmetic mean of crown length / tree height ratio (AM_K) [%]	0.587	0.744	24.944 ***
Deadwood volume (MOD) [m³/ha]	0.491	0.889	9.062 ***
Coverage of grasses (PK_T) [%]	0.469	0.932	5.314 **
Coverage of mosses and lichens (PK_M) [%]	0.465	0.940	4.608 **
Aggregation index (R)	0.458	0.953	3.580 **
Coefficient of variation of tree diameter (CV_D1.3) [%]	0.442	0.988	0.862

Table 14. Statistic characteristics of the discriminant model.

In order to explain the classification graphically, the canonical analysis was applied to the data set. Fig. 9. shows the position of the groups of the sample plots with the same forest naturalness class and their approximate borders. From this figure it is obvious that the marginal categories of naturalness class have the highest probability of correct classification because their overlap with the neighbouring class is the smallest.

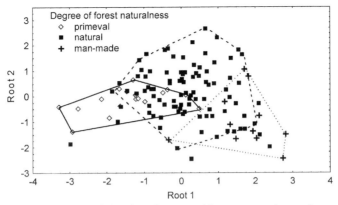

Fig. 9. Graphical interpretation of the classification of forest naturalness class with the discriminant model using canonical analysis.

Additive model

The partial indicators in the additive model are the same as in the discriminant model, i.e. the arithmetic mean of the ratio between crown length and tree height (AM_K), the deadwood volume (MOD), the coverage of grasses (PK_T), the coverage of mosses and lichens (PK_M), the aggregation index (R), and the coefficient of variation of tree diameters (CV_D1.3). The significance of the model was tested by singlefactor analysis of variance. The analysis revealed significant differences between the average values of IISP of the forest naturalness class (the whole model F(2, 150) = 21.849***, Tukey test). Figure 10. presents the graphical interpretation of the model. The range of IISP values was divided between the forest naturalness classes using the weighted approach, taking into account the error ranges of the average values of IISP and the percentiles of the values in every forest naturalness class. The objects, e.g. the stands, with the IISP values exceeding the value of 267 represent primeval forests; the IISP values in the range from 182 to 267 indicate that the forests are natural, while the values of IISP below 182 classify the objects as man-made forests.

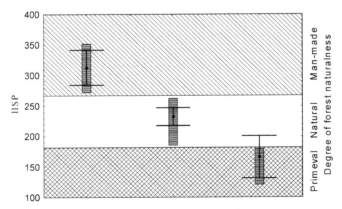

Fig. 10. Intervals of the integrated indicator of forest naturalness (IISP) specified for the three degrees of forest naturalness (primeval, natural, man-made forests); Legend: ▤ percentile 26–7 74% = 48% of values, ⊥ 95% confidence interval (1.96 × standard error).

The correctness of the model classification was determined on the base of the categorisation of individual plots into the forest naturalness class. The overall correctness of the classification using IISP is 63.4%. The individual forest naturalness classes 1, 2, and 3 were correctly classified in 74%, 56%, and 89% of cases, respectively.

Comparison of the models

The results of the classification of the forest naturalness degree indicate that both variants of the classification model have a similar probability of the correct classification of the assessed object into the forest naturalness class. The discriminant model behaves better, since its probability of correct classification is by approximately 11% higher than the probability of the additive model. Higher efficiency of the discriminant model is evident mainly in the proportion of correct classifications in 1st and 2nd forest naturalness classes. From the point of practical applicability, the additive model is simpler to use, but considering the current capacity of computers, it is also not difficult to apply the discriminant model in the form of a small computer program.

6. Conclusion

Because of enhancing requirements of public which are laid on forests the area of forests with prevailing social and ecological functions has been increasing in the last decades. Their important parts are forests in spruce vegetation zone. Therefore it is important to achieve status in which maximum fulfilment of mentioned functions through permanent existence of stable and healthy forest with corresponding stand structure is secured. Regeneration, improvement or maintenance of self-regulating ability of such forests should be essential. From these reasons forest management planning and subsequently also forestry operation should come out from appraisal and assessment of the class of natural structure conservation in carrying out of management measures. Naturalness class as indicator of natural structure conservation has to be a decisive criterion for determining need and urgency of the respective management measures.

For ensuring this approach there were chosen the most suitable indicators for quantification of stand structure status in primeval and natural forests which are characteristic with considerable degree of age, spatial, diameter and height diversity. Further there was collected and evaluated vast experimental material with objective of derivation of management targets – mainly target stand structure, target stocking and target tree species composition and criteria on identification of the basic naturalness classes – primeval forests, natural forests and man-made forests. The experimental material was collected in different natural, site and stand conditions from 122 permanent research plots in all significant groups of forest site types and altitudinal zones in the scope of spruce vegetation zone. Outcomes resulting from the evaluated experimental material confirmed statistically significant differences of forest status in various naturalness classes and actual development stages.

For needs of frameworking planning there were worked out: *differentiated tree species composition* by groups of forest site types and more detailed by altitudinal zones in the scope of spruce vegetation zone; *outlook target structure* derived following results of the primeval forests analysis and *available target structure* derived following results of the natural forest analysis and *target stocking* which was derived as optimum stocking on the basis of harmonising the requirements for fulfilment ecological functions, ensuring static stability and conditions for natural regeneration. The most suitable status of mentioned requirements was observed in stocking 0,7. There were derived differentiated rotation periods in dependence on the naturalness classes: 150 years for man-made forests, 200 years for natural forests and 250-300 years for primeval forests. However in primeval forests it is understood merely as a symbolic rotation period resulting from life-cycle of Norway spruce in respective natural preconditions. There were also identified the natural, site and stand preconditions of spruce vegetation zone in which forest stands could be utilised for more intensive commercial exploitation.

Likewise as the management targets and the basic decisions also management principles are differentiated in dependence on the naturalness classes. Basically it can be stated that only forests classified in the first naturalness class (primeval forests) can be left without any measures. In such stands concerning their structure, natural regeneration, health conditions self-regulating processes are usually in progress. Forest stands that do not meet these criteria – mostly man-made, even-aged, vertically and horizontally little-differentiated forests, but also natural forests with various development stages whose natural regeneration ability is insufficient – require concrete reconstruction measures. These stands can not be left for self-regulating because there is not possible to secure their required utility in acceptable temporal horizon.

Further there were proposed procedures for finding out and evaluating the forest status. They include the indicators and classification systems of evaluating the stand structure, status and conditions of natural regeneration, static stability, health conditions, determining the naturalness classes and evaluating ecological stability. Listed data are important for determining the need and urgency of respective measures.

Antoher very important outcome of this research is elaboration the methodology for the evaluation of forest naturalness on the base of the selected indicators of tree species and structural diversity. As we already stated, the knowledge about the naturalness of forest ecosystems is of great importance. Its objective assessment is essential in the decision-making process dealing with forest utilisation and subsequent forest management. Further more forest naturalness is the most significant and widely applied criterion for the forest evaluation from the viewpoint of nature conservation, and serves as a key tool in analyses and as a support in planning nature conservation measures. The currently proposed methodology, if applied within the practical forest management, can lead to the improvement of ecological stability of forests and landscape. Although the approach has already included several aspects of forest naturalness, it can be further enhanced by taking into account other components, e.g genetic diversity. The coupling of the model with statistical inventory and GIS tools can enable the creation of detailed maps of naturalness of forest ecosystems. Such information can further improve planning and practical application of nature conservation measures.

The developed classification model is easily applicable in practice and its application does not require intensive material and technical background. The applicability of the model for the classification of the forest naturalness classes has already been successfully tested on independent data (see Merganic and others 2009). **The method is applicable outside SVZ** or even outside Slovakia. In any other conditions, appropriate indicators of forest naturalness need to be selected, data need to be gathered, and the model needs to be re-parameterised. The coupling of the model with statistical inventory and GIS tools can enable the creation of detailed maps of naturalness of forest ecosystems. Such information is important for planning as well as for practical application of nature conservation measures. The model is a powerful tool for objectifying the assessment and the evaluation of the development of forest ecosystems within monitoring schemes.

7. Acknowledgments

The authors express their gratitude to the whole scientific and technical staff of Forest Research Institute in Zvolen, who participated in the collection of data in permanent research plots established within the scope of the project "Research of the methods of mountain forest management following the principle of sustainable development" in the years 1999–2002. We mainly acknowledge Ing. Jozef Vladovič, PhD. and Ing. Vladimír Šebeň, PhD. for their cooperation in the selection and the primary classification of selected permanent research plots into forest naturalness classes.

8. References

Assmann, E., 1961: Waldertragskunde. Munchen Bonn Wien, BLW Verlgsgesellschaft, 490 s

Bartha, D., Ódor, P., Horváth, T., Timár, G., Kenderes, K., Standovár, T., Boloni, J., Szmorad, F., Bodonczi, L., Aszalós, R., 2006: Relationship of tree stand heterogeneity and forest naturalness. Acta Silv. Lign. Hung 2: 7–22

Bublinec, E., Pichler, V., 2001: Slovenské pralesy - diverzita a ochrana. Ústav ekológie lesa SAV vo Zvolene, Zvolen, p 200

Clark, P.J., Evans, F.C., 1954: Distance to nearest neighbour as a measure of spatial relationship in populations. Ecology 35: 445–453

Cooley, W.W., Lohnes, P.R., 1971: Multivariate data analysis. John Wiley & Sons Inc, New York, p 400

Chernick, M.R., 2008: Bootstrap methods: a guide for practitioners and researchers, 2nd edn. John Wiley & Sons, Inc., Hoboken, New Jersey, 369 pp, ISBN: 978-0-471-75621-7

Faith, J. - Grék, J., 1975: Výskum metodiky prevádzkových cieľov a ich stanovenie. Záverečná správa, VÚLH Zvolen, 224 s

Faith, J. - Grék, J., 1979: Cieľové zakmenenie porastov. Záverečná správa, VÚLH Zvolen, 78 s

FAO (Food and Agriculture Organization) (2007) Specification of national reporting tables for FRA 2010. Working paper 135, pp 20–22. http://www.fao.org/forestry/media/6496/1/0/. Accessed 10th Nov 2008

Fleischer, P., 1999: Súčasný stav lesa v TANAP-e ako východisko pre hodnotenie ekologickej stability na príklade spoločenstva Smrekovcových smrečín. Dizertačná práca, TU Zvolen, 107s., 24 príloh

Frank, T. (ed), 2000: Természet – erdö - gazdálkodás. [Nature – forest – management] MME és Pro-silva Hungária Egyesü let, Eger, pp 116–118

Glončák, P., 2007: Hodnotenie prirodzenosti lesných porastov na základe typologických jednotiek (príklad z ochranného pásma Badínskeho pralesa). In: Hrubá V, Štykar J (eds) Geobiocenologie a její aplikace. Geobiocenologické spisy 11. MZLU, Brno, pp 39–46

Greguš, Ct., 1976: Hospodárska úprava maloplošného rúbaňového lesa. Príroda, Bratislava, 304 s

Greguš, Ct., 1989: Plánovanie ťažieb v ochranných lesoch. TÚ 3/1989, Lesoprojekt Zvolen, 55 s

Halaj, J., 1973: Porastové veličiny na meranie hustoty a zakmenenia porastov., Lesnictví 19, č. 10, s. 835 - 853

Halaj, J., 1985: Kritické zakmenenie porastov podľa nových rastových tabuliek. Lesnícky časopis, 31, č. 4, s. 267-276

Hančinský, L., 1972: Lesné typy Slovenska. Príroda Bratislava, 307 s

Hoerr, W., 1993: The concept of naturalness in environmental discourse. Natural Areas Journal 13(1): 29–32

Huberty, C.J., 1994: Applied discriminant analysis. John Wiley & Sons Canada, Ltd, New York, p 496

Jankovič, J., 2002: In: Moravčík, M., a kol.: Výskum metód obhospodarovania horských lesov na princípe trvalo udržateľného rozvoja (Správa pre priebežnú oponentúru). VTP č. 2730, LVÚ Zvolen, 2000, Ms. 214 s

Kamenský, m. a kol.: Pestovanie horských lesov na princípe trvalo udržateľného rozvoja. ZS, LVÚ Zvolen, 2002, 283 s.

Konôpka, J., 2002:. In: Moravčík, M., a kol.: Výskum metód obhospodarovania horských lesov na princípe trvalo udržateľného rozvoja (Správa pre priebežnú oponentúru). VTP č. 2730, LVÚ Zvolen, 2000, Ms. 214 s

Korpeľ, Š., 1978: Pralesy Slovenska. Veda Bratislava, 332 s

Korpeľ, Š., 1978: Štruktúra, vývoj a prirodzená reprodukcia prírodných lesov (pralesov) v 7. smrekovom vegetačnom stupni na Slovensku. Záverečná správa, LF VŠLD Zvolen, 158 s

Korpeľ, Š., 1979: Zásady pestovných opatrení pri obhospodarovaný porastov TANAP. In: Lesné hospodárstvo TANAP. Knižnica Zborníka TANAP, 7, s. 341- 366

Korpeľ, Š., 1980: Vývoj a štruktúra prírodných lesov Slovenska vo vzťahu k protilavínovej funkcii. Acta facultatis forestalis Zvolen, XXII: 9-39

Korpeľ, Š., 1989: Pralesy Slovenska. Veda Bratislava, 329 s

Korpeľ, Š., 1990: Štruktúra, vývin a dynamika zmien prírodných porastov trvalých výskumných plôch v TANAP-e. Zborník prác o TANAP, č.30, s. 43 – 86

Korpeľ, Š., 1992: Dynamické zmeny štruktúry, vývoj a produkčné pomery prírodných lesov pri hornej hranici lesa vo Vysokých Tatrách. Zborník prác o TANAP, 32, s. 245 – 272

Korpeľ, Š. – Saniga, M., 1993: Výberný hospodársky spôsob. VŠZ – LF Praha a Matice lesnická Písek, 128 s

Korpeľ, Š., 1995: Zásady a vhodné metódy pestovania horských lesov. In: Ott, E. a kol.: Pestovanie horských lesov Švajčiarska a Slovenska. Zvolen, ÚVVP LVH SR, 127 s

Mayer, H. – Ott, E., 1991: Gebirgswaldbau – Schutzwaldpflege. G. Fischer Verl. Stuttgart, 2. Aufl., 587 s

McComb, W., Lindenmayer, D., 1999: Dying, dead and down trees. In: Hunter ML Jr (ed) Maintaining biodiversity in forests ecosystems. Cambridge University press, Cambridge, UK, pp 335–372

MCPFE (Ministerial Conference on the Protection of Forests in Europe), 2002: Improved pan-European indicators for sustainable forest management as adopted by the MCPFE expert level meeting. http://www.unece.org/timber/docs/stats-25/supp/WA2-2. pdf. Accessed 6th June 2005

MCPFE (Ministerial Conference on the Protection of Forests in Europe), 2003: Annex 2 to Vienna resolution 4, MCPFE assessment guidelines for protected and protective forests and other wooded land in Europe. In: Fourth ministerial conference on the protection of forests in Europe. Conference proceedings 28–30 April 2003, Vienna, Austria. pp 216–219. http://5th. mcpfe.org/files/u1/vienna_resolution_v4.pdf. Accessed 6th June 2005

Merganič, J., Šmelko, Š., 2004: Quantification of tree species diversity in forest stands – model BIODIVERSS. European Journal of Forest Research 123: 157–165

Merganič,J. 2008: Proposal and derivation of the integrated indicator of forest naturalness and development of the classification model of forest naturalness degree. Partial report of the solution of the project „Research, classification, and application of forest functions in landscape"., FORIM, 18p., 06.11.2011, Available from http://www.forim.sk/index_soubory/Merganic_2008_Prirodzenost.pdf

Merganič, J., Moravčík, M., Merganičová, K., Vorčák, J., 2009: Validating the classification model of forest naturalness degree using the data from the nature reserve Babia Hora. European Journal ofForest Research (submitted)

Midriak, R., 1994: Funkcie a funkčný potenciál lesov. In: Vološčuk, I. : Tatranský národný park. Biosferická rezervácia. Správa TANAP, s.500 – 509

Míchal, I.., 1983: Dynamika přírodního lesa. V. Přírodní smrčiny. Živa, č.5, s. 166 – 170

Moravčík, M., a kol., 2000: Výskum metód obhospodarovania horských lesov na princípe trvalo udržateľného rozvoja (Správa pre priebežnú oponentúru). VTP č. 2730, LVÚ Zvolen, Ms. 214 s

Moravčík, M., Konôpka, B., Janský, L., 2003: Management of high mountain forests in the western carpathians, Slovak republic: research results and perspectives. Mountain Research and Development Journal 23(4): 383–386

Moravčík, M. et al., 2005: Zásady a postupy hospodárskej úpravy a obhospodarovania horských lesov smrekového vegetačného stupňa. Lesnícke štúdie č. 58. Lesnícky výskumný ústav Zvolen. ISBN 80-88853-91-5, 140 s

Moravčík, M., 2007a: Derivation of target structure for forests of Norway spruce vegetation zone in Slovakia. Journal of Forest Science 53(6) :267–277

Moravčík, M., 2007b: Derivation of target stocking for forests of Norway spruce vegetation zone in Slovakia. Journal of Forest Science 53(8): 352–358

Moravčík, M., Sarvašová, Z., Merganič, J., Schwarz, M., 2010: Forest Naturalness: Criterion for Decision Support in Designation and Management of Protected Forest Areas. Environmental Management. Springer Science+Business Media, LLC 2010 DOI 10.1007/s00267-010-9506-2

Mráz, I., 2001: Revízia ochranných lesov a lesov osobitného určenia. In: Financovanie 2001 Lesy – Drevo, Zborník z konferencie s medzinárodnou účasťou, TU vo Zvolene, s. 105 – 107

Müller-Starck, G. (ed), 1996: Biodiversität und nachhaltige Forstwirtschaft. Ecomed Verlagsgesellschaft, Landsberg, p 340

Peterken, G. F., 1996: Natural woodland: ecology and conservation in northern temperate regions. University Press, Cambridge, p 522

Plíva, K., 2000: Trvale udržitelné obhospodařování lesú podle souború lesních typú. Ministerstvo zemědelství, 34 s

Polák, P., Saxa, A. (eds), 2005: Priaznivý stav biotopov a druhov európskeho významu. Štátna ochrana prírody SR, Banská Bystrica, p 736

Pretzsch, H., 1996: Strukturvielfalt als Ergebnis Waldbaulichen Handels. Allgemeine Forst- und Jagdzeitung 167: 213–221

Saniga, M., 2000: Pestovanie lesa. Technická univerzita Zvolen, 247 s

Scherzinger, W., 1996: Naturschutz im Wald: Qualitätsziele einer dynamischen Waldentwicklung. Eugen Ulmer, Stuttgart, p 447

Schmidt, P., 1997: Naturnahe Waldbewirtschaftung – Ein gemeinsames Anliegen von Naturschutz und Forstwirtschaft? Naturschutz und Landschaftplanung 29(3): 75–82

Šmelko, Š., 1990: Zisťovanie stavu lesa kombináciou odhadu a merania dedrometrických veličín. Vedecké a pedagogické aktuality 6/1990. ES VŠLD Zvolen, 88 s

Šmelko, Š., Fabrika, M., 2007: Evaluation of qualitative attributes of forest ecosystems by means of numerical quantifiers. Journal of Forest Science 53(12):529–537

Šmídt, J., 2002: Metodika hodnotenia prirodzenosti lesov v Národnom parku Muránska Planina. In: Uhrin M (ed) Výskum a ochrana prírody Muránskej planiny 3. Správa NP Muránska planina, Bratislava & Revúca, pp 119–123

StatSoft 1996: STATISTICA for Windows. Tulsa, OK. http://www.statsoft.com. Accessed 27 Oct 2008

Turok, J., 1990: Vývoj, štruktúra a regenerácia prírodných lesov smrekovcových smrečín. ZPTNP, Martin, č. 30, s. 179 – 226

Turok, J., 1991: Vývoj, štruktúra a regenerácia prírodných lesov jarabinových smrečín (Sorbeto- Piceetum). Zborník prác o TANAP-e, 31, s. 119 – 159

Vacek a kol., 2003: Horské lesy České republiky. Praha, MZ ČR, 320 s

Viewegh, J., Hokr, J., 2003: Přesná typologická mapa – důležitý podklad pro hospodářská opatření v rezervacích. Příklad z části NPR Břehyně – Pecopala. In Geobiocenologické spisy, svazek č. 7. Zemědělská a lesnická univerzita v Brně Mendelova, Brno, pp 255–259

Vološčuk, I., 1970: Produkcia a zakmenenie porastov smrekového lesného vegetačného stupňa. Les, 26, č.1, s. 2 - 8, č.2, s. 50 – 57

Zlatník, A., 1956: Nástin lesnické typologie na biogeocenologickém základě a rozšíření československých lesů podle skupin lesních typů., In: Polanský, B. a kol. : Pěstení lesů III., SZNP Praha, s. 317 - 401

Zlatník, A., 1957: Poznámky k původnímu složení a typologickému zařazení tatranských lesů. Sborník Vysoké školy zemědelské a lesnické v Brně, řada C, s. 227 – 228

Zlatník, A., 1959: Přehled slovenských lesú podle skupin lesních typú. Lesnická fakulta vyskoké školy zemědelské v Brně. Spisy vědecké laboratoře biogeocenologie a typologie lesa, Brno, 195 s

Zlatník, A., 1976: Lesnická fytocenologie. SZN Praha, 495s

Yu, Ch.H., 2003: Resampling methods: concepts, applications, and justification. Practical Assessment, Research & Evaluation 8(19). http://PAREonline.net/getvn.asp?v=8&n=19. Accessed 2nd September 2009

12

Forest Transportation Systems as a Key Factor in Quality Management of Forest Ecosystems

Tibor Pentek and Tomislav Poršinsky
University of Zagreb/Forestry Faculty
Croatia

1. Introduction

Forests and forested land cover 24,018 km² in the Republic of Croatia, which accounts for 42% of its land surface. Its greatest part consists of state-owned forests (2,018,987 ha or 75.10%), which are, just as other forests owned by state institutions (87,930 ha or 3.30%), managed by the company »Hrvatske Šume« Ltd. Zagreb, through its 16 Forest Administration Units, distributed throughout the territory of the Republic of Croatia. A significantly smaller forest surface is in private ownership (581,770 ha or 21.60%).

It is indisputable that the forest transportation system is a required and above all necessary precondition in today's modern, technologically advanced, rational, economical, ecologically orientated, environmentally-friendly management of forest ecosystems, based on biodiversity, natural forests and income sustainability.

Type, amount and layout of all forest transportation system components have to be carefully planned in order to establish a truly optimal forest transportation system within a forest. The optimal quality of a primary forest transportation system is estimated from economic, technical-technological, environmental (ecological-esthetical) and sociological point of view, and it is necessary to achieve harmony among all the mentioned evaluation criteria, as well as reach the level of overall optimization (all evaluation criteria have to be brought within the limits of acceptability).

Each evaluation criterion of the optimal quality of the existing transportation system is composed of complex dominant influential factors (which combine close and mutually dependent simple dominant influential factors).

1.1 Environmental component of establishing the forest transportation system

As it was mentioned previously, the forest transportation system is an indispensable and obligatory component in quality management of forest ecosystems. Forest roads, both primary and secondary, are still a foreign body in a forest, therefore, in their planning, design, construction and maintenance (repairs) and possible reconstruction, we should take into account the minimal disruption of the laws, relations and balance existing in a forest ecosystem. When individual primary forest roads or secondary forest roads of permanent character (skid roads) are no longer necessary for forest ecosystem management, or in time they have lost most of its function because of which they had been built, they should be closed down, disbanded, and the surface on which they had been built should be restored to its previous purpose (productive forest land) by technical and biological methods. The procedure of removing the unnecessary (redundant) forest roads and restoring the habitat is neither cheap nor short, but it is necessary and, in the end, cost-effective in the long run.

Evaluation criteria for optimization network of truck forest roads	Complex dominant influential factors
The first priority evaluation level of truck forest road network optimality	
Economical criteria	Suitability of soil for truck forest road construction
	Terrain slope
	Existing traffic infrastructure
	Hydrographic network
Technical-technological criteria	Purpose of forests and forest land
	Forest owner
	Harvesting volume quality
	Applied technology and means of work
The second priority evaluation level of truck forest road network optimality	
Environmental-ecological criteria	Protected areas and buildings
	Protective areas and landscapes
	Danger of forest fire
Sociological criteria	Access to villages and hamlets
	Access to farm buildings
	Access to hunting lodges and weekend-houses
	Access to tourist and recreation buildings

Fig. 1. Structure of evaluation criteria for the optimal quality of the primary forest transport system estimation.

In the overall procedure of establishing each of the components of an optimal forest transportation system, the forest experts-designers should be guided with the idea of achieving an undivided whole consisting of constituents-habitat-forest road. In order to make that possible, the most important rules for the establishment of environmentally-ecologically-esthetically suitable truck forest roads in the planning and designing stage will be stated below:

While planning the environmentally-ecologically-esthetically suitable truck forest roads, the following remarks should be adhered to.

- Forest opening should be based on the *Studies of Primary and Secondary Forest Opening* of certain forest areas made according to a scientific-expert principle, because this is the only way to produce comprehensive, generally acceptable solutions.

- The production of a *Study of Primary and Secondary Forest Opening* should include all physical and legal entities (private forest owners, local government and self-government units, relevant ministries and institutions, etc.) which find it in their interest to participate in forestry planning (in this case, in planning truck forest roads).

- Bounded areas (protective zones) should be laid around the waterway network at a distance of at least 50 m.

- Truck forest road routes should not be laid in the immediate vicinity of accumulations and water-protective areas, unless it is absolutely necessary or contrary to the laws and sub-acts. Furthermore, lakes, surface waterways of permanent and periodic character, as well as the areas around water surfaces should be avoided.

- Low-land truck forest roads in the immediate vicinity of waterways may influence on the disruption of the underground water system, which would alter the microclimatic conditions of the site, resulting in a physiological decline of trees, lower quality of harvesting volume, and in the end, dieback and degradation of forests. The communication among underground water (even by means of artificially constructed technical objects) should be ensured.

- Because of rough ground conditions and required communication, it is necessary to set a truck forest road crossing over the waterway, subject to prior water management use approval, waterways should be crossed in their upper flow, and the road should not in any way influence on the change of the waterway direction and course. From the ecological and technical point of view, bridges would be preferred over roads, but they are expensive and non-profitable when it comes to low flow waterways.

- It is good to prefer soft construction material categories (soil with smaller or greater share of rock and softer rock), because on the one hand, earthwork expenses are low, and on the other, these categories have qualities sufficient for forming a quality truck forest road structure. The key factor in categorizing construction land favorability is its internal strength and hardness, or the share of rocky material of certain hardness in the overall amount of soil.

- Non-bearing base substratum, or soils having poor bearing capacity or none at all, should be avoided, because the construction procedure requires the use of one of the stabilization methods (soil improvement) or delivery of significant quantity of rock material in forming the truck forest road structure. Furthermore, such soils are quite often erosive, which increases the subsequent costs of truck forest road maintenance, as well as repair of possible damage and disturbing the forest ecosystem balance.

- Various categories of average terrain inclination are not equally favorable for truck forest road construction; therefore, it might be concluded that it is more favorable to construct truck forest roads on terrains with gentler average inclination than those with greater average inclination. Average terrain inclination has a direct influence on the transversal terrain inclination (terrain inclination perpendicular to the truck forest road axis), which dictates the terrain configuration at the truck forest road cross-section and has an influence on the amount of excavation and material filling during construction.

- In average terrain inclination of over 70%, it is very difficult to make stable excavation and embankment slopes in truck forest road construction of a normal cross section profile of fill slopes, so the danger of erosive processes is quite serious. In addition, in the process of construction, the level line elevation must be lowered deep below the terrain elevation at least for the roadway width, in order to make a normal cross-section profile of the fill slope (or the technology of embankment construction is used on inclined terrains with the use of excavators). Therefore, forest road construction on inclined terrains, particularly on slopes with an inclination of over 70%, is very expensive and demanding, it significantly encroaches upon the forest ecosystem and causes possible significant damage on the same, so truck forest road construction is not recommended on the slopes with an inclination of over 70%.

The following recommendations are stated for the design of environmentally-ecologically-esthetically suitable truck forest roads:

- The main truck forest road projects should be produced by authorized independent designers – foresters, who will gain the mentioned title by taking a professional examination at the *Chamber of Forestry*, and later on design the proscribed number of truck forest road kilometers, first with the guidance of a senior authorized designer, and only then independently.
- It is necessary to have good *Technical Requirements for Truck Forest Roads* in the framework of which the basic components of truck forest road main design content should be defined, with a detailed analysis of each sub-appendix in order to achieve uniformity and standardized quality of produced projects.
- A professional, qualified committee for the revision of produced projects should be constituted within the Chamber of Forestry, which would ensure the credibility and quality of technical documentation prior to starting a construction business.
- Integrating the truck forest road level line into the existing longitudinal profile of the terrain is one of the more sensitive and more responsible jobs of a designer in the very procedure of project production. The positionally (horizontally) fixed route (in the direction of x and y coordinates) should be defined in the direction of the z-axis as well. Balance should be found between the efforts to integrate the route as well as possible into its surroundings and satisfying minimal proscribed technical requirements of a certain truck forest road category.
- From the point of view of minimal costs of subsequent upper structure maintenance, the most favorable solution would be to design the truck forest route level line on an inclination of 4 to 6%. This is also a favorable longitudinal inclination for traffic operation. On level line inclinations greater than 8%, besides other drainage structures, it is necessary to make soakaways (transversal drainage ditches over the truck forest road structure at an angle of 30° on the longitudinal road axis).
- Maximal values of longitudinal inclinations of truck forest road level line should remain within permitted limits, while the same should be applied only on the most difficult parts of the route and on as short a distance as possible. The consequence of the use of great road inclinations in longitudinal direction is a significant problem which leads to harmful erosive effects of water on the upper structure and its washing away. This automatically requires high maintenance costs and investment into the construction of drainage elements.
- The maximal permitted longitudinal level line slopes are connected with the truck forest road category and the terrain configuration opened by a road. Before, between and after the maximal longitudinal slopes, smaller level line slopes are integrated, in order to provide a relief to the vehicle engine, as well as to reduce the strength of erosive influence of water.

- The distance between oppositely directed vertical curve apex should be at least 60 m (the exact value depends on the value of level line grade change points) in order to ensure a straight level line stroke of minimally 35 m between the end of the previous and the beginning of the following vertical curve arc (it would be better if we could achieve greater distances). We are often torn between the wish to follow the ground conditions as well as gaining minimal costs (greater distance between vertical curve apex are proportional to the earthwork volume and its value), and simultaneously respect the rules described in this passage.

- Considering the level line slopes, the grade change points become round in concave or convex vertical curve arches. The following factors must be decisive in the choice of vertical curve radius: safety against grinding of the bottom part of the vehicle against the truck forest road surface, safety against lifting the wheels off the road under the influence of the centrifugal force, and sufficient visibility of the road in case two vehicles meet at the vertical grade change point. The minimal radius of concave vertical curves amounts to 200 m, while the least radius of convex vertical curves amounts to 400 m.

- The reduction of the level line longitudinal slope at steep and step-like terrains is often possible only by increasing the earthwork volume. This is more expensive at the beginning, but in the end, it is a much better solution because the overall costs in the truck forest road depreciation period are definitely reduced when compared with maintaining greater level line slopes with fewer initial construction costs.

- The truck forest road route level line should be laid in a way as to avoid the deep cuttings and high embankments (over 3 m) because, besides representing an aggression to the environment, they also have little esthetic value. The so-called „dead sections" of truck forest roads render it more difficult to perform works of forest harvesting, and other primary or secondary forest roads cannot be connected to them. The same are justified in the case of the necessary crossing of truck forest road route over prominent ridges and larger lowlands when the terrain configuration simply does not enable a different, better solution.

- In designing level line, account should be taken of the land mass distribution diagram. A quality distribution of land masses presupposes equal amounts of excavation and embankment at a distance of up to 50 m, with at least oscillations of the cube profile ordinate curve as possible around the x-axis. This puts the cheaper side material transport into the limelight, thus avoiding the longitudinal material transport at longer sections.

- Often not even the use of the greatest allowed level line slopes can negotiate a certain altitude difference, which is why truck forest road constructive elements are made, characteristic, above all, for mountainous and hilly areas – switchback. When they are unavoidable, switchbacks should be laid on locations of milder transversal terrain slopes (up to 40%), in order to avoid greater works on the terrain and the necessity of constructing retaining walls, but also for reducing danger of stimulating erosion processes.

- In lowland areas, special attention should be paid to raising the truck forest road route level line, owing to drainage above the surrounding terrain, that is, above the level of the highest water. The level line slope must amount to a minimum of 0.5%, because in smaller level line slopes there is more damage on the upper structure due to the interaction between vehicles and water retaining on the roadway.

- Minimal diameters of horizontal circular arches amount to 20 m and they should be avoided, because they have an indirect influence on the truck forest road width by widening the road in the curves, reducing the safety of operation speed and having a negative influence on the safety of traffic. However, the fact is that we often follow the terrain owing to the road construction cost-effectiveness in the hilly and mountainous area, which of course, has an influence on laying the route with smaller diameters. Therefore, it is necessary to find an optimal compromise in such terrains between the well-integrated truck forest roads into the terrain contours on the one hand, and the safety of traffic, minimal diameters of horizontal curves, cost-effectiveness of construction and other relevant environmental, technical, financial, ecological and social factors, on the other.

- Passing areas (full or partial) are built on straight lines, outside vertical curves, on smaller longitudinal level line slopes and points with good visibility, often on the excavation side of fill slopes (due to stability), at a distance of 200 to 500 m. Their task is to ensure the possibility of evasion of two vehicles moving in opposite directions, as truck forest roads are made with one lane.

- Taking into consideration everything that was said before, a passing area should be located, according to the designer's evaluation and perhaps according to the collected samples, where there is material, which by using certain contemporary, environmentally friendly work technologies and technical means, might be used for the building into the upper structure. That is a soft and averagely tough rocky material, in which excavations can be done by means of a hydraulic hammer excavator (without the use of explosives). In this way, we avoid the opening of stone material borrow pits (quarry) in a forest, thus completely reducing or decreasing the need for the supply and delivery of stone material of a certain granulation from often very distant quarries (the overall costs of procurement and remote transport of the stone material do not incur the total expenses of truck forest road construction).

- Drainage ditches of trapeze or triangle shape (grader ditches) are always at the internal (excavating) fill slope side. Water from the drainage ditches can be drained into the surrounding terrain, and in order for the ditches to preserve their function, they need to be regularly maintained.

- Pipe culverts should be built on the crossing of truck forest road routes over small capacity surface waterways, which, considering their cross-section might be round or square, and they are most often made of reinforced concrete. The diameter dimension is determined with regard to the calculated water flow. Each culvert, regardless of its dimension, must be set in an appropriate way at sufficient depth under the level line elevation (so that during traffic operation they would not break).

- The junction of truck forest and asphalt public roads (crossroads) should be made in accordance with the proscribed Technical Requirements. The approach from a truck forest road into an asphalt public road should be elevated between 2% and 4%, so that during precipitations the stone material would not be washed away from the truck forest road into the public road, thus endangering traffic. If it is not possible to follow the proscribed inclinations, then the end of the truck forest road superstructure (20 m in length) should be made with concrete or asphalt carpet.

- The retaining walls are used for the stabilization of embankment slope, shortening the length of embankment and reducing the volume of material that should be fitted into the embankment on steep terrains. There are several forms of retaining walls, but regarding all of them, attention should be paid to regular dimensioning and fitting into the environment (which will best be achieved by using autochthonous stone material, which besides their functionality will provide the retaining walls with a more prominent esthetic function). Similarly, lining walls built on the excavation side of the fill slope and serving for the repair of excavation slopes (they do not carry the traffic burden).

- Owing to the evaluated construction material category, while forming cross-sections, it is necessary to make the excavation slopes and the embankment slopes at a certain inclination. Besides reducing the possibility of material sliding down to the surface of the roadway and endangering traffic (at the excavation side), as well as material sliding down the embankment, damaging trees at the lower side of the truck forest road and stimulating erosion processes, it also achieves an esthetic effect and a more pleasant surroundings for the drivers. Besides the mentioned ones, there are other technical recovery methods of excavation and embankment inclinations, which should definitely be combined with biological methods of recovery, because that is the only way to achieve the ultimate effect.

2. Research area

2.1 Classification of forest transportation systems

Forest transportation systems may be classified into primary, secondary and special-purpose forest transportation systems. Public roads are a generally common good owned by the Republic of Croatia, and according to the Public Roads Act (Official Gazette no. 180/04, 82/06, 138/06, 146/08, 152/08, 38/09, 124/09, 153/09, 73/10 and 91/10), depending on social, traffic and economic significance, they may be: motorways, state roads, county roads and local roads.

Primary forest transportation system consists of all categories of truck forest roads, as well as public roads which may be used for forestry operations (these are often lower level public roads – county roads and local roads). Forest roads are permanent construction facilities, enabling constant motorcar traffic for the completion of tasks anticipated by the Management Plan. With their construction, the amount of productive land in a forest is permanently reduced (except in the case of their closing down and revitalizing sites). They consist of a lower structure and an upper structure with all technical characteristics of a road. They may be divided on the basis of several criteria.

The components of secondary forest transportation system are secondary forest roads: skid roads, skid trails and cable yarder corridors. Their main purpose is timber extraction from the bunching point to roadside landing (primary timber transport) and, occasionally, completing assignments anticipated by the Management Plan. From the roadside landing to the ultimate user, timber may be transported by constructed transportation systems (forest and public roads, as well as railroads) or waterways (rivers, lakes, seas, oceans).

Skid roads are construction facilities of permanent character (except in the case of their closing down and revitalizing sites), built only with a lower structure. They are associated with sloping terrains, heavier material construction categories and the presence of surface obstacles.

Skid trails are secondary forest roads of temporary character, made by cutting a route through a forest, possible extraction of stumps and a repeated passing of a timber extraction machine (skidder, forwarder) on the same route. They are characteristic of flat terrains, lighter material construction categories and the absence of surface obstacles (easily passable terrains).

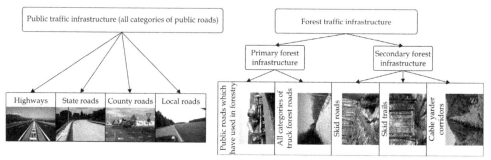

Fig. 2. Classification of forest transportation system.

2.2 Stages of establishing an optimal forest transportation system network

Establishing an optimal network of forest roads in the field has to unfold through the following operational stages: planning, designing, construction with supervision and maintenance/repairing (*Pentek et al. 2004*). These stages are mutually related and dependent, they should be performed in the order as they are stated, taking into account the unfeasibility of each of the operational stages in case the previous one has not been completed in a satisfying manner.

Besides the mentioned, always present stages of forest transportation system optimization, there are occasionally two additional operational stages: stage of forest roads reconstruction (in order to increase their standards when it comes to forest roads, or turning skid roads into forest roads) and the stage of forest road removing/restoring (besides the revitalization/restoration of sites, or restoring the function and form of a site as close as possible to what it had been before the road was built).

2.3 Planning of forest roads

A comprehensive planning of forest roads is the first, initial and unavoidable stage of establishing an optimal forest transportation system network in the field. In the last 20 years, GIS (Geographical Information System) is applied in primary and secondary forest opening, in combination with other contemporary technologies (*Pentek, 2007a*). The data required for making a quality GIS of a research area are collected from the following sources (*Pentek, 2002*): thematic maps, computer databases, written databases, Management Plans, field measurements, field observations and notes, other sources, arithmetic and logical operations with the data from previously mentioned sources.

The result of a contemporary approach to forest opening are the Studies of Forest Opening (primary and secondary), made for the period of 10 (20) years, after which they are renewed or revised. These documents are considered as a part of tactical planning in forestry. Tactical plans provide answers to the question of what to do in order to achieve the set goals of strategic planning and what decisions are necessary for that (*Kangas and Kangas, 2002*), and they are made for a shorter period of time (depending on the circumstances, they cover a period of 5 to 20 years); they result in a list of measures (operations or interventions) planned to be done within the following time period.

2.3.1 A study of primary forest opening

Every good Study of Primary Forest Opening should contain the following data:

- for the existing forest transportation system:
  - a complete (updated) cadastre of the existing primary forest transportation system,
  - a complete (updated) cadastre of the existing secondary forest transportation system,
  - the existing primary and secondary road density (m/ha),
  - the existing mean distance of timber extraction for each particular compartment (m),
  - the target primary road density and the target (planned) mean distance of timber extraction calculated from it,
  - numerical, graphical and pictorial (map) results of the analysis of the existing relative primary openness;
- for the improved primary forest transportation system:
  - numerical, graphical and pictorial (map) results of the analysis of the existing relative primary openness for the improved primary forest transportation system,
  - primary road density of the improved primary forest transportation system (m/ha),
  - mean distance of timber extraction for each particular compartment (m),
  - conceptual route of planned truck forest roads (defined by the coordinates of route break-points),
  - category of each conceptual truck forest road route,
  - cost component (anticipated expenses) and economic justification for the construction of each conceptual truck forest road route,
  - dynamics of the construction of the overall (optimal) future primary forest transportation system network, aligned with the proscribed works in the Management Plan,
  - dynamics of the maintenance of the overall (optimal) future primary forest transportation system network,
  - other data significant for any of the stages of establishing the optimal primary forest transportation system network.

2.3.2 Primary road density

Primary road density represents the sum of lengths of all components of primary forest transportation system (which influence on the openness of the respective area) divided with the surface on which the respective roads are located. It is expressed in m/ha or km/1000 ha. *Šikić et al.* (1989) defined the fundamental criteria on the basis of which a certain road, or a particular part of it, is taken into consideration when calculating primary road density.

Pentek (2002) distinguishes five basic variants of primary road density:

- The existing primary road density – calculated for the existing (real) primary forest transportation network of a certain forest area, often a Management Unit,
- Minimum required primary road density – set for a greater forest area, in most cases related to a relief area, used in the strategic planning of forest-management area as a minimum goal which should be reached within a given time period for a more rational forest management,
- Planned primary road density – it is also set for a greater forest area (relief area), and serves as a marker within a defined time period in the strategic planning of forest-management area and in making long-range plans of primary forest transportation system construction,
- Target primary road density – most often defined for a management unit area and represents the final goal of primary road density of a certain forest area; it is closely connected with the methods and procedures of timber harvesting, as well as morphological relief characteristics in a specific Management Unit; it is used in the context of tactical planning and making Studies of Primary Forest Opening,
- Optimal primary road density – calculated by applying a known method of primary forest transportation system optimization; it is related to the management unit surface area and is most often based on the model of minimum overall cost of timber harvesting.

Relief area of the Republic Croatia	Minimum required road density	Planned road density 2010th	Planned road density 2020th
	km/1000 ha		
Low-land area (Flat terrain)	7.00	15.00	It is not the subject of research
Hilly area	12.00	20.00	25.00
Mountainous area	15.00	25.00	30.00
Karst area	No data	10.00	15.00

Table 1. Minimum required (*Šikić et al. 1989*)., planned 2010th(*Anon., 1997*) and planned 2020th *(Pentek et al. 2007a, Pentek et al. 2011)* primary road density for different relief areas in Croatia.

2.4 Truck forest roads designing

To design a certain truck forest road (*Pentek, 2010b*) means to conceptualize it, describe it and present it arithmetically and graphically. Only a completely finished main truck forest road design may be analyzed, and construction may ensue after its approval.

Designing truck forest roads consists of collecting general and technical data, as well as route layout and design creation. The first stage of designing includes collecting general and technical data, which present a basis for making a feasibility study. Route layout (field measurement) and design creation (office data processing and print-out of results) represents a designing components which combines all field and office route layout work, making of investment program, as well as conceptual, general and main forest road design.

Truck forest road layout is performed by way of a direct layout. The result of the planning stage, observed from the level of a single truck forest road, is a larger number of projected zero line variants (at least three) on forest-management contour maps with the scale 1:10000 or even better 1:5000 in digital form. These are the so called conceptual layouts of the future truck forest road.

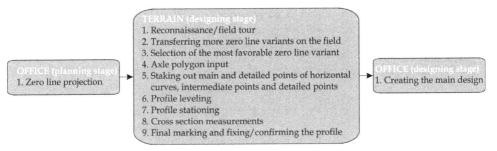

Fig. 3. Direct layout – basic operation stages.

Two procedures may be applied in truck forest road designing (*Pentek, 2010*), according to the valid regulations – short procedure: implies the making of the conceptual and main truck forest road design; and full or complete procedure: encompasses the making of conceptual, general and main truck forest road design.

Conceptual design – this design deals with the conceptual layouts of truck forest roads, along with the making of technical and economic study. It is made on contour maps with the scale 1:5000, 1:10000, 1:25000 and 1:50000. Maps need to have a marked management division – borders Management Units, compartments and sub-compartments as well as a complete cadastre of primary forest transportation system (if possible, also the cadastre of secondary forest roads). Important components of the maps are permanent and occasional waterways. It is necessary to be acquainted with the Management Plan, as well as study the information about growing stock, harvesting volume (allowable cut) and timber assortment structure of an individual compartment and a harvesting plan. The contour map has to contain more zero line variants for each truck forest road. At the same time, attention should be paid to key points, which have to be connected by the future route, as well as to the position and capacity of landings, branching of skid roads, etc. All zero line variants are transferred to the field, the most favourable one is chosen, and then follows the making of a rough cost estimate and a technical and economic study for it.

General design – made on the basis of previously made and approved conceptual design. Tachometry measurement is made around the operative polygon of conceptual route (in order to make a contour plan). Then, a contour plan is made in the office, the axle polygon is integrated into the zero line, horizontal curves of the selected radius are determined and drawn in. A longitudinal section is drawn on the basis of a situation plan, and cross sections are made based on a round level line in order to make a report of land work cubage. General design provides more realistic technical and economic indicators on the future truck forest road than the conceptual design.

Main design – made on the basis of the conceptual and general design, or just on the conceptual design. This is the most comprehensive design, which represents the basis for commencing the construction procedure.

• Court Register Certificate	• Declaration of conformity of project documentation	• Drawn longitudinal section
• Certificate on the Appointment of an Architect	• Calculation of mechanical resistance and stability	• Normal cross sections
• Certificate on Architect Authorization	• Technical description of forest road layout	• Drawn cross sections
• Ownership Certificate and a copy of cadastre plan	• Position plan	• Land mass cubage and a quantity statement
• Certificate of Title and a land register transcript	• Detailed position plan	• Mass-haul diagram
• Record of zero line handover	• Print-out of longitudinal section	• Bill of quantities
• Record of staked-out route handover	• Data on horizontal and vertical curves	• Estimate – priced bill of quantities
• Fire protection document		

Table 2. Basic components of the main truck forest road design.

2.5 Truck forest road construction with supervision

Following the designing stage, there is the construction stage with supervision, representing the greatest expense in the overall process of making a new truck forest road. In the Republic of Croatia (*Pentek, 2010*), the construction procedure, in the widest sense of the word, is performed through the following operational stages:

- carrying out the public tender procedure and the selection of the most favorable tenderer,
- signing the Implementation of Works Contract and reporting the work-site to relevant institutions,
- record on possession of site,
- renewal of construction stake-out of truck forest road axle layout,
- execution of works of truck forest road construction,
- permanent and occasional work supervision,
- taking-over certificate.

The works during truck forest road construction (*Pentek, 2010*) are divided into several main groups: preparatory works, works on lower structure, improvement of the soil with various stabilization methods, works on facilities of underground and surface drainage, works on incline/slope stabilization, works on upper structure and other works.

Until the middle of the 1990s, the construction of truck forest roads in all relief conditions was performed by means of dozers, while explosives and pneumatic hammers were used on rocky soils. Nowadays, in general, dozers are used in lowlands and for lighter material construction categories, while excavators fitted with hydraulic hammers are used on sloping terrains and rocky media. Explosives are applied only for the toughest rocks where the use of a hydraulic hammer would not yield satisfying results.

The choice of technology to be used in the construction of a certain truck forest road depends on the following: relief characteristics of the terrain where the works are performed, material construction categories on the truck forest road route, economic indicators, availability of construction machinery and equipment, valid regulations in the area of forestry, civil engineering, protection of nature and the environment, as well as other influential factors.

Lowland area (flat terrain)	Hilly and mountainous area (inclined terrain)
• non-bearing and poor capacity soil (necessary soil stabilization)	• heavy material construction categories (sometimes the use of explosives required)
• lack of rocky material on the forest road route	• large transversal inclinations of grounds
• distance of a quarry from the site (high cost of stone material transport)	• deep cuttings and high embankments (construction of retaining and revetment walls)
• developed hydrographic network – constant large flow waterways (necessary construction of bridges)	• danger of stimulating erosion processes
• high level of underground water (construction of embankment in order to raise the forest road level line, construction of drainage ditches)	• necessity of using larger longitudinal level line slopes
• significant oscillations of water levels in a forest stand (construction of overflow channels),	• necessity of using minimum horizontal curves radius
• avoiding the compartmentalization of forest areas (necessary construction of culverts)	• danger of sudden rush of water on the forest road route (construction of drainage ditches, culverts and soakaways)

Table 3. Most important problems encountered during truck forest road construction.

The construction of truck forest roads should be approached in a very responsible and professional manner, at the same time trying to the utmost to minimize their harmful (negative) influence on the forest ecosystem. Soil compaction and reduction of airiness, interruption of waterways, reduced biological activity in the soil, erosion, floods, landslide sites, etc., represent just some of the consequences of the construction and use of truck forest roads. A good way of achieving a balance between the forest ecosystem and truck forest roads is through high expertise and constant professional improvement of forest road designers, developed awareness of maintaining the forest ecosystem and knowledge of its functioning as a whole, good Technical Requirements for Forest Roads, valid and consistently respected regulations, good and thoroughly respected procedures of establishing optimal primary forest transportation system network (according to defined operational stages), and a selection of ecologically suitable construction technologies and construction supervision on more levels.

Expert supervision of truck forest roads construction on more levels is necessary in order to ensure the adherence to project documentation (main design of truck forest road), or transferring the vision and conception of forest roads designers from paper into the forest ecosystem. It is recommended to have designer's supervision because, besides guaranteeing expertise and a good knowledge of project documentation of a certain forest road, this kind of supervision provides good and fast elimination of possible vagueness or disputable situations.

Constant and occasional work supervision is performed on more levels:

- the construction site superintendent controls the machinist and keeps the Engineering Log and Engineering Record on a daily basis,
- the supervising engineer controls the contractors (machinist and construction site superintendent) and signs the Engineering Log and Engineering Record on a daily basis,
- occasionally and if necessary, the main supervising engineer controls the works,
- the final works control is performed by the Committee for Handover of Works.

2.6 Truck forest roads maintenance

Maintenance of truck forest roads represents a series of construction-technical procedures, which should be performed regularly in order to keep the roads in their original condition, in which they may complete all the tasks proscribed by the Management Plan. Construction cost and costs of truck forest road maintenance in the period of its depreciation (25 – 40 years, depending on the authors and calculation method) constitute the overall costs of truck forest road management.

As a rule (in normal weather conditions, the usual regime of usage and similar site and stand conditions), well-built truck forest roads require lower maintenance costs during the depreciation period than those truck forest roads in the construction of which the costs were cut down at the expense of quality or the works were in a hurry (each work needs to be realized within a certain time factor). In the end, the overall costs of well-built and well-kept truck forest roads are considerably lower than those of badly and quickly built roads (which often in certain periods of the year, during rough weather, cannot complete their tasks).

2.6.1 Forest road maintenance types

According to the frequency and regularity of performing maintenance works, there are several types of maintenance:

- Regular maintenance – consists of constant visiting and inspections of truck forest roads, as well as establishing possible defects and damages. Besides recording damages, we should also determine the measures for their elimination, define the time for the performance of works, necessary material, machinery and the number of workers, as well as calculate the costs of repair. The following works form a part of regular maintenance: cleaning of drainage ditches, culverts and other drainage facilities, cleaning of road upper structure, maintenance of incline/slope and road shoulders, mowing grass, maintenance of plants, etc.
- **Investment maintenance** – implies larger works on the earth road structure, replacement of damaged and worn-out culverts and drainages, repair of retaining and revetment walls, etc.
- **Periodic maintenance** – related to a certain period, season or particular circumstances (e.g. snow cleaning, works after sudden floods, etc.).

According to the component of a truck forest road being maintained, road maintenance may be divided into:

- lower structure maintenance:
  - maintenance of the earth road structure,
  - maintenance of the surface and underground drainage system,
  - maintenance of retaining and revetment walls,
  - maintenance of cuttings and embankment slopes,
  - vegetation maintenance (also includes road shoulder maintenance),
  - bridge maintenance;
- upper structure maintenance.

3. Research goals and methods

3.1 Research goal

Research Goals are defined by these encompassed and logical units:

- classification of Management Units (MU) and Forest Administration Units (FAU) into relief categories,
- establish the existing primary road density (Management Units, FAUs and relief categories),
- calculate the length of the planned truck forest road network (in FAUs and in relief areas, with the purpose of achieving the planned primary road density until 2010 and 2020),
- calculate the construction cost of the planned truck forest road network (in FAUs and in relief areas, with the purpose of achieving the planned primary road density until 2010 and 2020),
- suggest guidelines for further primary opening of the forests of the Republic of Croatia.

3.2 Research methods

3.2.1 Classification of MUs and FAUs into relief categories

There are four categories of relief areas: lowland area, hilly area, mountainous area and karst area. According to Management Plans, each Management Unit is located within its relief category. The surfaces of each relief category on the level of FAUs were summed up in order to calculate first the absolute, and then the percentage share of each relief category in the overall FAU surface.

3.2.2 Establish the existing primary road density

The existing primary road density by Management Units will be determined on the basis of the cadastre of primary forest transportation system, constituted on the level of the company »Hrvatske Šume« Ltd. Zagreb, as it was on 31st December 2009. This is followed by the collection of all data from all Management Units of the same relief category on the level of FAUs and the overall research area.

3.2.3 Calculate the length of the planned truck forest road network

The difference between the existing and the planned primary road density (year 2010 and 2020) of an individual FAU and its surface gives the overall length of planned truck forest roads which ought to be built. It is assumed that all future truck forest roads will be a part of the calculation of openness with their entire length. Planned primary road density in 2020 for lowland relief area has not been calculated; lowland area has not been the subject of research neither the length nor the cost of planned truck forest road network in 2020.

3.2.4 Calculate the construction cost of the planned truck forest road network

Cost analysis of the new planned truck forest road network will be constructed according to technical characteristics of forest roads proscribed by the valid Technical Requirements for Industrial Roads (*Šikić et al., 1989*) and planned costs of truck forest road construction in various relief categories by the company »Hrvatske Šume« Ltd. Zagreb.

3.2.5 Suggest guidelines for further primary opening of the forests in the republic of Croatia

The dynamics and the priorities of the development of the existing primary forest transportation system network will be defined through a detailed analysis of the need for the construction of primary forest transportation system for the following 20 years, upon examination of the past dynamics of truck forest road construction, taking into consideration the financial, productional, organizational, expert and technical-technological capacities of the company »Hrvatske Šume« Ltd. Zagreb.

4. Research area

Research was conducted on the area of 15 Forest Administration Units which are a part of the company »Hrvatske Šume« Ltd. Zagreb. Owing to the lack of data, FAU Split was not included in the research.

Fig. 4. Location and area of each Forest Administration Unit.

5. Research results

5.1 Classification of FAUs into relief categories and determining the existing primary road density

On the basis of the conducted relief area classification (Fig. 5.), FAUs have been grouped into relief categories for the purpose of easy reference and result comparability.

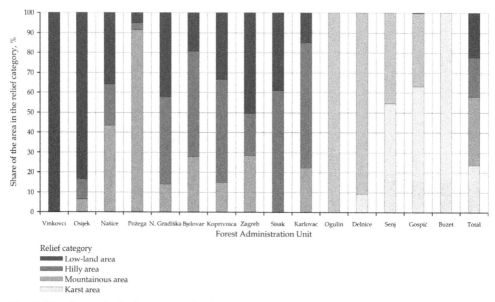

Fig. 5. The share of relief area in each of the FAU.

In the overall forest surface under research (1,442,140 ha), the lowland area accounts for 322,320 ha (22.35%), the hilly area for 282,560 ha (19.59%), the mountainous area for 497,830 ha (34.52%) and the karst area for 339,430 ha (23.54%). Only three FAUs are located with their entire surface in one relief category (FAU Vinkovci in lowland area, FAU Ogulin in the mountainous area and FAU Buzet in the karst area). The areas under management by the other FAUs stretch throughout two (four FAUs) or three (eight FAUs) relief categories.

In the lowland area, the average existing primary road density amounts to 8.85 m/ha, in the hilly area 11.26 m/ha, in the mountainous area 15.64 m/ha and in the karst area 7.63 m/ha.

By comparing the existing density of the primary forest transportation system by a particular relief area category, we may conclude the following: in the lowland area, the greatest primary road density exists in the FAU Karlovac (16.08 m/ha), and the least in FAU Osijek (4.37 m/ha); in the hilly area, FAU Koprivnica (17.02 m/ha) has the best primary road density, while FAU Sisak (6.88 m/ha) has the lowest classical primary openness; in the mountainous area, the highest degree of classical primary openness is in the FAU Našice (23.27 m/ha), and the lowest in FAU Gospić (10.32 m/ha); in the karst area the greatest priary road density is present in FAU Delnice (12.47 /ha), and the lowest in FAU Gospić (5.76 ha).

5.2 Calculate the length of planned truck forest road network in 2010 and 2020

The primary road density in 2010 and 2020 was calculated according to relief categories and the overall surface of each FAU. The length of truck forest roads to be built until the expiration of the planned period was set for both variants of primary road density. The results are shown in the Tables 5 and 6.

The data from the Tables 4 and 5 show that, although we stepped out of 2010, primary road density defined as planned for that year (Table 1) has not been achieved in any FAU in the whole area. In FAU Karlovac (lowland) and in FAU Delnice and Senj (karst) the planned primary road density for 2010 was exceeded (FAU Karlovac built 13.13 km, FAU Delnice 21.97 km, and FAU Senj 82.36 km of truck forest roads which influence on primary road density more than was planned).

Forest Administration Unit	Total area of FAU	Low-land area		Hilly area		Mountainous area		Karst area	
		Area	RD	Area	RD	Area	RD	Area	RD
	1000 ha	1000 ha	km/1000 ha	1000 ha	km/1000 ha	1000 ha	km/1000ha	1000 ha	km/1000 ha
VINKOVCI	72.37	72.37	6.84	-	-	-	-	-	-
OSIJEK	62.83	52.41	4.37	6.35	6.99	4.07	11.06	-	-
NAŠICE	82.95	29.73	12.70	17.26	16.84	35.96	23.27	-	-
POŽEGA	51.23	2.60	5.14	1.70	12.76	46.93	15.59	-	-
NOVA GRADIŠKA	73.57	31.07	8.73	32.21	8.52	10.29	11.70	-	-
BJELOVAR	131.83	25.42	11.92	69.84	11.68	36.57	11.84	-	-
KOPRIVNICA	62.37	20.79	13.74	32.29	17.02	9.29	17.50	-	-
ZAGREB	81.52	41.13	10.88	17.21	13.26	23.18	15.90	-	-
SISAK	87.99	34.28	6.68	53.71	6.88	-	-	-	-
KARLOVAC	82.45	12.11	16.08	51.99	11.32	18.35	12.02	-	-
OGULIN	59.58	-	-	-	-	59.58	14.11	-	-
DELNICE	96.31	-	-	-	-	87.41	22.55	8.90	12.47
SENJ	112.19	-	-	-	-	51.01	17.03	61.18	11.35
GOSPIĆ	312.67	0.41	12.10	-	-	115.19	10.32	197.07	5.76
BUZET	72.28	-	-	-	-	-	-	72.28	9.01
Total/Average	1,442.14	322.32	8.85	282.56	11.26	497.83	15.64	339.43	7.63

Table 4. Existing primary road density (RD) by relief areas in each FAU.

The greatest volume of new truck forest road construction should be carried out in FAU Gospić (2,529.04 km). According to relief areas, truck forest roads should be constructed mostly in FAU Vinkovci (590.36 km) in the lowland area, in FAU Sisak (704.66 km) in the hilly area in the mountainous area (1,691.35 km) and in the karst area (836.50 km) in FAU Gospić.

Planned primary road density for 2010 in the lowland area, in comparison with the existing primary road density in the same relief area, implicates the need of very intensive interventions of truck forest road construction (especially in the area of FAU Vinkovci and Osijek).

The obtained results should be observed in the context of historic guidelines in lowland forest management in FAU Vinkovci and Osijek (compartments of a symmetrical quadrangular shape with dimensions 750x750 m with a regular pattern of secondary forest roads, so called »šljukarica«, with the mutual distance among the middle of the passages (axle) from 37.5 m (*Posarić, 2007*)), but also in the sense of new (today accepted) technologies of timber harvesting in Croatian low-land forests.

Forest Administration Unit	Length of TFR									
	Low-land area		Hilly area		Mountainous area		Karst area		Total	
	km	km/1000 ha	km	km/1000 ha	km	km/1000 ha	km	km/1000 ha	km	km/1000 ha
VINKOVCI	590.36	8.16	-	-	-	-	-	-	590.36	8.16
OSIJEK	557.21	10.63	82.64	13.01	56.75	13.94	-	-	696.60	11.09
NAŠICE	68.41	2.30	54.47	3.16	62.21	1.73	-	-	185.09	2.23
POŽEGA	25.64	9.86	12.31	7.24	441.60	9.41	-	-	479.55	9.36
NOVA GRADIŠKA	194.70	6.27	369.93	11.48	136.81	13.30	-	-	701.44	9.53
BJELOVAR	78.39	3.08	581.08	8.32	481.14	13.16	-	-	1,140.61	8.65
KOPRIVNICA	26.19	1.26	96.12	2.98	69.64	7.50	-	-	191.95	3.08
ZAGREB	169.36	4.12	116.04	6.74	211.03	9.10	-	-	496.43	6.09
SISAK	285.38	8.32	704.66	13.12	-	-	-	-	990.04	11.25
KARLOVAC	0.00 (+13.13)*	-	451.44	8.68	238.12	12.98	-	-	689.56	8.36
OGULIN	-	-	-	-	648.54	10.89	-	-	648.54	10.89
DELNICE	-	-	-	-	213.99	2.45	0.00 (+21.97)*	-	213.99	2.22
SENJ	-	-	-	-	406.62	7.97	0.00 (+82.36)*	-	406.62	3.62
GOSPIĆ	1.19	-	-	-	1,691.35	14.68	836.50	4.24	2,529.04	8.09
BUZET	-		-	-	-	-	71.52	0.99	71.52	0.99
Total	1,996.83	6.20	2,468.69	8.74	4,657.80	9.36	908.02	2.68	10,031.34	6.96

* FAU Karlovac (low-land area), Delnice and Senj (karst area) built more roads than planned so the need to build by 2010. in these FAU is 0.00 km.

Table 5. Required length of truck forest roads that need to be built to achieve the planned primary road density for 2010 by the FAU and relief categories.

Forest Administration Unit	Length of TFR							
	Hilly area		Mountainous area		Karst area		Total	
	km	km/1000 ha	km	km/1000 ha	km	km/1000 ha	km	km/1000 ha
OSIJEK	114.39	18.01	77.10	18.94	-	-	191.49	3.05
NAŠICE	140.77	8.16	242.01	6.73	-	-	382.78	4.61
POŽEGA	20.81	12.24	676.25	14.41	-	-	697.06	13.61
NOVA GRADIŠKA	530.98	16.48	188.26	18.30	-	-	719.24	9.78
BJELOVAR	930.28	13.32	663.99	18.16	-	-	1,594.27	12.09
KOPRIVNICA	257.57	7.98	116.09	12.50	-	-	373.66	5.99
ZAGREB	202.09	11.74	326.93	14.10	-	-	529.02	6.49
SISAK	973.21	18.12	-	-	-	-	973.21	11.06
KARLOVAC	711.39	13.68	329.87	17.98	-	-	1,041.26	12.63
OGULIN	-	-	946.44	15.89	-	-	946.44	15.89
DELNICE	-	-	651.04	7.45	0.56*	0.06	651.60	6.77
SENJ	-	-	661.67	12.97	141.18*	2.31	802.85	7.16
GOSPIĆ	-	-	2,267.30	19.68	1,821.85	9.24	4,089.15	13.08
BUZET	-	-	-	-	432.92	5.99	432.92	5.99
Total	3,881.49	13.74	7,146.95	14.36	2,396.51	7.06	13,424.95	9.31

Values are reduced by the length of more constructed forest roads in the FAU Delnice and Senj by 2010 shown in Table 5.

Table 6. Required length of truck forest roads that need to be built to achieve the planned primary road density for 2020 by the FAU and relief categories.

According to the openness plan for 2020, the greatest volume of new truck forest road construction should be carried out in FAU Gospić (4,089.15 km). Analyzing the relief areas, the construction of most truck forest roads will be required: in the hilly area in FAU Sisak (973.21 km), in the mountainous area (2,267.30 km) and in the karst area (1,821.85 km) in FAU Gospić.

Forest Administration Unit	Current (existing):		Planned 2010th:		Planned 2020th: (without low-land area)	
	RD	Length of TFR	RD	Length of new TFR	RD	Length of new TFR
	km/1000 ha	km	km/1000 ha	km	km/1000 ha	km
VINKOVCI	6.84	495.19	15.00	590.36		
OSIJEK	5.07	318.30	16.15	696.60	26.95	191.49
NAŠICE	18.14	1,505.06	20.38	185.09	28.38	382.78
POŽEGA	14.97	766.70	24.33	479.55	29.83	697.06
NOVA GRADIŠKA	9.05	666.06	18.59	701.44	26.21	719.24
BJELOVAR	11.77	1,551.74	20.42	1,140.61	26.72	1,594.27
KOPRIVNICA	16.00	997.95	19.08	191.95	26.12	373.66
ZAGREB	12.81	1,044.22	18.90	496.43	27.87	529.02
SISAK	6.80	598.36	18.05	990.04	25.00	973.21
KARLOVAC	12.17	1,003.77	20.38	689.56	26.30	1,041.26
OGULIN	14.11	840.96	25.00	648.54	30.00	946.44
DELNICE	21.62	2,082.23	23.61	213.99	28.61	651.60
SENJ	13.93	1,562.79	16.82	406.62	21.82	802.85
GOSPIĆ	7.44	2,327.56	15.53	2,529.04	20.53	4,089.15
BUZET	9.01	651,28	10.00	71.52	15.00	432.92
Total/Average	11.38	16,412.17	18.25	10,031.34	24.19	13,424.95

Table 7. Existing and planned primary road density for 2010 and 2020 by the FAU and the length of truck forest roads that need to be built.

Relief category	Length of TFR	Existing TFR density	Planned length of TFR 2010th	Planned TFR density (until 2010)	Planned length of TFR 2020th	Planned TFR density (until 2020)
	km	km/1000 ha	km	km/1000 ha	km	km/1000 ha
Low-land area	2,851.10	8.85	1,983.70	15.00		
Hilly area	3,182.51	11.26	2,468.69	20.00	3,881.49 (1,412.80)	25.00
Mountainous area	7,787.95	15.64	4,657.80	25.00	7,146.95 (2,489.15)	30.00
Karst area	2,590.61	7.63	803.69	10.00	2,500.84 (1,697.15)	15.00
Total/Average	16,412.17	11.38	10,031.34	18.25	13,424.95 (3,393.61)	24.19

() Length of forest roads that need to be constructed in 2010-2020 to achieve the planned openness 2020th.

Table 8. Existing and planned primary road density for 2010 and 2020 in different relief areas and length of truck forest roads that need to be built.

5.3 Calculation of the cost of planned truck forest road construction in 2010 and 2020

On the basis of planned costs of truck forest road construction for each relief area (*Anon., 2010*): lowland area (500,000.00 HRK/km), hilly area (350,000.00 HRK/km), mountainous area (250,000.00 HRK/km) and karst area (225,000.00 HRK/km) and the applicable Technical Requirements for Economic Roads (*Šikić et al. 1989*), the total price was calculated for all truck forest roads which need to be built for achieving the planned primary road density in 2010 and 2020.

Forest Administration Unit	The construction costs of a new truck forest roads network, €			
	Planned by 2010		Planned by 2020 (without low-land area)	
	kn	€*	kn	€*
VINKOVCI	295,180,000.00	40,056,234.51		
OSIJEK	321,716,500.00	43,657,265.30	59,311,500.00	8,048,632.54
NAŠICE	68,822,000.00	9,339,217.33	109,772,000.00	14,896,175.13
POŽEGA	127,528,500.00	17,305,750.74	176,346,000.00	23,930,336.51
NOVA	261,028,000.00	35,421,772.42	232,908,000.00	31,605,859.03
BJELOVAR	362,858,000.00	49,240,209.85	491,595,500.00	66,710,023.15
KOPRIVNICA	64,147,000.00	8,704,814.94	119,172,000.00	16,171,764.95
ZAGREB	178,051,500.00	24,161,774.64	152,464,000.00	20,689,524.15
SISAK	389,321,000.00	52,831,266.61	340,623,500.00	46,222,964.96
KARLOVAC	217,534,000.00	29,519,591.16	331,454,000.00	44,978,654.23
OGULIN	162,135,000.00	22,001,888.96	236,610,000.00	32,108,224.30
DELNICE	53,497,500.00	7,259,666.66	162,886,000.00	22,103,800.44
SENJ	101,655,000.00	13,794,689.75	197,183,000.00	26,757,939.19
GOSPIĆ	611,645,000.00	83,000,865.77	976,741,250.00	132,544,808.49
BUZET	16,092,000.00	2,183,701.22	97,407,000.00	13,218,231.71
Total	3,231,211,000.00	438,478,709.86	3,684,473,750.00	499,986,938.77

* Middle exchange rate of euro in the Croatian National Bank on day 13.11.2010 (1 € = 7.36914 kn).

Table 9. Construction costs of the planned network of truck forest roads in 2010 and 2020 according to the current Technical Requirements.

In order to achieve the planned primary road density in 2010, at the level of »Hrvatske Šume« Ltd. Zagreb, according to the applicable Technical Requirements, it is necessary to invest HRK 3,231,211,000, and for achieving the planned primary classical openness in 2020 it is necessary to invest HRK 3,684,473,750 (without the lowland relief area).

Fig. 7. shows the dynamics of the construction of lower and upper truck forest road structure for the period from 2004 - 2009 by Forest Administration Units. During the six observed years, a total of 1,295.86 km of lower structure and 1,438.64 km of upper structure were built, or on average 215.98 km of lower structure and 239.77 km of upper structure of truck forest roads per year. The greatest amount of lower and upper structure were built in 2006 (246.69 km and 307.49 km), and the least in 2009. The difference between the constructed lengths of lower and upper structures indicate that truck forest roads are not always constructed all at once, but the construction procedure extends over two or more years. Most often the reasons are the following: lack of financial means, unfavourable weather conditions at the end of the year, construction technology or simply, because that is how it was conceived by the construction plan.

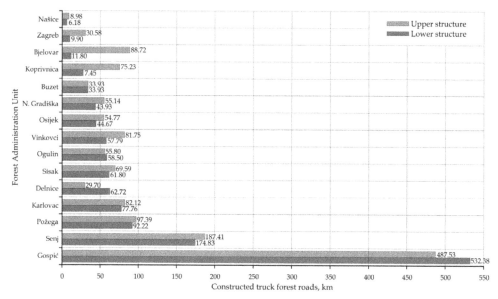

Fig. 6. Construction of the lower and upper structure of the TFR for the period from 2004 to 2009 by the FAU-s.

6. Discussion

A significant diversity of Croatian forestry from the point of view of terrain orthography (lowland, hilly, mountainous and karst), site and stand characteristics, as well as the way of forest management (regular, selection), but also the ways of forest opening in certain orthographic areas, or the degree of forest road density, indicates the need of good planning in the entire forestry department, as well as in timber harvesting works, or forest opening. Quality and reliable planning, in this case of forest roads, guarantees rationalization (a collection of procedures for achieving savings in business) in a part of forestry production.

Planning forest roads is a starting, unavoidable and very important stage of establishing an optimal forest road network in the field. Planning may be strategic, tactical and operative. On the level of strategic planning, we use the so called descriptive (primary) classification of terrains, which describes a terrain according to measurable characteristics and divides it into categories, independently of the applied timber harvesting systems.

Planned values of primary road density in 2010 and 2020 on the level of relief categories represent only guidelines which should be followed, but should not (and must not) be strictly adhered to in the final design of primary forest transportation system network. It is recommended to re-examine and, if necessary, redefine the values of planned primary road density by relief areas (Table 1), at the same time recognizing all the factors which have an influence on the calculation of planned primary forest transportation system density.

At the lower, also more accurate, planning stage (tactical planning), it is possible to deviate from the values defined at the strategic level, both in positive and negative, but acceptable intervals. Target (optimal/best possible) primary road density is defined on the level of a Management Unit; Studies of Primary Forest Opening are made at this level.

The accuracy of planning is the greatest and suits the best to the actual condition at the lowest planning level, but it also requires the most precise and the most complete data and information. In operative planning, sometimes even the division into Management Units is not accurate enough, because within the same Management Unit there may be two or more (very rarely) relief categories. This planning level presupposes a purposeful (functional or secondary) terrain classification, which associates the possibility of application of potential and suitable harvesting systems with classes of terrain factors. The harvesting system is determined by procedures, method of timber processing (cut-to-length, half-tree, full-tree, tree-length), as well as machines and tools used in the harvesting of a cut-block. The selection (use) of a timber extraction device (skidder, forwarder, adapted farm tractor – AFT, AFT with semitrailer, cable yarder) in the light of the influence of terrain (relief categories) and stand factors, as well as the level of primary and secondary forest openness is the most important determinant of the entire harvesting system.

By analyzing the existing primary road density by FAUs and by relief categories conclusively with 31st December 2009 at the researched area, it is concluded that the planned primary road density has not been achieved on the greatest part of state-owned forests. Moreover, in the better part of the forests, not even the minimal necessary primary road density has been reached, specifically, in the lowland area in 4 out of 11 FAUs, in the hilly area in 5 out of 9 FAUs, in the mountainous area in 6 out of 12 FAUs, which has a significant negative influence on the quality, efficiency and rationality of management in these insufficiently opened forests.

With the average annual intensity of truck forest road construction, based on the data from 2004 – 2009 (taking into consideration the average constructed length of the lower truck forest road structure of about 216 km/y), it would take 47 years to achieve the planned openness in 2010, and 63 years to achieve the same for 2020 (without the construction of truck forest roads in lowland area).

It should be considered to extend the time period for the attempt to achieve the planned values of primary forest transportation system density. At the same time, the construction of truck forest roads should be intensified, and raised from the present 216 km/y to at least 600-800 km/y. Thus, the planned values of primary road density for 2010 (counting with the construction of 700 km/y of truck forest roads) might be achieved in about 15 years, and for

2020 in about 20 years (without the construction of truck forest roads in the lowland area). Because of that, the objective financial, expert and infrastructural capacities of the company »Hrvatske Šume« Ltd. should be acknowledged. Besides the existing sources of financing, the means from current business operations and the means from the fund for generally beneficial forest functions, it is necessary to search for other sources of financing of all the stages for establishing an optimal primary forest transportation system network, e.g. EU funds, etc.

There is a significant gap among the primary road density of forest areas which belong to the same relief category. In the future, while planning investments into the rebuilding and development of primary forest transportation system, the priority orientation of financial means into the less opened forest areas should be taken into consideration, all until there is a balance of primary road density on the level of the overall relief area.

The values of primary road density (existing, as well as planned for 2010 and 2020) will be compared with the average values of the existing primary road density in the forests of the Republic of Austria (*Stampfer 2011* according to *Austrian Forest Inventory 1992/96*). In order to make the comparison complete, the basic characteristics of Austrian forests and harvesting works are provided:

- forest ownership (48.3% are private forests with the surface of less than 200 ha, 22.4% are private forests with the surface of more than 200 ha, 15.7% are state-owned forests, and 15,6% are forests owned by other forest owners – community forests, communal forests and provincial forests),
- inclination of the terrain where forests grow (more than 22% of the forests grow on inclined terrains with an inclination of more than 60%, and 39% of forests on terrains inclined 30-60%; other forests are situated on inclinations up to 30%),
- particularities (way) of forest management,
- generally useful forest functions,
- harvesting systems (procedures, method of timber processing, machines and tools used):
  - applied device for timber cutting and processing (82.41% of harvesting volume is cut and processed with a chainsaw, and 17.59% with a harvester),
  - means used for timber extraction (skidder extracts 53.1%, forwarder 26.8, cable yarder 14.2%, manually 4.6%, horse-power 0.4% and other ways – e.g. helicopter 0.9% of the overall annual allowable cut).

Forest ovners	Road density km/1000 ha
Smale scale forest owners (< 200 ha)	49.1
Private Companies (> 200 ha)	41.8
Federal Austrian Forests	33.7
Average	**45.0**

Table 10. The existing primary road density in the Republic of Austria (*Stampfer 2011* according to *Austrian Forest Inventory 1992/96*).

In smaller forest properties, where less modern machinery is used for timber extraction (mostly AFTs and AFTs with semitrailers); there is a denser primary forest transportation system network owing to the rationalization of overall timber harvesting costs.

The existing primary road density in the Republic of Austria (in Austrian state forests) is far greater than the planned openness in the hilly area of the Republic of Croatia for 2020; of course, there is an even greater difference when comparing with the planned openness for 2010, while the greatest differences exist when comparing with the existing primary road density. There are Management Units in Croatian state forests whose primary road density is on the level of those in Austrian state forests, but these are rare. As an example, there are three Management Units situated in selection forests of Gorski Kotar, owned by the state but managed by the University of Zagreb Faculty of Forestry. The primary road density in those units amounts to between 32 and 36 km/1000 ha, with the average values of mean timber extraction distance of 150 m.

Habsburg (1970), Sanktjohanser (1971) and Piest (1974) agree that the optimal primary road density for the needs of forest exploitation varies between 17 and 30 m/ha, depending on the terrain and site characteristics, while the optimal primary road density for rational overall forest management is a bit greater. The suggested values correspond very well to the planned primary road density in Croatia for 2010 and 2020, while there are slight deviations in lowland forests and karst area forests. Considering the time distance and the development of the entire timber harvesting system, as well as the total forest ecosystem management that has been established in the meantime, the recommended primary forest transportation system density by the three above mentioned authors should be taken with a grain of salt.

7. Concluding remarks

The planned values of primary road density in 2010 and 2020 on the level of various relief categories of the Republic of Croatia, besides being the guidelines for strategic planning in the Republic of Croatia, may also be used as landmarks in primary forest road planning on a strategic level in countries of similar orographic, site and stand conditions, as well as the ways of forest management. The existing primary road density should certainly be taken into consideration, and in accordance with the financial, professional and infrastructural resources of a certain country and its forestry, the deadlines for achieving the planned values of primary road density, annual intensity and construction priorities should be defined.

The more developed countries and countries with a long forestry tradition, which could have invested significant financial means permanently and systematically into the primary openness of their forests during the last few decades, are expected to have a better primary road density than the Republic of Croatia (which could have started with a more systematic and more intensive forest opening only after being proclaimed independent in the 1990s), and therefore, less need for primary classical openness in the future (with the purpose of achieving planned primary road density) and probably less differences in the existing primary road density of equal or similar (comparable) forest areas, that is, they have a uniform existing primary road density of the same relief categories. A multiple use of truck forest roads, by first of all users outside forestry (e.g. tourism), contributes to greater density and better quality of truck forest roads.

Operative planning, as the lowest and the most accurate level of forest road planning, requires a purposeful analysis of the terrain, connecting the possibility of applying potential and suitable timber harvesting systems with terrain factor analysis. At this planning level, timber harvesting systems have a significant influence on the shape and density of the truck forest roads, but even more on the on the shape and density (and existence, in general) of the secondary forest road network. The application of certain timber harvesting systems is, besides the terrain factors, conditioned by the degree of technological growth (technological awareness), resulting in utilization (the possibility of using) the most up-to-date means of timber extraction, and connected with that, the procedures and methods of timber processing. The selection of a timber extraction system is often under the influence of traditional forestry values of a certain country.

This paper describes and applies the methodology, but it can serve as a starting point for making a case study in any European and non-European country. Individual differences (specific qualities) of a certain country should be recognized and integrated into the modified methodology in a proper way to make the research results achieve an expected high level.

The data about the primary road density does not say much about the quality of spatial distribution of primary forest transportation system components. For better understanding of the real value of primary road density, it is always necessary to present primary classical openness coupled with the average timber extraction distance, or the mean distance of access to the endangered forest area, in the case of forest fire-prevention roads in the karst area. A clear insight into the real, quantitative (amount of primary forest roads) and qualitative (spatial coverage with primary forest roads) parameters of primary forest transportation system may only be achieved by a parallel consideration of primary road density and mean timber extraction distance.

8. References

Anon. (1997). Austrian Forest Inventory 1992/96.
Anon. (1979. Izvješće o problematici gradnje i održavanja šumskih i protupožarnih prometnica i stanju otvorenosti šuma (Report on the issue of construction and maintenance of truck forest economic roads and forest fire-prevention roads and the condition of forest openness), J.P. Hrvatske šume, Zagreb, pp. 1-11.
Anon. (1997b). Prijedlog metodologije izrade katastra šumskih i protupožarnih prometnica na području J.P. Hrvatske šume (Proposition of the methodology of making truck forest economic and forest fire-prevention roads cadastre on the area company Hrvatske šume), J.P. Hrvatske šume, Zagreb, pp. 1-14.
Anon. (2006). Šumskogospodarska osnova područja Republike Hrvatske, razdoblje 2006-2015 godina (Management plan basis of the area of the Republic of Croatia, period 2006 - 2015 year).
Anon. (2010). Izvješće o izgradnji donjeg i gornjeg ustroja šumskih cesta na području HŠ d.o.o. Zagreb za razdoblje 2004 – 2009 (Report on the construction of lower and

upper truck forest road structure in the area of Hrvatske Šume Ltd. Zagreb for the period of 2004 – 2009).

Chung, W., J. Stückelberge, K. Aruga, T. W. Cundy (2008). Forest road network design using a trade-off analysis between skidding and road construction costs. *Canadian Journal of Forest Research* 38(3): pp. (439-448). ISSN 0045-5067.

Habsburg, U. (1970). Sind Knickschlepper und Forststrassen Gegensätze ? Betrachtungen über den Einfluss der Rückemethoden auf den Wegeabstand. (Are folding draggers and forest roads contradictions ? Reflections on the impact of the extraction method on the distance between roads.). Allgemine Forstzeitung.

Kangas&Kangas, A. (2002). Multiple criteria decision support methods in forest management. An overview and comparative analysis. In. Pukkala, T. (ed.). *Multi objective forest management.* (37-70).

Nevečerel, H., Pentek, T., Pičman, D., Stankić, I. (2007). Traffic load of forest roads as a criterion for their categorization – GIS analysis, *Croatian Journal of Forest Engineering*, 28(1): pp. (27-38). ISSN 1845-5719.

Pentek, T. (1998). Šumske protupožarne ceste kao posebna kategorija šumskih cesta i čimbenici koji utječu na njihov razmještaj u prostoru (Forest fire prevention roads as a special category of forest roads and factors that influence their distribution in space). *Glasnik za šumske pokuse.* 35: pp (93-141).

Pentek, T., Pičman, D. (2001). Šumske protupožarne prometnice – osnovne zadaće, planiranje i prostorni raspored (Forest fire-prevention roads – basic tasks, planning and lay-out), Znanstvena knjiga, pp. 545-554.

Pentek, T. (2002). Računalni modeli optimizacije mreže šumskih cesta s obzirom na dominantne utjecajne čimbenike (Computer models of forest road network optimization regarding dominant influencing factors). Disertacija, Šumarski fakultet Sveučilišta u Zagrebu, pp. 1-271.

Pentek, T., Pičman, D. (2003). Uloga šumskih prometnica pri gospodarenju šumama na kršu s posebnim osvrtom na Senjsku Dragu (The role of forest roads in forest management on karst with the special reference to Senjska Draga., *Šumarski list,* vol. 127 (suplement), pp. (65-78): ISSN 0373-1332

Pentek, T., Pičman, D., Krpan, A., Poršinsky, T. (2003). Inventory of primary and secondary forest communications by the use of GPS in Croatian mountainous forest, Proceedings of Austro 2003 (FORMEC) *CD/DVD MEDIJ - High Tech Forest Operations for Mountainous Terrain,* Schlaegl, Austrija, 5-9.10.2003., pp. (1-12).

Pentek, T., Pičman, D., Nevečerel, H. (2004). Srednja udaljenost privlačenja drva (The mean timber skidding distance). *Šumarski list* 127(9-10): pp. (545-558): ISSN 0373-1332.

Pentek, T., Pičman, D., Nevečerel, H. (2004). Environmental-ecological component of forest road planning and designing. Proceedings of International scientific conference: *Forest constructions and ameliorations in relation to the natural environment,* Technical University in Zvolen, Slovakia, 16th – 17th September 2004. *CD/DVD MEDIJ,* pp. (94-102).

Pentek, T., Pičman, D., Potočnik, I., Dvorščak, P., Nevečerel, H. (2005). Analysis of an existing forest road network, *Croatian Journal of Forest Engineering*, 26(1). pp. (39-50): ISSN 1845-5719.

Pentek, T., Pičman D., Nevečerel, H. (2005b). Planiranje šumskih prometnica – postojeća situacija, determiniranje problema i smjernice budućeg djelovanja (Planning of forest roads – current status, identifying problems and trends of future activities), *Nova mehanizacija šumarstva*, 26(1), pp. 55-63. ISSN 1845-8815

Pentek, T., Pičman, D., Nevečerel, H. (2006a). Uspostava optimalne mreže šumskih cesta na terenu – smjernice unapređenja pojedine faze rada (Establishing the optimum forest road network on the terrain – guidelines for improving individual work stages), *Glasnik za šumske pokuse*, Posebno izdanje 5, pp. (647-663).

Pentek, T., Pičman, D., Nevečerel, H. (2006b). Definiranje faza postupka optimiziranja mreže ŠC-a sa dizajniranim dijagramima toka podataka (Defining procedure stages of forest road network optimizing with designed of data flow diagrams), *Glasnik za šumske pokuse*, Posebno izdanje 5, pp. (665-677).

Pentek, T., Nevečerel, H., Pičman, D., Poršinsky, T. (2007a). Forest road network in the Republic of Croatia – status and perspectives, *Croatian Journal of Forest Engineering*, 28 (1), pp. (93-106): ISSN 1845-5719.

Pentek, T., Nevečerel, H., Poršinsky, T., Horvat, D., Šušnjar, M., Zečić, Ž. (2007b). Quality planning of forest road network – precondition of building and maintenance cost rationalisation, *Proceeding of Austro2007/FORMEC´07: Meeting the Needs of Tomorrows´ Forests – New Developments in Forest Engineering*, BOKU, Vienna, Austria, 07-11.10.2007.; *CD ROM*.

Pentek, T., Poršinsky, T., Šušnjar, M., Stankić, I., Nevečerel, H., Šporčić, M. (2008). Environmentally sound harvesting technologies in comercial forests in the area of Northern Velebit – Functional terrain classification, *Periodicum Biologorum*, 110 (2), pp. (127-135): ISSN 0031-5362.

Pentek, T., Pičman, D., Nevečerel, H., Lepoglavec, K., Poršinsky, T. (2008). Road network quality of the management unit Piščetak – GIS analysis; *Proceedings of the 3rd international scientific conference FORTECHENVI 2008* / Skoupy, Alois ; Machal, Pavel ; Marecek, Lukas (ur.). Brno : Mendel University of Agriculture and Forestry, 2008. pp. (45-53).

Pentek, T., Nevečerel, H., Poršinsky, T., Pičman, D., Lepoglavec, K., Pičman, Potočnik, I. (2008). Methodology for development of secondary forest traffic infrastructure cadastre, *Croatian Journal of Forest Engineering*, 29(1), pp. (75-83): ISSN 1845-5719.

Pentek, T. (2010a). Predavanja-prezentacije iz nastavnog predmeta Otvaranje šuma (Lecture-presentation from the subject Opening up of forests), pptx prezentacije (1-10).

Pentek, T. (2010b). Predavanja-prezentacije iz nastavnog predmeta Šumske prometnice (Lecture-presentation from the subject Forest roads), pptx prezentacije (1-13).

Pentek, T., Pičman, D., Nevečerel, H., Lepoglavec, K., Papa, I., Potočnik, I. (2011). Primarno otvaranje šuma različitih reljefnih područja Republike Hrvatske (Primary forest opening of different relief areas in the Republic of Croatia), *Croatian Journal of Forest Engineering*, 32(1), pp. (401-416): ISSN 1845-5719.

Pičman, D., Pentek, T. (1996). Čimbenici koji utječu na opravdanost izgradnje mreže šumskih prometnica (Factors affecting the validity of a forest road network building). Savjetovanje »Skrb za hrvatske šume od 1846. do 1996.«, Znanstvena knjiga 2 »Zaštita šuma i pridobivanje drva«, pp. (293-300).

Pičman, D., Pentek, T., Družić, M. (1997). Utjecaj troškova izgradnje i održavanja šumskih cesta na njihovu optimalnu gustoću u nizinskim šumama Hrvatske (The impact of construction and maintenance costs of forest roads on their optimal density in the lowland forests of Croatia), *Mehanizacija šumarstva*, 22 (2), pp. (95-101): ISSN 0352-5406.

Pičman, D., Pentek, T. (1998). Rašclamba troškova izgradnje šumskih protupožarnih cesta i mogućnosti njihova smanjenja (Analysis of construction costs of forest fire-prevention roads and possibilities of reducing them), *Mehanizacija šumarstva*, 23(3-4), Zagreb, pp. (129-137): ISSN 0352-5406.

Pičman, D., Pentek, T. (1998). Relativna otvorenost šumskoga područja i njena primjena pri izgradnji šumskih protupožarnih prometnica (Relative opness of forest area and its use in the construction of the forest fire-prevention roads), *Šumarski list*, 122(1-2), pp. (19-30): ISSN 0373-1332.

Pičman, D., Pentek, T. (1998). Rašclamba normalnog poprečnog profila šumske protupožarne ceste i iznalaženje troškovno povoljnijih modela. (Analysis of a normal cross section of a forest fire-prevention road and finding expenditure favourable models). *Šumarski list*, 122(5-6), pp. (235-243): ISSN 0373-1332.

Pičman, D., Pentek, T., Poršinsky, T. (2001). Relation between Forest Roads and Extraction Machines in Sustainable Forest Management, FAO/ECE/ILO & IUFRO *Workshop on "New Trends in Wood Harvesting with Cable Systems for Sustainable Forest Management in the Mountains"*, Osiach, Austrija, 18-24.06. Workshop Proceedings, June 2001, pp. 185-191.

Pičman, D., Pentek, T., Poršinsky, T. (2002). Application of modern technologies (GIS, GPS) in making methodological studies on the primary open of hilly-mountain forests. *Forest Information Technology 2002 – International Congress and Exhibition*, 3-4 September, 2002 Helsinki, Finland. Proceedings pp. 1-10.

Pičman, D., Pentek, T., Nevečerel, H. (2006). Otvaranje šuma šumskim cestama – odabir potencijalnih lokacija trasa budućih šumskih cesta (Forest opening by forest roads – choosing the potential locations of the future forest road routes), *Glasnik za šumske pokuse*, Posebno izdanje 5, pp. (617-633).

Pičman, D., Pentek, T., Nevečerel, H. (2006). Katastar šumskih prometnica – postojeće stanje, metodologija izradbe i polučene koristi (Forest road cadastre – the present condition, the working methodology and obtained uses), *Glasnik za šumske pokuse*, Posebno izdanje 5, pp. (635-646).

Piest, K. (1974). Einfüsse auf Walderschliessung und Wegegestaltung, (Impacts on forest development and road design), *Forsttecchnische Informationen*, Nr. 3, pp. (27-30).

Posarić, D. (2007). Vodič za revirničke poslove (Guide to District Forestry Works), Hrvatske šume d.o.o. Zagreb, pp. (1-225).

Potočnik, I. (1998). The multiple use of roads and their classification. *Proceedings of the Seminar on environmentally sound forest roads and wood transport* : Sinaia, Romania, 17-22 June 1996. Rome: Food and agriculture organization of the United Nations, pp. (103-108).

Potočnik, I. (1998). The environment in planning a forest road network. *Proceedings: Environmental Forest Science:* Kyoto, Japan on 19-22 October 1998., pp. 67-74.

Potočnik, I., Pentek, T., Pičman, D. (2005). Impact of traffic characteristics on forest roads due to forest management, *Croatian Journal of Forest Engineering*, 26(1), pp. (51-57): ISSN 1845-5719.

Sanktjohanser, L. (1971). Zur Frage der optimalen Wegendichte in Gabirgswaldungen. Forstwissenschaftliches (On the question of the optimum road density in mountain forests), Centralblatt, Nr. 3. pp. (142-153).

Šikić D., B. Babić, D. Topolnik, I. Knežević, D. Božičević, Ž. Švabe, I. Piria, S. Sever (1989). Tehnički uvjeti za gospodarske ceste (Technical Requirements for Economic Roads), Znanstveni savjet za promet Jugoslavenske akademije znanosti i umjetnosti, pp. 1-78.

Permissions

The contributors of this book come from diverse backgrounds, making this book a truly international effort. This book will bring forth new frontiers with its revolutionizing research information and detailed analysis of the nascent developments around the world.

We would like to thank Dr. Juan A. Blanco and Dr. Yueh-Hsin Lo, for lending their expertise to make the book truly unique. They have played a crucial role in the development of this book. Without their invaluable contribution this book wouldn't have been possible. They have made vital efforts to compile up to date information on the varied aspects of this subject to make this book a valuable addition to the collection of many professionals and students.

This book was conceptualized with the vision of imparting up-to-date information and advanced data in this field. To ensure the same, a matchless editorial board was set up. Every individual on the board went through rigorous rounds of assessment to prove their worth. After which they invested a large part of their time researching and compiling the most relevant data for our readers. Conferences and sessions were held from time to time between the editorial board and the contributing authors to present the data in the most comprehensible form. The editorial team has worked tirelessly to provide valuable and valid information to help people across the globe.

Every chapter published in this book has been scrutinized by our experts. Their significance has been extensively debated. The topics covered herein carry significant findings which will fuel the growth of the discipline. They may even be implemented as practical applications or may be referred to as a beginning point for another development. Chapters in this book were first published by InTech; hereby published with permission under the Creative Commons Attribution License or equivalent.

The editorial board has been involved in producing this book since its inception. They have spent rigorous hours researching and exploring the diverse topics which have resulted in the successful publishing of this book. They have passed on their knowledge of decades through this book. To expedite this challenging task, the publisher supported the team at every step. A small team of assistant editors was also appointed to further simplify the editing procedure and attain best results for the readers.

Our editorial team has been hand-picked from every corner of the world. Their multi-ethnicity adds dynamic inputs to the discussions which result in innovative outcomes. These outcomes are then further discussed with the researchers and contributors who give their valuable feedback and opinion regarding the same. The feedback is then collaborated with the researches and they are edited in a comprehensive manner to aid the understanding of the subject.

Apart from the editorial board, the designing team has also invested a significant amount of their time in understanding the subject and creating the most relevant covers. They scrutinized every image to scout for the most suitable representation of the subject and create an appropriate cover for the book.

The publishing team has been involved in this book since its early stages. They were actively engaged in every process, be it collecting the data, connecting with the contributors or procuring relevant information. The team has been an ardent support to the editorial, designing and production team. Their endless efforts to recruit the best for this project, has resulted in the accomplishment of this book. They are a veteran in the field of academics and their pool of knowledge is as vast as their experience in printing. Their expertise and guidance has proved useful at every step. Their uncompromising quality standards have made this book an exceptional effort. Their encouragement from time to time has been an inspiration for everyone.

The publisher and the editorial board hope that this book will prove to be a valuable piece of knowledge for researchers, students, practitioners and scholars across the globe.

List of Contributors

Maria de Lourdes Pinheiro Ruivo, Cristine Bastos Amarante and Maria Lucia Jardim Macambira
Museu Paraense Emilio Goeldi, Brazil

Antonio Pereira Junior and Quezia Leandro Moura
Program of Post-Graduate in Environmental Sciences/Federal University of Pará, Brazil

Keila Chistina Bernardes
Post-Graduate Program in Agronomy/ Federal Rural University of Amazonia, Brazil

Ayu Toyota
Institute of Soil Biology, Biology Centre, Academy of Sciences of Czech Republic, České Budějovice Soil Ecology Research Group, Czech Republic
Yokohama National University, Yokohama, Japan

Nobuhiro Kaneko
Soil Ecology Research Group, Yokohama National University, Yokohama, Japan

Masamichi Takahashi and Keizo Hirai
Forestry and Forest Products Research Institute, Tsukuba, Japan

Dokrak Marod
Faculty of Forestry, Kasetsart University, Bangkok, Thailand

Samreong Panuthai
National Parks, Wildlife and Plant Conservation Department, Bangkok, Thailand

Su-Jin Kim, Hyung Tae Choi, and Chunghwa Lee
Division of Forest Water & Soil Conservation, South Korea

Kyongha Kim
Division of Forest Disaster Management, Department of Forest Conservation, Korea Forest Research Institute, South Korea

Anna Augustyniuk-Kram
Institute of Ecology and Bioethics, Cardinal Stefan Wyszyński University, Warszawa, Polish Academy of Sciences Centre for Ecological, Poland
Research in Dziekanów Leśny, Łomianki, Poland

Karol J. Kram
Polish Academy of Sciences Centre for Ecological Research in Dziekanów Leśny, Łomianki, Poland

Mohammed Mahabubur Rahman
United Graduate School of Agricultural Science, Ehime University, Matsuyama, Ehime, Japan
Education and Research Center for Subtropical Field Science, Faculty of Agriculture, Kochi University, Kochi, Japan

Rahman Md. Motiur
Silvacom Ltd., Edmonton, Canada

K.K. Nkongolo and M. Mehes-Smith
Department of Biology, Canada
Biomolecular Science Program, Laurentian University, Sudbury, Canada

R. Narendrula, S. Dobrzeniecka, K. Vandeligt, M. Ranger and P. Beckett
Department of Biology, Canada

Irina Likhanova and Inna Archegova
Institute of Biology Komi SC UrD RAS, Russia

Yueh-Hsin Lo and Biing T. Guan
National Taiwan University, Taiwan

Yi-Ching Lin
Tunghai University, Taiwan

Juan A. Blanco
University or British Columbia, Canada

Chih-Wei Yu
New Taipei City Government, Taiwan

Murat Demir
Istanbul University, Faculty of Forestry, Department of Forest Construction and Transportation, Turkey

Martin Moravčík, Zuzana Sarvašová and Miroslav Kovalčík
National Forest Centre – Forest Research Institute Zvolen, Slovak Republic

Ján Merganič
Czech University of Life Sciences in Prague, Faculty of Forestry, Wildlife and Wood Sciences, Department of Forest Management, Praha, Czech Republic
Forest Research, Inventory and Monitoring (FORIM), Železná Breznica, Slovak Republic

Tibor Pentek and Tomislav Poršinsky
University of Zagreb/Forestry Faculty, Croatia